Sustainable Communities Design Handbook

Sustainable Communities Design Handbook
Green Engineering, Architecture, and Technology

Woodrow W. Clark, II, Ph.D

ELSEVIER

Amsterdam • Boston • Heidelberg • London
New York • Oxford • Paris • San Diego
San Francisco • Singapore • Sydney • Tokyo
Butterworth-Heinemann is an imprint of Elsevier

Butterworth-Heinemann is an imprint of Elsevier
30 Corporate Drive, Suite 400, Burlington, MA 01803, USA
The Boulevard, Langford Lane, Kidlington, Oxford, OX5 1GB, UK

Library of Congress Cataloging-in-Publication Data
Sustainable communities design handbook: green engineering, architecture,
and technology/Woodrow Clark II, author and editor.
   p. cm.
   Includes bibliographical references.
ISBN: 978-0-12-810204-6
   1. Environmental policy. 2. Sustainable development. 3. Green technology.
   4. Sustainable architecture. 5. Sustainable engineering.   I. Clark, Woodrow W.   II. Title.
GE170.S875 2010
   333.79—dc22   2010012045

British Library Cataloguing-in-Publication Data
A catalogue record for this book is available from the British Library.

For information on all Butterworth-Heinemann publications
visit our Web site at www.elsevierdirect.com

Printed in the United States of America
10 11 12 13 14   10 9 8 7 6 5 4 3 2 1

CONTENTS

Advance Praise   *xi*
About the Editor   *xiii*
Contributors   *xv*
Preface   *xvii*

1. Introduction   1
Woodrow W. Clark, II
References   8

2. The Third Industrial Revolution   9
Woodrow W. Clark, II and Grant Cooke
2.1 Introduction   9
2.2 How Do Communities and Nations Move Ahead?   13
2.3 What Is a Renewable Energy Power Source?   16
References   22

3. Public Policy and Leadership: "We Have Met the Enemy and He Is Us"   23
Scott G. McNall
3.1 The Energy Efficiency Gap   26
3.2 People Do Not Make Rational Choices   27
3.3 Human Beings Want to Make Sense of the World   28
3.4 People Can Worry About Only a Limited Set of Factors at Any Given Time   32
3.5 People Focus on Short-Term, Not Long-Term, Threats   32
References   41

4. Achieving More with Less: The State of Energy Conservation and Energy Efficiency   45
Michael F. Hoexter
4.1 Introduction   45
4.2 Electrical Energy Efficiency: Generating "Negawatts"   49
4.3 Utility Revenue Decoupling and Energy Efficiency   50
4.4 Green Design: Guiding Natural Energy Flows   51
4.5 Near-Zero, Net-Zero, and Plus-Energy Buildings   53
4.6 Electricity and Energy Efficiency Retrofits of Existing Buildings   55
4.7 Efficient Lighting: Efficient Fluorescents, Induction Lighting, and LEDs   56
4.8 Heat Pumps: Ground Source and BTES Linked   56
4.9 Key Technologies for More Energy Efficient, Eventually Carbon-Neutral
    Living   58

4.10 Quality Assurance and Certification in Energy Efficiency   59
4.11 Energy Efficiency in Transport: Short-Term and Long-Term Solutions   59
4.12 Price Signals and Energy Efficiency   62
4.13 Conclusion: Energy Efficiency in the United States Today   62
References   62

5. Renewable Energy   **65**
Joe Katenbacher
5.1 Introduction: Energy Use in the United States   65
5.2 Wind   67
5.3 Solar   70
5.4 Geothermal   73
5.5 Biopower   76
5.6 Marine   78
5.7 Advanced Renewables Deployment   78
5.8 Summary   80

6. How Energy Conservation Fits in an Existing Facilities Master Plan: A Case Study   **83**
Alison Gangl and Ben Johnson
6.1 Introduction   83
6.2 Getting Started   84
6.3 Investment Grade Audit   86
6.4 Installation   92
6.5 Measurement and Verification   95
6.6 Conclusion   97
References   97

7. Life-Cycle Analysis: The Economic Analysis of Demand-Side Programs and Projects in California   **99**
Don Schultz, Woodrow W. Clark, II, and Arnie Sowell with the California State Government Green Accounting Team (2001–2003) of forty active members
7.1 Preface   100
7.2 Introduction   101
7.3 Basic Methodology   103
7.4 Participant Test   110
7.5 Ratepayer Impact Measure Test   115
7.6 Total Resource Cost Test   121
7.7 Program Administrator Cost Test   126
Reference   129

Appendix a Inputs to Equations and Documentation   129
Appendix b Summary of Equations and Glossary of Symbols   131
Appendix c Derivation of RIM Life-Cycle Revenue Impact Formula   136

8. **Life-Cycle and Cost-Benefit Analyses of Renewable Energy: The Case of Solar Power Systems**   **139**

Thomas Pastore and Maria Ignatova
8.1 Introduction   139
8.2 A Successful Energy Plan   140
8.3 Due Diligence Procedures   140
8.4 Life-Cycle Analysis of a PV System from a Financial Perspective   141
8.5 Financing Structures   143
8.6 Measuring Savings from a PV System   144
8.7 Conclusion   148

9. **Public Buildings and Institutions: Solar Power as a Solution: Legal Mechanisms for Sustainable Buildings**   **149**

Douglas N. Yeoman
9.1 Alternative Energy Policy   149
9.2 Legal Mechanisms Facilitating Development of Alternative Energy Sources   150
9.3 Treatment of Environmental Incentives   161

10. **Seven Principles for Interconnectivity: Achieving Sustainability in Design and Construction**   **165**

Christine S. E. Magar
10.1 Introduction   165
10.2 Some of Today's Most Influential Sustainable Design Maxims   166
10.3 A New Set of Principles   168
10.4 Principle 1. Building Independence   170
10.5 Principle 2. Building Natural Form   171
10.6 Principle 3. Building Service   172
10.7 Principle 4. Building Interconnectivity   172
10.8 Principle 5. Building Adaptability   173
10.9 Principle 6. Building Performance   174
10.10 Principle 7. Building Independence   175

11. **The Los Angeles Community College District: Establishing a Net-Zero Energy Campus**   **181**

Calvin Lee Kwan and Andrew Hoffmann
11.1 Introduction   181
11.2 Background   182

11.3 Goals and Objectives   184
11.4 Importance of This Study   184
11.5 Renewable Energy Options   185
11.6 LACC Current Situation   188
11.7 Projected 2015 Campus Energy Demand and Consumption   201
11.8 LACC Solar Insolation   202
11.9 Solar PV Array and Setup   204
11.10 Discussion   209
11.11 Conclusion   210
Appendix a Map of LA City College Campus Indicating
            Previous, Current, and Planned Renovations / Construction   212
Appendix b LADWP Energy Rates as of October 1, 2009 – Specific for
            LACC Operations   213
References   214

12. Transformational Relationship of Renewable Energies and the Smart Grid

217

A. J. Jin

12.1 Introduction   217
12.2 Solar Electricity Systems and their Relationship with the Grid   219
12.3 Wind Power   223
12.4 Data Response and Power Transmission Lines   226
12.5 The Smart Grid and a Market Solution   228
Acknowledgments   231
References   231

13. Clarifying American Values through Sustainable Agriculture

233

Sierra Flanigan

13.1 The Murphy Apple Orchard   233
13.2 The SAD Truth   242
13.3 A New Beginning   243
13.4 Intercampus Produce Exchange   244
13.5 Appreciating the Value of Food   245

14. Climate Change Mitigation from a Bottom-up Community Approach

247

Poul Alberg Østergaard and Henrik Lund

14.1 Introduction   247
14.2 Energy Demand in Fredrikshavn   250
14.3 Current Energy System in Fredrikshavn   253
14.4 Energy Resources and Energy Scenario for Fredrikshavn   256

14.5 Energy System Integration in Fredrikshavn   259
14.6 Public Involvement   263
References   264

15. Conclusions: Toward a Global Sustainable Future   267
Woodrow W. Clark, II
15.1 The Third Industrial Revolution in the United States: Second
      Decade of the 21st Century   267
15.2 "Green" Careers and Businesses   269
15.3 Public Policy: Local, State, and National   270
15.4 Sustainable and Smart, Agile Communities   272
Reference   275

Appendix A.   California Standard Practice Manual: Economic
                Analysis of Demand-Side Programs and Projects   277
Woodrow W. Clark, II, Arnie Sowell, and Don Schultz
For California Governor Gray Davis
      A.1 Basic Methodology   278
      A.2 Participant Test   286
      A.3 The Ratepayer Impact Measure Test   291
      A.4 Total Resource Cost Test   296
      A.5 Program Administration Cost Test   302
      Appendix a Inputs to Equations and Documentation   304
      Appendix b Summary of Equations and Glossary of Symbols   306
      Appendix c Derivation of RIM Life-Cycle Revenue
                  Impact Formula   311

Appendix B.   Request for Qualifications   313
Los Angeles Community College District with
lead author Woodrow W. Clark, II
      B.1 Purpose and Scope of Request for Qualificatons   315
      B.2 Schedule of Events   319
      B.3 RFQ Instructions and General Provisions   320
      B.4 Scope of Services, Work, and Deliverables   329
      B.5 Qualification Requirements   341
      B.6 Evaluation and Selection Criteria   346
      B.7 Content and Format of Response to RFQ   352

Appendix C.   Los Angeles Community College District
                Energy Program   367
Los Angeles Community College District with
lead author Woodrow W. Clark, II
      C.1 Section One: Energy Strategy   370

C.2 Section Two: Central Plant   372
C.3 Section Three: Demand Side Management/Performance
     Contracting   374
C.4 Section Four: Renewable Energy Technologie   377
C.5 Section Five: Campus Plans   388
C.6 Section Six: Financial Structure   399
C.7 Section Seven: Power Purchase Agreement   401
C.8 Section Eight: Finance Partners   404
C.9 Section Nine: Feed-in Tariff Analysis/RECs   406
C.10 Section Ten: Education/Jobs   410

Index   **415**

Sustainable Communities Design Handbook

After authoring over a dozen books throughout my career, I appreciate and admire such a book as *Sustainable Communities Design Handbook*, since it identifies and brings applied skill sets in line with actual needs for our communities of any kind to become sustainable. This book represents a landmark for others to follow. Communities of all kinds need the tools that the book discusses in order to stop climate change.

Jeremy Rifkin
Founder/CEO, Foundation on Economic Trends
Bethesda, Maryland

My entire career has been dedicated to understanding, operating, and now regulating the energy sector. In many ways, both Edison (as CEO for a decade), in the private energy sector (started an energy company) and now CPUC (as Chair for another decade) has meant that I needed some knowledge about each of the chapters in the book. From law to economics to technologies with engineering and designing, in order to make communities sustainable there a number of different skill sets. The *Handbook* should be on everyone's shelf or computer as a reference, a guide with tools and inspiration that sustainable communities are and have been achievable.

Mike Peevey
Chair, California Public Utilities Commission
San Francisco, California

Woody and I have known each other for over two decades. We have been in the trenches together working on bringing renewable energy systems to local on-site use for generating power. Our first meetings in the 1990s began with the UN IPCC's Third Assessment Report and the Special Report on Technology Transfer. Since then, while going our different ways, we have stayed in constant touch. The *Sustainable Communities Design Handbook* represents yet another milestone in his career, but even more significantly an advance in the field of systems sustainability. In his latest work the field now has a guidebook that documents and explains the use of academic skills in the actual real world in order to stop and reverse climate change. This applied academic handbook to mitigate global warming/cooling is long overdue and needed by everyone. Let there be more.

Dan Kammen, Ph.D
Class of 1935, Distinguished Professor of Energy, and
Founding Director, Renewable and Appropriate Energy Laboratory
University of California, Berkeley
Berkeley, California

Woodrow W. Clark, II, MA³, Ph.D.

Dr. Clark was Senior Policy Advisor to California Governor Gray Davis (2000–2003) for renewable energy reliability. He was formerly Manager of Strategic Planning/Implementation at Lawrence Livermore National Laboratory. He is a qualitative economist who is the Managing Director of Clark Strategic Partners, in Los Angeles, CA.

Grant Cooke, MJ

Cooke is the founder of Sustainable Energy Associates, an energy efficiency engineering firm in northern California. Originally a journalist, he is an entrepreneur and marketing professional with 30 years of management, business development, sales, and communications experience. He joined his first start-up in 2005 and helped it catch the surge that evolved into California's $3 billion energy efficiency industry. With his strong business and communications skills, Cooke helped the company grow to 30 employees and $100 million in projects in less than five years. As a pioneer in the energy efficiency industry, he helped shape California's energy efficiency program as it evolved into the world's largest, including some of the nation's premier energy efficiency programs. Besides his work in energy efficiency, he opened up key business areas such as water conservation, energy renewables, and sustainability. His degrees are from the University of California.

Sierra Flanigan

Sierra is a recent college graduate and a young aspiring agent for change. She works as the Campus Programs Coordinator for EcoMotion Inc., an energy and environmental consulting firm. Sierra and her colleagues assist schools of all kinds to broaden and strengthen campus sustainability, a concept and practice near and dear to her heart.

Alison Gangl, CRM

Alison has been with Schneider Electric for five years, working closely with schools and colleges in Washington and California providing energy conservation projects through performance contracting.

Michael F. Hoexter, Ph.D.

Terraverde Consulting.

Andrew Hoffmann

Technical Associate, Energy and Engineering Group, Los Angeles Community College District.

Maria Ignatova, CFA

Ms. Maria Ignatova is a manager at Sanli Pastore & Hill, Inc. Ms. Ignatova is a Chartered Financial Analyst (CFA) Charterholder, and received her Bachelor in Commerce and Business Administration (BComm). She has been a consultant on numerous large scale renewable energy projects since joining the firm in 2006.

A.J. Jin, Ph.D.

Dr. AJ Jin has passion for bringing renewable energy products into the market at high efficiency and low cost. He founded Suzhou Renewable Energies Technology Ltd. and has invented a solar electricity product, the portable direct solar thermal electricity generator.

Ben Johnson, PE, LEED AP, CEM

Ben has worked for Schneider Electric for five years; in 2007, he helped launch their performance contracting presence in California, and currently leads the development and engineering operations for the state.

Calvin Lee Kwan
Environmental Science and Engineering Program, School of Public Health, University of California, Los Angeles.

Henrik Lund
Aalborg University, Denmark.

Christine S.E. Magar, AIA, LEED AP
Since joining the USGBC National LEED Steering Committee in 2001, Christine has been instrumental in steering the development of the LEED Green Building rating system. While the organization's LEED Core & Shell rating system was being developed she was vice-chair of Core & Shell Committee. She was also chair of the AIA-LA Committee on the Environment, and has served on the LAUSD High Performance Schools Working Group, CHPS Technical Committee, City of Seattle's Green Building Team, and the City of LA Sustainable Design Task Force. Her current role as President of Greenform, a Los Angeles-based woman business enterprise, allows her to remain dedicated to managing people, ideas, and decision-making in the area of sustainable design. Christine directs Greenform's efforts as a sustainability consultant, assisting clients by identifying and implementing sustainability goals for their organization and building a portfolio from a holistic and integrated point of view.

Scott G. McNall
Executive Director, The Institute for Sustainable Development, California State University, Chico.

Poul Alberg Østergaard
Aalborg University, Denmark.

Thomas Pastore, ASA, CFA, CMA, MBA

Mr. Thomas E. Pastore is Chief Executive Officer and co-founder of Sanli Pastore & Hill, Inc. Mr. Pastore is an Accredited Senior Appraiser (ASA), Business Valuation Discipline, of the American Society of Appraisers, a Chartered Financial Analyst (CFA) Charterholder, a Certified Management Accountant (CMA), and received his Masters in Business Administration (MBA). He has valued over 2,000 businesses during his career, including numerous energy and clean technology companies. He regularly testifies in court as an expert witness. Mr. Pastore frequently speaks on business valuation to professional organizations.

Don Schultz
Dr. Schultz was an Analyst for the California Office of Ratepayer Advocates for over 20 years. He had been an advisor to the California Energy Commission when it was first formed in the mid-1970s and participated in the early *Standard Practices Manual* released by the California Public Utilities Commission and the California Energy Commission. He retired from California state service in 2009.

Arnie Sowell
Mr. Sowell was the Assistant Deputy Secretary for the California Consumer Affairs Agency. He had been a project analyst for the California State Legislature and advisor to elected officials on revenue, budget, and management issues. Currently, he is the Policy Advisor to California State Senate President, Karen Bass.

Douglas N. Yeoman
Parker & Covert LLP, Tustin, California.

PREFACE

This book reflects at least a decade (1990s into the 21st Century) of my applied work in the environmental and energy sectors, both of which are critical in understanding and making sustainable development for any community.

For any author, the creation of a book and what needs to be done to finish it in a timely manner are significant. And, in this case, authoring a book does not pay anything, certainly not enough for the time that is spent on it. However, what most people do not understand is that the creation of a book takes time and is thoroughly reviewed by the publisher. Elsevier Press, with whom I have had a decade of peer review relationships for a book (*Agile Energy Systems*, about the global lessons learned from the California energy crisis, 2004) and several journals in the energy sector to which I have been a contributor as well as associate editor and special edition editor.

The peer review process was extensive for this book. Sustainable development, as a field, has developed and hence created the need for mechanisms or tools in which to implement sustainable buildings, communities, and regions as well as nation-states. A key issue is how to define *sustainable development* (which comes in Chapters 1 and 2) and identify what these mechanisms are. Given my experience, the mechanisms include technologies (storage devices and wireless smart grids), standards (such as Leadership in Energy, Environmental Design from the U.S. Green Building Council), economics and accounting, including finance, that are created as legal contracts (power purchase agreements and feed-in tariffs) as well as architecture and design for buildings and their surroundings, including transportation, water, and waste systems.

This book contains chapters on each of these topics. I purposely avoided chapters on each topic but rather have them include the application of the mechanisms. After Chapter 1 introduces the book, Chapter 2 discusses the Third Industrial Revolution. In other words, the world today is moving rapidly from the Second Industrial Revolution of fossil fuels and nuclear power to renewable energy, new technologies, and smart communities. This dramatic industrial change affects and guides the "paradigm shift" that sustainable development represents.

Hence, Chapters 3–11 reflect these basic concerns as to applied mechanisms and tools covering the setting of public policy (Scott McNall), technologies (Alison Gangl, Ben Johnson, Calvin Kwan, Andrew Hoffman, and I), design and architecture (Christine Magar), economics and accounting (Tom Pastore, Arnie Sowell, Don Schultz, and I) to legal and contract areas (Doug Yeoman) and how the communities are connected through smart grids (Jerry Jin).

In particular, however, I wanted to "push the envelope." Therefore there is a chapter (13) from a recent college graduate (Sierra Flannigan) on her experiences as a student creating viable, organic agricultural products. I also include a chapter (14) on a city in Denmark by Henrik Lund and Poul Alberg Østergaard so the reader can see how a nation like Denmark has gotten ahead of the sustainable development curve to power an entire city on 100% renewable energy. These examples are critical for understanding on how society needs to both empower itself (especially the youth of today) and have successful examples that work in the real world.

I thank, in particular, Ken McCombs, from the publisher's office at Elsevier. He has been both an inspiration and motivation for this book. Ken not only came through after I had to respond to six reviewers and answer their questions but also kept after his staff and me to perform. Like me, Ken sees this book as a standard manual, which has useful tools for all practitioners, but also will be available online and as a series in the future. Indeed, I hope so. This topic will change, as it should. The chapters will need to be revised, as they should. And the content will expand, as it should. The biggest issue will be to measure the positive impact on our environment and communities. That will generate a whole new area of literature, data, and analyses. All of this is needed, *not* later but *now*.

In that context, I am dedicating this book to my wife Andrea and our son Paxton. Without their total support and encouragement (hard for a two year old, but he has certainly done that) the book would not have been completed. I especially want this book to help my son in his future life, because my generation (the Baby Boomers) left him with a world that is becoming increasingly environmentally polluted and economically dysfunctional. The solution is in sustainable developed communities—that is, ones that are transformed, rebuilt, or even created so as *not* to violate or harm their environment or others around them or far away. What we do locally, affects others globally. We need to stop now and implement the solutions that exist and will be created in the near future.

Woodrow W. Clark, II

CHAPTER 1

Introduction

Woodrow W. Clark, II, Ph.D

Contents

References   8

Books are often difficult for a number of reasons. This one is an exception. The topic of sustainable development has been part of my lexicon for over two decades. While the terms were first used in the late 1980s in the Brundltand Report for the United Nations, they were and still are the subject of much debate and, to some extent, controversy. The reasons are complex but the terms are now common and in constant use, particularly with the wide release and acceptance of Al Gore's film, *An Inconvenient Truth*, which won Gore both a Nobel Peace Prize (along with the members of the UN Intergovernmental Panel on Climate Change, to which this author was a corecipient) and an Academy Award in 2007. Each of those honors alone was a remarkable accomplishment, but the deeper value was the global awareness of climate change and the need for solutions.

My last book, *Sustainable Communities* (Clark, 2009a), was a series of case studies in sustainability that were actually implemented. The communities ranged from a public elementary school and community colleges in the nonprofit world to city governments, corporations, and businesses from around the world. The idea was to provide models to understand the breadth and depth of sustainability as a concept and mechanism for action.

Sustainable development thus has become an acceptable policy initiative and programmatic implementation strategy. However, the "devil is still in the details" as to what *sustainable development* means. On the one hand, some businesses and governments see sustainable development as a strategy for building more homes, office buildings, and large communities. To acknowledge that communities were sustainable, groups formed to provide scores and credit points. The most popular one in the United States is the USGBC (U.S. Green Building Council), which developed a scoring mechanism called *Leadership in Energy and Environmental Design* (LEED), whose basic score starts with being "accredited" goes to its highest, being "platinum."

Sustainable Communities Design Handbook
ISBN: 978-1-85617-804-4, DOI: 10.1016/B978-1-85617-804-4.00001-X
1

Without going into details, the USGBC has become a world leader in certifying sustainable buildings. There is now an effort and pilot programs to do the same for communities and even cities. A number of other organizations in the last decade have done something similar in the United States, including the Climate Action Registry, originally founded to be California-centric but now is national and global. Because of the energy crisis that hit California in 2000, a number of energy efficient and conservation programs were created, including Flex Your Power, which still exists a decade later. These programs and subsequent actions taken by the California Energy Commission and the California Public Utility Commission established the state as a world leader in energy conservation and renewable power generation.

Meanwhile, the U.S. government also established programs (e.g., Energy Star) that ranked the energy output (hence, savings from conservation) of appliances and equipment and the national energy laboratories (e.g., National Renewable Energy Laboratory, NREL) for establishing the rankings of electric power derived from sunshine, wind, geothermal energy, and other renewable energy resources. Also international organizations have been formed for similar rankings and scores of EU and other nations.

Sustainable Communities Design Handbook is a book long in the making and long overdue. Basically, the book maps out what communities need to do when thinking about how to protect their environment while repairing, building, or expanding.

At this point, there is no need to review Chapters 2 and 3, "The Third Industrial Revolution" and "Public Policy and Leadership: 'We Have Met the Enemy and He Is Us'," respectively. The best way to tell the story about sustainable development is by way of examples. In this book, colleges often illustrate the case of communities, since they are often self-contained communities with all the attributes therein, from residential to recreational areas that include the use of basic systems from water and energy to transportation and telecommunication.

The Third Industrial Revolution (3IR) is a concept that puts these issues into a broader picture about society and its industrial growth. The 3IR is an industrial revolution that uses renewable energy, power, and fuel generation along with storage and new technologies, including the interconnection of communities into "smart girds." The 3IR started in the European Union and Asia and is only now coming to the United States. It was remarkably different from the Second Industrial Revolution (2IR) of fossil fuels and nuclear power. Jeremy Rifkin, the environmental economist, saw this

3IR coming and has started a series of groups to support and implement it. Now communities throughout the world are participating and receiving advice and plans to implement.

"Public Policy and Leadership" is the subject of Chapter 3 by Scott McNall from Chico State University in California. The chapter is significant in that *no* plans or programs are done without a prior public policy decision being made. Be the officials elected or appointed, leadership over government programs is critical. Today, however, leadership among corporate decision makers is equally significant. Most companies will make decisions based on profit and loss, but the political arena and economic concerns over how to handle global warming is critical for most corporate leaders. The bottom line tends to always be there and provide the baseline or bar to prevent sustainable development programs.

A key component to this book is the perspective that communities must be more efficient to conserve their use of energy, as well as use renewable energy to generate that which they need. In Chapter 4, Michael Hoexter covers the issue of conserving energy with "Achieving More with Less: The State of Energy Conservation and Energy Efficiency." Then, in Chapter 5, Joe Kantenbacher discusses and analyzes this concept in critical detail in "Renewable Energy."

Chapter 6, "How Energy Conservation Fits in an Existing Facilities Master Plan: A Case Study," on the use of an optimization energy plan for college campuses, is useful in that it provides a working model on how to understand, apply, and analyze technologies for the use in localized areas, such as college campuses. The same optimization code can be used for non-profit organizations as well as government. With some modifications, the optimization may also be very useful for companies and businesses, especially those that have clusters or groups of buildings in one area.

That topic of finance and accounting is explored in "Life-Cycle Analysis" by Don Schultz, Woodrow Clark, and Arnie Sowell. Chapter 7 walks through the process and the appendices to that chapter provide the spreadsheets and accounting mechanisms for people and businesses to follow. However, Chapter 8 by Tom Pastore and Maria Ignatova provides the actual analyses for the accounting of sustainable development costs. Life-cycle and cost-benefit analyses are critical to organization decision making. Even more significantly, rebates, incentives, and tax benefits need to be factored in for technologies and programs.

The big issue is often the bottom line for businesses and this is increasingly so for governments and nonprofits in the 2009–2010 economic "de-recession." What is important about Chapters 7 and 8 (as well as

Appendix A) on economics and accounting are that they point out and demonstrate the shift from cost-benefit analysis to "life-cycle analysis" and how this economic change has helped bring innovations, emerging technologies, and the 3IR into the market sooner than normally expected in the conventional "market economy."

The economic change has, however, also come with new economic and legal programs, which have helped in the longer term financing of innovation and the 3IR. A key area is the use of power purchase agreements (PPAs), which provide financing to an organization from 15 to 25 years. Such long-term or "life-cycle" legal and financial commitments have helped solar systems be installed at a faster rate, along with government tax and grant incentives, than in the last decade. Hence, 2010 appears to be moving rapidly into more renewable energy (especially solar) to be installed than in prior years. This process also reduces the price of solar systems, due in part to a growing number of new solar manufacturing companies.

The result is that solar systems are now almost cost competitive with regular energy generation systems. The same economic phenomena came with wind, when a combination of long-term financing mechanisms and government tax and incentive programs reduced the costs of wind turbines to below that of natural gas power generation by the end of the first decade of the 21st century.

Hence, the financial and accounting economics for renewable energy power generation for the 3IR is moving away from a PPA long-term finance mechanism. This does *not* mean that the cost-benefit analysis (short-term in quarters or even 2–3 years) works now. Soon the shift might be to a more traditional economic and accounting model. Now, however, two models have become part of the 3IR to which Chapter 8 refers and several remaining chapters mention. The two emerging models are feed-in tariffs and regular lease.

Feed-in tariffs (FiTs) charge a higher energy purchase rate to consumers but also allow consumers to sell their power to the central grid supplier or other energy customers. The FiT was started and achieved success in Germany in the early 1990s. It has since been expanded and revised. Spain then started a national program in the mid-2000s that appeared to be too aggressive, with overbuilding of solar plants and systems. The net result in both Germany and Spain was higher employment and job creation in the solar sector, and Germany became the number 1 nation in solar manufacturing. Now other nations and communities are starting FiT programs.

The second emerging model is a regular lease. However, the costs are high, although for a shorter time. The costs of solar and other renewable energies are coming down, so that, according to one "old economic" model, all renewable power generation might be factored into the regular operational costs (like heating, air conditioning, electricity, and plumbing) for buildings (homes, offices, storage, etc.). These systems can then be part of the total cost for a mortgage of any building and applied to different or more expansive areas, like college campuses or shopping malls, that have clusters of buildings.

Chapter 9 "Public Buildings and Institutions: Solar Power as a Solution" by Douglas Yeoman explores the PPA and also some of the newer models. The legal section, however, also covers the need for construction contracts, liabilities, warrantees, and insurance. What is important is to know about such legal documents, which can be templates and models for other programs, buildings, and clusters.

"Seven Principles for Achieving Sustainability in Design and Construction" by Christine Magar, Chapter 10, covers why communities need designs and plans to provide direction, goals, and measurable objectives. Magar is a certified architect and has LEED advanced placement. Her work has often involved communities and clusters of buildings. This criterion is important, since most communities include more than one building and must also include other infrastructures, like transportation, energy, water, and waste as well as, increasingly today, wireless and telecommunication. All of this is part of the 3IR and helps provide the basis for smart grids.

Chapters 11 through 13 get into the actual cases or examples of how the 3IR works. Chapter 11 covers agile sustainable communities, with the case of "The Los Angeles Community College District." The authors, Calvin Kwan and Andrew Hoffmann, have brought extensive technical, finance, and policy experience to these colleges. Some of the information is rooted in Appendix B, which uses the same database for the technical and financial optimization for the same colleges.

Chapter 12 then looks into the smart grid infrastructures and local grid applications. Jerry Jin provides some examples and plans from a business perspective in smart communities and the girds that connect them. Then Chapter 13, which looks at sustainable agriculture and the abundance of human resource potential, is by by Sierra Flannigan, whose just-completed work at a small college is a good case in point on how local communities, and students in particular, can bring about sustainable communities in terms of food and recycling.

In the end, however, it is important to identify and find communities that have been sustainable and represent what the 3IR is about. Denmark (where I was a Fulbright fellow in 1994, then visiting professor for the next six years) has been a leader in this regard in terms of national policies, financing, and operational programs (Clark, 2009b; Clark and Lund, 2006 and 2008). Lund in particular has been tracking the renewable energy and wind manufacturing industry that started in Denmark through a merger of several Danish wind turbine companies (Lund and Clark, 2002). Vestas is today the largest, most dominant wind turbine manufacturing company in the world.

Hence, Chapter 14 on "Climate Change Mitigation from a Bottom-up Community Approach" by Poul Alberg Østergaard and Henrik Lund (academic practioners from Aalborg University in Denmark) looks at Frederikshavn, a small city in the northern Jutland region of Denmark. This city is a main transportation and shipping hub for northern Demark with western Sweden and southern Norway. Because of that, the city is aware and very aggressive in becoming sustainable, in addition to controlling energy and fuels from the 2IR.

The three appendices to the book are well worth using as references. Appendix A presents the *California Standard Practices Manual* (CSPM), of which I was a coauthor while in state government. Basically it is an economic model for doing life-cycle analysis on projects, containing guidelines and formulas. The CSPM was published in 2002 but was originally created in the mid-1980s for doing cost analyses on projects for the California Public Utility Commission (CPUC). However, it was not revised until 2002. The CSPM is now used extensively in California government project finance. It remains the guiding model for doing economic data projections, analysis, and evaluative outcomes.

However, even with decision makers providing leadership and public policy, the main question remains, what new technologies and scientific planning need to be in place before change can occur? Even more significantly, how does an organization or company know when it is the right time to "try something new" from the 3IR? The answer is that there is no perfect time. These are subjects explored in Appendices B and C, which respectively concern public requests for proposals from colleges to vendors to implement renewable energy systems, and then how these sustainable systems can work in community colleges. Most construction contractors and builders use the same tried and true technologies, or whatever is on their shelf from their last client.

Often all these traditional or conventional technologies are from the 2IR and therefore inappropriate for sustainable development. The "stranded costs" alone will be long lasting (20–30 years) and delay the 3IR for another two to three decades. This is unacceptable, given the need to mitigate global warming and climate change.

Finally, the last chapter has conclusions for the next generation of people, companies, and governments about sustainability. I wrote this chapter because today's problems with global warming and climate change are directly the fault of my generation. We "baby boomers" lived off our land and exploited others for decades. We are now paying the price, because the world is indeed "round"—not "flat" as some "populist" economic commentators would argue—hence the atmospheric and other pollution that we cause in one part of the world, travels to our part of the globe, too. In fact, this group's current defense of the "flat" world idea in economic and business terms is equally as wrong and shortsighted.

Consider the economic crisis that did *not* hit just the United States or even the European Union and Asia. Rather, the "de-recession" has affected everyone. Money has been misplaced and even stolen from New York City to London, from Paris to Rome, and around the world again to Tokyo, from Seoul to Sydney, New Delhi, and now to Beijing, Shanghai, and Hong Kong. Everyone must be careful.

The point is that the last chapter tries to provide the details from the Third Industrial Revolution perspective in terms of what this means for jobs, careers, and the future of our planet, which we are handing over to young people. Without doubt, the problems are enormous and will need time, money, and people to solve them. But the clock is ticking and the environment is becoming worse daily around the world.

For those who said that Katrina and the hurricanes of the mid-decade in the 21st century were just a short-term and normal weather change, which would correct itself, the data and facts prove them wrong. In the United States alone, the number and level of hurricanes have doubled. The number and impact of tornados have become more intense and damaging by a factor of 2.

These are not statements from one more of those crazy scientists, academics, or economists on the far Left. *No.* The facts exist. Insurance companies are the bottom line here and their rates and even reluctance to insure certain communities, regions, or areas of the United States has expanded. Some countries and their cities, as well as industries, can no longer get insurance. In short, our world is at risk. We do not have any time left to debate it. We *all* must act now.

REFERENCES

Clark II, W.W. (Ed.), 2009a. Sustainable Communities: Case Studies. Springer Press.

Clark II, W.W., 2009b. Analysis: 100% renewable energy. In: Lund, H. (Ed.), Renewable Energy Systems. Elsevier, Burlington, MA, pp. 129–159. Chapter 6.

Clark II, W.W., Lund, H., June 2008. Integrated technologies for sustainable stationary and mobile energy infrastructures. Utility Policy J. 16 (2), 130–140.

Clark II, W.W., Lund, H., December 2006. Sustainable development in practice. J. Clean. Prod. (15), 253–258.

Lund, H., Clark II, W.W., November 2002. Management of fluctuations in wind power and CHP: comparing two possible Danish strategies. Energy Policy 5 (27), 471–483.

CHAPTER *2*

The Third Industrial Revolution

Woodrow W. Clark, II, Ph.D and Grant Cooke, MJ

Contents

2.1	Introduction	9
2.2	How Do Communities and Nations Move Ahead?	13
2.3	What is a Renewable Energy Power Source?	16
References		22

2.1 INTRODUCTION

Countries, states, regions, and communities of all kinds must eliminate their dependence on fossil fuels, stop their renewed focus on nuclear power, and reverse the pace of environmental degradation. To do so, communities must embrace the reality of smart grids, emerging storage technologies, and renewable energy generation. For example, the United States must leapfrog into the Third Industrial Revolution, which started two decades ago in Europe and Asia, in particular Japan and South Korea, by creating sustainable, agile smart communities that are energy independent by the use of renewable power and hence carbon neutral.

Time is passing quickly for the United States and other nations, while Europe and Asia, now with China, have been developing sustainable, energy-independent communities for the last two decades. As a nation and the leader of democracy for two centuries, the United States must examine its own "roots" and provide the future direction for humanity. The Third Industrial Revolution (3IR) is now strongly embedded in other nations, so that these countries are no longer dependent on fossil fuels or nuclear power, which defined the Second Industrial Revolution (2IR). The 3IR primarily generates stationary power and creates fuel from renewable energy sources.

Nations must take these actions now to create and implement the 3IR for themselves. They need to reduce the energy dependency on the Middle East, a geopolitical region whose instability constantly threatens national security and keeps nations from focusing on crucial domestic issues such as health care, financial reform, and innovation as well as the planetary

Sustainable Communities Design Handbook
ISBN: 978-1-85617-804-4, DOI: 10.1016/B978-1-85617-804-4.00002-1
9

environmental crisis. Becoming involved and part of the 3IR must be recognized and implemented by nations and communities sooner rather than later. The United States, for example, must go beyond the 2IR, with its massive and inefficient fossil fuel generation and environmental degradation to move rapidly into the 3IR, with its community-centric and environmentally friendly renewable energy generation. Europe and Japan have already done so for the last two decades. Where is America? Clark (2009) illustrated this point in his book, *Sustainable Communities*, which gives examples from all over the world.

Social and economic forces are coming together as the nation ponders its sustainable future. Now, with global warming and climate change having an impact on everyone's daily lives, can anyone wait any longer? On an economic front, the world is battling the most severe economic turndown since the Great Depression of the 1930s. Nationwide, states are reeling with the loss of tax and real estate development revenue. California has been "bankrupted" by its governor, whose efforts to balance a shattered budget are subject to serious questions.

California is the world's eighth largest economy. Nine years ago, the state was the world's sixth largest economy and held the distinction as number 7 from 2003 to 2008. However, in mid-2008, the recession started. The basic result of the California budget signed in September 2009 was to handicap the entire state, from its public education and welfare systems to basic needs such as fire, police, water, energy, waste, transportation, and prisons. In the Midwest, the American auto industry, once the nation's pride as the leader of the global manufacturing sector, is on life support from the federal government. The era of the V8 and the megaton SUV is fading in the rearview mirror, as it should have been a decade ago. Now General Motors is renamed by the general public and federal government decision makers, as Government Motors.

Americans are wondering what their vital interests in any international arena should be. The world's oil supplies have peaked and are rapidly declining. M. King Hubbert, a Shell Oil geophysicist, observed in a startling prediction, first made in 1949, that the fossil fuel era would be of very short duration (Hubbert 1949). In 1956, Hubbert predicted that U.S. oil production would peak in about 1970 then decline. At the time, he was scoffed at, but in hindsight, he proved remarkably accurate.

Just as the world's oil and natural gas supplies have peaked, there is renewed interest in nuclear power. This, too, is a false hope. The U.S. Department of Energy (see Figure 2.1) reported a key set of figures

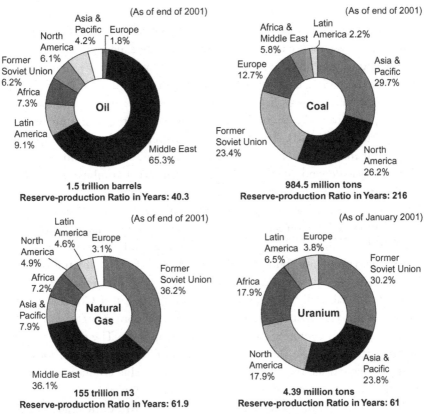

Figure 2.1 Global fuel reserves and production resources (Source: U.S. Department of Energy Information Service.)

documenting the declining and limited supplies of gas, oil, and coal. These surprising statistics, which bode badly for the nuclear industry, show that only 61 years of uranium remain.

While America's domestic oil supplies peaked in the 1970s and international oil supplies will peak some time around the early part of the 21st century (estimates are now at 2030) and with demand rising from newly developed nations, pushing for more oil and gas with tax breaks or even land options are the wrong policies and certainly not part of the 3IR. When these measures and others related to "balancing a budget" in the short term are then implemented, this means future generations will be paying taxes for years at triple or quadruple the original costs. This one-year "fix" to balance budgets or justify expenses is misguided, wasteful, and economically crippling on our children and grandchildren. Any new funds and resources must be focused on renewable energy generation and related technologies for storage and waste, transportation, and related areas. If not, global political and social tensions will mount, since fossil fuels will become scarce and more expensive.

No nation can afford any more "oil wars" nor can any continue to deny that the nation needs to take a new path. Americans must come up with a national energy policy that makes sense, as the entire country must move rapidly from the Second Industrial Revolution that dominated the 20th century to the 3IR. This transition has already started in Europe and Asia, and it may be the "new world order" of the 21st and 22nd centuries. The 2IR was dependent on fossil fuels and internal combustion engines, along with massive infrastructures to support energy and transportation. The 3IR is focused on using renewable energy to power "smart" local communities, where on-site building-by-building renewable power and smart grids can monitor usage to conserve power and increase efficiency.

Europe, Japan, and South Korea have been in the 3IR for the last two decades. A large-scale effort is now underway in China. A recent report by the international think tank, the Climate Group, finds that China is rapidly gaining in the race to become the leader in development of energy technologies. America definitely has some catching up to do. The sooner it starts, the faster it can achieve the inherent benefits of a sustainable and localized energy-generated lifestyle, which focuses on sustainable communities while creating new companies, careers, and areas for employment.

In the 19th century, the United States started to be the leader in the 2IR. By the end of the 20th century, America was the world leader in innovation and entrepreneurship, so that by the new millennium (21st century), it was creating the historic advances in computerization and information technology. Now, that distinction as innovator and entrepreneurial dynamo is challenged, as the world seeks leadership in the battle to stop global warming and reverse climate change.

Germany is now the number 1 producer and installer of solar panels for homes, offices, and large open areas. Japan is now leading the world in auto manufacturing, since it started to make vehicles that are not damaging the environment and atmosphere. Other nations in the European Union, such as the Nordic countries and Spain, have been aggressively implementing policies and programs to become energy independent in four decades, and they are succeeding. See Figure 2.2 as to Denmark's accomplishments already. However, unlike other European Union nations, the Danish government is focused both on national policy and plans and local distributed systems as it moves ahead to implement the 3IR.

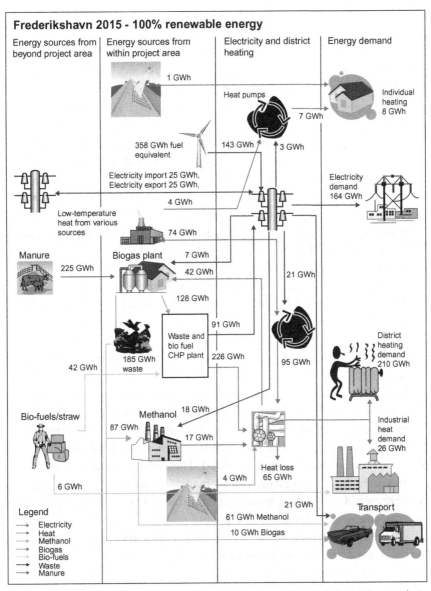

Figure 2.2 Flowchart of the proposed energy system in Frederikshavn, Denmark, in 2015.

2.2 HOW DO COMMUNITIES AND NATIONS MOVE AHEAD?

The place to start is to recognize that there is confusion in the national dialogue, as the nation waves good-bye to the 2IR and cast its eyes and focus toward the 3IR. Why? The confusion is exacerbated by the American media, with its digital communication systems that are ubiquitous and instantaneous yet are shallow and politically biased when dealing with significant issues. The public is besieged by the latest "buzz words" and concepts like *sustainability, renewable energy, green jobs and careers* along with *energy efficiency, conservation, greenhouse gases, global warming,* and *climate change.* All these words are without definition.

Basically, as Clark and Fast (2008) argued in their book *Qualitative Economics*, there must be definitions of concepts and ideas such as *clean.* For example, there is a qualitative and quantitative difference between "clean" and "green." By 2010, with the success of Al Gore's film *An Inconvenient Truth* making the public and policy makers aware of the problem of global warming, too many concepts were "greenwashed" and passed off as something they are not. The terms get tossed back and forth by scientists and politicians so that everyone thinks they know what they mean, until they try and use them in a sentence, then the conversation quickly becomes as painful as that of the 2007 South Carolina's Miss Teen Contestant YouTube video incident.

Even *The Economist* (July 2009) admitted that "Modern Economic Theory" had failed along these lines. As the journal put it, "economics is not a science." To help sort this out, Clark (2009), Clark and Fast (2008) created the field of "qualitative economics" to make distinctions between words, concepts, and even numbers that are often misused (See Figure 2.3 for an example of how audited data was misused by ENRON during the California energy crisis). The issue is that numbers, words, and ideas all too often are not defined or even discussed. The public and decision makers just use them. So, to companies and lobbyists, "clean" energy means the use of energy and fuels such as natural gas and diesel. These are fossil fuels and emanate gases and particulates that pollute the environment. These chemical wastes cause massive health and environmental problems. "Green," on the other hand, in the context of energy and fuel, means renewable energy from natural resources like wind, sun, geothermal energy, and ocean and tidal waves as well as the flow of water in rivers.

Whether America is ready or not, 3IR is at its doorstep, now. The huge amounts of federal stimulus money in 2009, about $250 billion, earmarked for energy conservation and renewable generation, coupled with crashing local government budgets (particularly in California and New York) will propel Americans to look in the direction of energy independence and sustainable activities and communities. In the small town of Benicia,

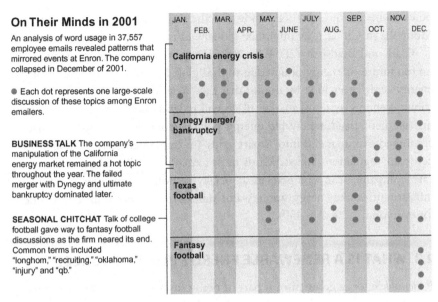

On Their Minds in 2001

An analysis of word usage in 37,557 employee emails revealed patterns that mirrored events at Enron. The company collapsed in December of 2001.

● Each dot represents one large-scale discussion of these topics among Enron emailers.

BUSINESS TALK The company's manipulation of the California energy market remained a hot topic throughout the year. The failed merger with Dynegy and ultimate bankruptcy dominated later.

SEASONAL CHITCHAT Talk of college football gave way to fantasy football discussions as the firm neared its end. Common terms included "longhorn," "recruiting," "oklahoma," "injury" and "qb."

Figure 2.3 Email surveillance data on ENRON.

California (population of 30,000), the city's $2 million annual energy bill represents about 5% of the budget. Eliminating that expense would allow the city to beef up safety personnel and community services or give the city a buffer for the leaner days ahead. Unfortunately, most of the federal stimulus funds for energy are focused only on efficiency and conservation.

While this is a start, renewable power generation is the core need for 3IR. Again, will America fall further behind? That is a distinct possibility, as the European Union, Japan, and now China become ever more aggressive in renewable energy generation and technologies.

Energy independence and subsequent elimination of energy bills are part of the potential benefits waiting as we make the transition into the 3IR. As soon as possible, America needs to give up freebasing fossil fuels and embrace a healthier community with intelligent development and greater community connectivity. What is crucial is that Americans, starting in local communities, must see the vision and take action. Almost every community has the renewable resources to make itself energy independent and carbon neutral. The United States must get started. Americans must come to an understanding and develop a national energy policy, then get out of the way and let America's historic innovation and entrepreneurship take over and "leapfrog" what other nations have done and are doing. Clark and Isherwood (2010) notes some of this in their study for the Asian Development on Inner Mongolia in China, now published globally in the *Utility Policy Journal.*

Clark and Bradshaw (2004), in *Agile Energy Systems,* the pioneering book on the future of energy policy due to the "global lessons from the California energy crisis," conclude by noting that the "new localized energy [read: distributed energy systems] market place will redefine how integrated resource management [read: renewable energy power generation and storage that is combined or integrated into "smart grids"] is implemented in a public market [read: regulations and standards must exit and be adhered to] where private companies can compete in a socially responsible manner [read: basic infrastructures like energy, water, waste, transportation, etc. must be provided for everyone]" (2004, p. 459).

2.3 WHAT IS A RENEWABLE ENERGY POWER SOURCE?

Renewable energy generation is part of being sustainable, one of those terms that everyone thinks they understand until forced to use it in conversation. Basically, it is a source of energy that is not carbon based and does not diminish; that is, it is the "gift that keeps on giving." For example, the sun is always shinning and the wind blows fairly consistently. Each needs some form of storage or feedback when the wind is not blowing or at nighttime when there is no sunshine. That is why these forms of energy generation are called *intermittent* and need technologies to provide for base load (around the clock power availability) energy generation.

The ocean is always there, with tides and water. The most common renewable energy sources are systems that use the wind, the sun, water, or a digestive process that changes waste into biomass and waste recycling for fuel generation. Other renewable sources include geothermal energy, "run of the river streams," and now, increasingly, bacteria and algae.

Wind generation is fairly straightforward and has been used as power sources for hundreds of years. A large propeller is placed in the path of the wind, the force of the wind turns it, a gear coupling interacts with a turbine and electricity is generated and captured. While ancient in form, there have been significant technological advances. The new generation of wind turbines are stronger, more efficient, quieter, and less expensive. Today, wind turbines are being installed in small communities and even smaller systems on rooftops as part of the natural flow of air over buildings.

Solar generation systems capture sunlight including ultraviolet radiation via solar cells (silicon). This process of passing sunlight through silicon creates a chemical reaction that generates a small amount of electricity. The process is described as a photovoltaic (PV) reaction and is at the core of solar panel systems. A second process uses sunlight to heat liquid (oil or water), which is converted to electricity. A number of communities are now looking into more and more "solar-concentrated" systems, where the sun is captured in heat tubes and used for heating and cooling. This is a great renewable technology for the use in water systems and buildings that have swimming pools.

Biomass is a remarkable chemical process that converts plant sugars (like corn) into gases (ethanol or methane), which are then burned or used to generate electricity. The process is referred to as *digestive* and it is not unlike an animal's digestive system. The ever-appealing feature of this process is that it can use abundant and seemingly unusable plant debris—rye grass, wood chips, weeds, grape sludge, almond hulls, and so forth.

Geothermal energy is power extracted from heat stored in the earth, which originates from the formation of the planet, radioactive decay of minerals, and solar energy absorbed at the surface. It has been used for space heating and bathing since ancient Roman times but is now better known for generating electricity. Worldwide, geothermal plants have the capacity to generate about 10 GW as of 2007 and, in practice, generate 0.3% of global electricity demand. In the last few years, engineers developed several remarkable devices, called *geothermal heat pumps*, *ground source heat pumps*, and *geo-exchangers*, that gather ground heat to provide heating for buildings in cold climates. Through a similar process, these devices can use ground sources for cooling buildings in hot climates. More and more communities with concentrations of buildings, like colleges, government centers, and shopping malls, are turning to geothermal systems.

Ocean and tidal waves generate power that was been pioneered by the French and the Irish with their revolutionary SeaGen tidal power system. The French have been generating power from the tides since 1966, and Electricité de France announced a large commercial scale tidal power system that will be big enough to generate 10 MW per year. America, particularly the Pacific coastline, is equally capable of producing massive amounts of energy with the right technology. Ocean power technologies vary, but the primary types are *wave power* conversion devices, which bob up and down with passing swells; *tidal power* devices, which use strong tidal variations to produce power; *ocean current* devices, which look like wind turbines and are placed below the water surface to take advantage of the power of ocean currents; and *ocean thermal energy conversion devices*, which extract energy from the differences in temperature between the ocean's shallow and deep waters.

Bacterial or microbial fuel cell energy generation, sounds too far out there, but Better Products (or British Petroleum) made a $500 million investment in the process, which is now being developed by researchers at the University of California, Berkeley, and the University of Illinois, Urbana. The process uses living, nonhazardous microbial fuel cell bacteria to generate electricity. Researchers envision small household power generators that look like aquariums but are filled with water and microscopic bacteria instead of fish. When the bacteria inside are fed, the power generator, referred to as a *bio-generator*, would produce electricity.

While all these power generation systems result in electricity, none is as cheap as current fossil fuels, such as coal, oil, and natural gas. Nor were fossil fuels cheap when they started in the late 1890s, forming the basis for the 2IR. To maximize the renewable power efficiency, renewable energies need to be integrated as linked or bundled supply sources according to the natural physical characteristics of the area where they exist. Further, these intermittent power generation resources are greatly enhanced by storage devices, since the sun is not always shining (especially at night) or the wind blowing.

Hence, there is a need for storage devices, either natural, like a salt formation, or artificial, like a battery—new, advanced batteries and fuel cell programs are now coming out in California and through the U.S. Department of Energy. Once the electricity is collected, the storage device allows regulating the distribution to optimize the process. The government support for the 2IR in terms of tax incentives, funds, and even land must be repeated for the 3IR. The incentives for the 2IR must be reduced and applied to the 3IR. This is called *tax shifting* and has been very successful in other areas, so that there is little or no additional tax on the consumers. There is no need for further debate or delay.

As people become familiar with the concepts and the fact that renewable energy technology is not as cheap today as fossil-fuel systems, Americans will begin to understand the economic move and change for local renewable energy generation and distribution of power (see Figure 2.4) from central grid power plants. The Industrial Revolution that developed central grid power plants was significant at the time for the coordination and costs for generating power supplies to communities. Nonetheless, this Second Industrial Revolution meant that the price of fossil fuels for power plants was reduced over time. At least three or four decades were needed to achieve that goal. Substantial evidence and the series of laws at the turn of the 20th century document how central power plants, along with fuel supplies, became economic monopolies that then controlled the fuel

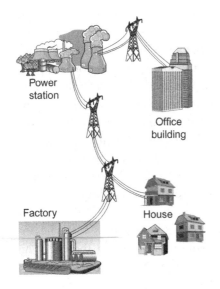

Energy Systems are the Changing Nature of Power Generation

Conventional electrical power grid (left)

Centralized power stations generate electricity and transmit via power lines, pipelines or other means over long distances to customers in their homes, and businesses in factories, offices and public buildings such as fire stations, city, county state city offices, court houses and others like hospitals, colleges, etc.

The New Agile Energy System (next slide)

Many small on-site generating facilities, including those based on renewable (green) energy sources such as wind, bio-mass and solar power, are efficiently utilized and coordinated using real-time monitoring, control systems along with power generation and storage.

Offices, hospitals, shopping malls, campuses, and other facilities generate their own power and sell the excess back to the grid. Electric and hydrogen-powered cars can act as generators when not in use. Energy storage technologies provide base load in supply from intermittent wind and solar power generation sources.

Agile "green" power generation provides energy security while it reduces transmission losses, operating costs and the environmental impact on global warming.

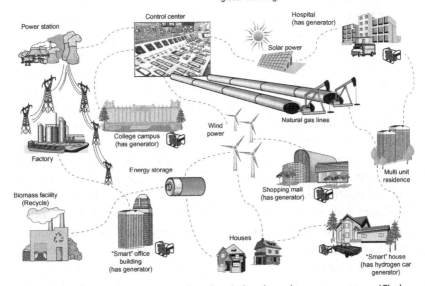

Figure 2.4 From central power grid to local distributed power systems (Clark and Eisenberg, 2008).

(primarily turning to fossil fuels such as coal, oil, and gas) supplies and hence large manufacturing and industrial markets. Despite litigation over the next four to five decades, these large fuel suppliers and power generators remain the global dominate economic business organizations.

While some fossil fuels, like coal, are still cheap today, they are the major American and global atmospheric polluters. If the human and environmental impacts of coal were calculated into its costs, then the real cost of coal energy generation for power would soar. The Third Industrial Revolution needs the same sort of economic, tax, and funding support or incentives that the Second Industrial Revolution received over a century ago. This is a key action point for all American communities, regions, and states. And these financial actions will enhance sustainability and reduce global warming as a result.

The result of the 2IR was the creation, operation, and maintenance of big centralized fossil-based power plants, as Figure 2.4 illustrates. They had to be powerful to withstand the degradation over the vast distribution of a central-powered grid system. At each conversion from AC to DC, electricity loses some of its capacity. However, there is so much of it at the beginning that it does not matter several thousand miles away at the end. This results in the loss of efficiency in transmission over power lines as well as the constant need for repairs and upgrades.

This is not so in the case of environmentally sound renewable systems. For the best results, they need local renewable power generation and distribution systems, "smart local and on-site grids" that do not travel far and lose none of the electricity to inefficiencies. The other way to do it is to hook into a transmission line. This way, the system is additive to existing energy distribution, so that the transmission line acts as a "battery" for the renewable energy that needs storage. Some have equated this to a model of the Internet, where there is no one area for control over data (or, in this case, power), rather it is spread out and localized.

Energy independence will not happen tomorrow, just like the SUV and the carbon-intensive economy did not become social and political realities overnight. America spent a trillion dollars on the Iraq war, and it will probably cost at least that much to turn America into the leader of the Third Industrial Revolution. However, national survival and international political leadership compel us to quickly surpass what has begun in parts of Europe, Japan, and China.

Fortunately, some in America are taking the first step. Consider California, where in the early part of the 21st century, the world's largest energy efficiency program was implemented. The state taxes the utilities' ratepayers and pushes that money back into making business and facilities more efficient. California is putting about $3 billion into the 2010–2012 energy efficiency cycle, with energy savings targets for the years 2012–2020 of over 4500 MW, the equivalent of nine major power plants.

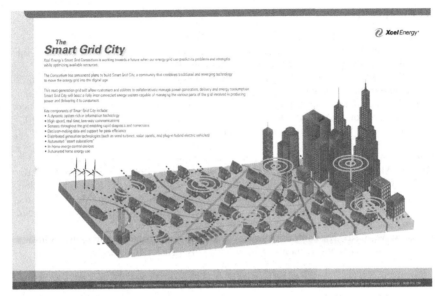

Figure 2.5 Smart grid that includes local distributed power and renewable energy generation (Source: Xcel Corporation, 2009).

New York City, which is struggling to hold onto its leadership in the financial world, is facing severe capacity issues, particularly in Manhattan. Taking a page from California, New York embarked on a similar state-policy-directed energy efficiency effort. Other states like Pennsylvania, New Jersey, Illinois, and Missouri are coming along. But the heavy coal burning states in the Midwest are in denial mode and refuse to give up burning coal, probably since the rancid and toxic residue is blowing east and not spoiling their own environments.

While energy efficiency is a first and important step, complete energy independence is within our technological grasp. A third generation of renewable technologies is coming and it is much better—lighter, thinner, stronger, and cheaper. Wind and solar power, coupled with highly efficient storage devices, smart grids, and local distribution systems are coming together (see Figure 2.5 for a graphic example). These independent power systems need to be integrated. What is lacking is the large national financing and political leadership to make the commitment and push America past the threshold into the 3IR. Sustainable development as a key component in the 3IR, like its predecessor, depends on this leadership and financial support.

REFERENCES

Clark II, W.W. (Ed.), 2009. Sustainable Communities. Springer Press, New York.
Clark II, W.W., Bradshaw, T., October 2004. Agile Energy Systems: Global Lessons from the California Energy Crisis. Elsevier Press, London, UK.
Clark II, W.W., Fast, M., 2008. Qualitative Economics: Toward a Science of Economics. Coxmoor Press, London, UK.
Clark II, W.W., Isherwood, W. (Ed.), 2010. Energy Instructions in the West: Lessons Learned for Inner Mongolia, PRC. (Asian Development Bank, 2007 Report). Utility Policy J.
The Economist, Special issue and cover on modern economic theory failing, July 17, 2009.
Hubbert, M. King, February 4 1949. Energy from Fossil Fuels. Science 109, 103–109.
Hubbert, M. King, (Chief Consultant, General Geology), June 1949. Nuclear Energy and the Fossil Fuels. Publication Number 95, Exploration and Production Research Division, Shell Development Company, Houston, TX. Presented at Spring Meeting of the Southern District, American Petroleum Institute, San Antonio, Texas, March 7–9, 1956.
Rifkin, J., 2006. The European Dream. Tarcher/Penguin, New York.
Woodrow W., Clark II., Larry Eisenberg, 2008. Agile Sustainable Communities: On-Site Renewable Energy Generation. Journal of Utility Policies 16(4), 262–274.

CHAPTER 3

Public Policy and Leadership
"We Have Met the Enemy and He Is Us"

Scott G. McNall

Contents

3.1	The Energy Efficiency Gap	26
3.2	People Do Not Make Rational Choices	27
3.3	Human Beings Want to Make Sense of the World	28
	Fear Causes People to Make Poor Decisions	29
	People Seek to Confirm What They Already Believe	29
3.4	People Can Worry About Only a Limited Set of Factors at Any Given Time	32
3.5	People Focus on Short-Term, Not Long-Term, Threats	32
	Most Consumption Behavior Results Not From Rational Deliberation but From Automatic Responses or Emotional Responses	33
	People Receive Conflicting Messages About Which Actions They Should Take	34
	People Need to Have Pathways to Efficiency Pointed Out to Them and They Need to Be Encouraged to Take Small Steps to Reduce Their Carbon Footprints	35
	People Follow Leaders	36
	Social Networks are Key to Change	37
	Information is More Readily Diffused if the Source is Regarded as Trustworthy and Reliable and is Factually Correct	38
	The Theory of Moral Hazards Derives from the Work of Behavioral Economists	38
	Our Use of Energy is Socially Constructed	39
	There is No Silver Bullet	41
	References	41

Walt Kelly, the creator of Pogo, first used the quote "We have met the enemy and he is us" on a poster for Earth Day in 1970. Pogo and Porky are portrayed in a swamp—their home—filled with trash. Pogo and Porky stand looking at the mess, and Pogo says to Porky, whom we do not see, "Yep, son, we have met the enemy and he is us." There may be no better quotation than this to begin a discussion about what we humans have done to our home on planet earth. The climate of the earth is warming because of increased concentrations of CO_2 in the atmosphere. In terms of human history the buildup started a relatively short time ago, with the beginning of the Industrial Revolution in the late 19th century. The literature on climate

Sustainable Communities Design Handbook
ISBN: 978-1-85617-804-4, DOI: 10.1016/B978-1-85617-804-4.00003-3

change is substantial, and there is substantial agreement that the impacts of rapid climate change will be devastating (Speth 2008).

A decade ago, Jason Shogren and Michael Toman (2000) noted that: "Having risen from relative obscurity as few as ten years ago, climate change now looms large among environmental policy issues. Its scope is global; the potential environmental and economic impacts are ubiquitous; the potential restrictions on human choices touch the most basic goals of people in all nations; and the sheer scope of the potential response—a shift away from using fossil fuels as the primary energy source in the modern economy—is daunting."

More recently (2009a), the Nobel-Prize-winning economist Paul Krugman noted that the planet is warming even faster than pessimists have expected. "The ice caps are shrinking, and arid zones are expanding at a terrifying rate. And, according to a number of recent studies, catastrophe—a rise in temperature so large as to be almost unthinkable—can no longer be considered a mere possibility. It is, instead, the most likely outcome if we continue along our present course."

The U.S. National Intelligence Council (2008) warned that we must prepare for a carbon-constrained world and shift away from the use of petroleum by developing fuels that do not use more energy to produce than they consume, consider building third generation nuclear power plants, developing new technologies so we can burn our substantial reserves of coal but still capture and sequester the carbon dioxide produced, and developing alternative energy courses such as wind and solar power. We must also, according to the National Intelligence Council, prepare to deal with the fact that only 1% of the world's water is drinkable. Energy, food, water, and climate change are the new horsemen of the apocalypse. How did this situation come about so quickly?

At the beginning of the Industrial Revolution, the concentration of carbon in the atmosphere was about 660 billion tons; it now stands at close to 880 billion and we add about 4.4 billion more tons each year. (We actually add more, but almost 40% is absorbed by the oceans and land vegetation.) Once carbon dioxide is added to the atmosphere it stays there for almost a century. The "safe" upper level is estimated to be around 935 billion tons (Holdren 2008). *Safe* does not mean good; it means the upper level at which scientists believe the planet's atmosphere is stable enough to sustain human life. But we are likely to reach this "safe" number in a decade (2020). Thus the urgency of those who say we must address these problems in the next ten years or it will be too late (Pearce 2007). The Kyoto Protocol of 2005, which was not signed by the United States, would have required us to begin immediately reducing greenhouse gases by up to 5.2% a year. The Waxman-Markley bill, which passed the House in 2009, and is still being debated in the Senate, would require a reduction of 17–20% in greenhouse gas emissions at 2005 levels by 2020 and an 83% reduction from 2005 levels by 2050. This is a huge challenge. No matter what we do, there is about a 2–3°C temperature rise in the pipeline, because of the length of time CO_2 stays in the atmosphere. If we do not act soon, we could be faced with increases of between 3.5–7.4°C by the end of century (Chandler 2009).

The solutions proposed for dealing with rapid climate change fall into two broad categories: *mitigation* and *adaptation*. As noted, even if we are able to roll back our release of carbon dioxide to 80% of 1990 levels by 2050 (and this would have to account for population growth and economic expansion[1]), the earth would still heat up by almost 2–3°C at the poles, resulting in the melting of polar icecaps and a rise in sea levels. Adaptation can take the form of building seawalls, as New Zealand is doing now, or moving the entire population of the Maldives before the country disappears under the waves of the Pacific.

Some writers (Dumanoski 2009) believe that humans have "not only pushed the Earth system well outside its normal operating range," they have caused such severe distortions in the earth's ecosystems and atmosphere that we now face conditions unlike anything we have seen in the past 200,000-year evolutionary history of humans; "or in a worse case, beyond anything encountered by our more distant ancestors over the past 5 million years." Dumanoski argues that we must, in fact, prepare survival strategies that include developing redundant economic and social systems, decoupled from a global economic system, so that localized groups can survive based on their own sustainable, and closed, economies.

Strategies for mitigating climate change focus on lowering the amount of carbon dioxide we send into the air. The list is long: use LED bulbs for light instead of incandescent or fluorescent bulbs, insulate homes and make them airtight, install more efficient cooling and heating systems,

---

[1] Many lay people do not understand that, when climate scientists speak of the need to lower current carbon emissions to reduce the level of carbon by 20% by 2020 or some other level, population and economic growth must be factored into the equation. This does not mean that we roll back population or eliminate all the roads, buildings, and other infrastructure built since 2005. It means that we must take the growth into account and still reduce the carbon levels by 80% by 2050. And, growing into the future, it means that we have to build homes that have no negative energy consequences or find ways to offset the increased carbon emissions.

paint roofs white, lighten the color of asphalt roadways, capture the sun's energy through the use of solar cells and thermal storage units, capture the wind, sequester carbon by growing algae from carbon dioxide emissions, and planting more trees and deep-rooted, drought-resistant grasses. Automobile companies are pushing the development of all-electric cars, although the manufacturing of lithium–ion batteries carries its own environmental hazards. There are other, more dramatic, ways to mitigate the crisis, but they require the reengineering of the planet. Geoengineering "solutions," whose full implications are not yet well understood, involve spraying sulfur aerosols into the atmosphere to cool down the earth, dumping limestone and iron ore into the ocean, shooting reflective materials into the sky, and suspending reflective mirrors in space, to prevent the sun's rays from reaching the earth (Fleming 2007; Holdren 2008).

3.1 THE ENERGY EFFICIENCY GAP

We noted that a 2–3°C warming already is in the pipeline and unless we act to reverse the level of carbon in the atmosphere, catastrophe is likely to follow. By 2020, we have to bring emissions down to 1990 levels. New emerging technologies (wave-generation machines) will take time to build and place on-site. Even if we start building nuclear power plants tomorrow, we would not be able to produce the clean energy we need in the next ten years. So, what can we do? We can stop using as much energy as we do. It is useful to remember that the United States, which constitutes only 5% of the world's population of 6+ billion people, each year consumes over 20% of the entire world's energy. We burn coal, petroleum, and natural gas, all of which contribute to global warming. What are we using all of this energy for? Thirty-one percent of *all electricity* generated in the United States is used to heat, cool, and ventilate often ridiculously large homes. More is used to power refrigerators, big-screen televisions, computers, microwaves, and nightlights (Mouwad and Galbraith 2009; U.S. Department of Energy 2009). Most of the oil we use (71%) goes to fuel cars and trucks. So, briefly then, we drive ourselves around and heat and cool ourselves using vast amounts of energy (coal and petroleum), which contribute to greenhouse gases, approximately 2.3 billion metric tons annually (Exxon Mobil 2009). If we used 10% less electricity in our homes, it would be the equivalent of taking 15 million cars off the road. We can, then, control our energy use and mitigate global warming by getting people to use less energy. We can and must develop programs based on what we know about why people do what they do (McMakin et al. 2002; Thompson 2002).

The energy efficiency gap refers to the fact that there are technologies that people do not use because they do not know about them or lack the information to act (Greenfield 2009; Koopmans and Velde 2001; Office of Technology Assessment 1993; Wilson and Dowlatabadi 2007).[2] Some are very simple: weather stripping doors and windows, adding more insulation in the attic, wrapping cold and hot water pipes, sealing all leaks in a home's duct system, installing high efficiency heating and cooling systems, and wrapping hot water tanks with bats of fiberglass. Some are more complicated or expensive. High efficiency doors and windows can provide significant savings on energy bills and often government tax incentives help offset the cost. The same is true with solar arrays. They are a good idea and save money; they are economically viable and have a relatively short payback period. There are, of course, economic constraints on what we can expect people to do to lower their carbon footprint. Nevertheless, we must do better in terms of energy consumption. We cannot meet the ambitious goals that would have us reduce our carbon emissions by 50% from 1990 levels by 2050 unless we get people to act. There are numerous campaigns by corporations, municipalities, universities, and colleges to get people to be wise stewards of scarce resources. How effective are these campaigns likely to be, given what we know about human behavior?

3.2 PEOPLE DO NOT MAKE RATIONAL CHOICES

Most people, most of the time, do *not* have sufficient information (about markets, financing, long-term consequences) to make rational choices about energy consumption (or to make rational choices about any number of issues; see Ariely 2008; MacCoun 2002; Perkins et al. 1983; Thaler and Sustein 2008). If I want to install energy-efficient doors and windows, a series of questions immediately follow: Who sells them? How much do they cost? How do they work? How will I know I am saving money? How long will it be before I start really saving money? Few of us know what the anticipated long-term price increases are going to be for local utilities and how that should be factored into decisions. If I want to put up a solar array, I want to know whether or not the technology is changing, and whether or not the price of the panels is expected to come down in the future. In

---

[2] Wilson and Dowlatabadi (2007) have done an exceptional job of exploring how different disciplines frame discussions of energy use and how, based on those frameworks, one would approach energy-reduction strategies.

the world of the Chicago School of Economics (see in particular the work of Becker 1993, and Lucas 2002) our "rational consumers" would carefully sort through options, costs, payback, technology, feasibility of the technology for their home or site, the benefits to their family or themselves, and enter into the marketplace prepared to negotiate the best possible deal. Our most recent example of why this just does not happen can be found in the subprime mortgage crisis. If we are to believe those who lost their homes or are now faced with payments they cannot make, they were not able to make rational decisions, because they did not understand the financing, they did not understand the real estate markets, and they usually did not understand the implications of their decisions. Rational actors exist only in the minds of those who believe we also have *rational markets*, in which information is shared and transparent, resulting in decisions that produce the greatest good for the greatest number of people (Evans 1997; Frank 2009; Levitt and Dubner 2005; Krugman 2009b). It is not happening.

The implications of this for climate change is that we cannot simply present information to people about energy choices, the problems associated with climate change, costs, amortization schedules, benefits to themselves or others, and expect the sheer volume of information to move them to action. "It's good for you and it's good for the planet" will not motivate people to act.

3.3 HUMAN BEINGS WANT TO MAKE SENSE OF THE WORLD

The sociologist, Max Weber (1978), argued that people seek explanations for the way in which the world works, particularly when something goes wrong. Consider almost any tragedy, personal or large scale. Personal tragedies often provoke questions about whether or not it was God's judgment. Others may seek to place blame or judge people. "My daughter died because the person who hit her was a drunk driver or a convicted child molester who should never have been released from prison." Immediately after the Twin Towers were struck in New York, explanations involving blame exploded across the airwaves.

Discussions about rapid climate change, given American's fondness for conspiracy theories, often involve claims that the idea of climate change must be a scheme to channel resources, wealth, and power to a small group of people (Dunlap et al. 2001). Science is, not for everyone, a sufficient explanation of how the world works, in part because they do not understand the science, but also because science is itself portrayed as a self-interested ideology on blogs and A.M. talk shows. Alternative ideas are more simple and sometimes compelling, such as the earth goes through cycles, it is caused by solar flares, and because it is cold in Afghanistan the globe cannot be warming. This suggests that, if change is to be effective, we must use the language used by, common to, and understood by specific audiences. If we want to enlist agriculturalists, we need to talk about stewardship, something understood by virtually every independent farmer. Many religious groups also believe that Man was supposed to take care of the earth and serve as a steward of God's creation. Many college students want to hear that the future belongs to them, and they need to step forward and help to implement change.

Fear Causes People to Make Poor Decisions

The military and police devote considerable training time to helping soldiers and officers develop "automatic" responses and overcome feelings of fear so they can do their jobs. Of course, their training also involves using their own judgment and taking the initiative, because situations change quickly and unpredictably. It takes a lot to overcome fear, and fear can often make people do the wrong thing (Polski 2009)..

Consider some of our approaches to public education campaigns. We may begin with an accounting of the coming apocalypse and go on to list the things that will happen if people do not act: They will have no water; they might not have food; they will run out of energy; and they will live in a world made by hand (Kunsler 2008). The animals we are fond of—polar bears, pandas, dolphins, Koala bears, the great apes—will be no more. The reality is that very few people are motivated to act by such campaigns, because they feel helpless in the face of this array of problems without one clear way forward. A sense of helplessness does not lead to action.

Fear also causes people to focus on how to solve a problem immediately, for themselves, their family, and the members of their social network (Dukas 2004). Fear can cause people to buy guns, ammunition, and retreat to compounds in isolated regions, rather than focusing on the long term. Climate change is a long-term problem. Many of the full effects sketched out will not happen for another 50 years, or longer. To get people to act, we need to show them what the benefits of action *are*, not what the long-term consequences are if they *do not* act.

People Seek to Confirm What They Already Believe

Once somebody has developed an explanation of a situation, they seek out information to confirm what they believe and simultaneously rejects mountains of information of solid evidence to the contrary (LePage 2009).

Leonard Mlodinow (*The Drunkard's Walk: How Randomness Rules Our Lives*, 2008) refers to this as *confirmation* bias. Let me take a personal example. Our local (Chico, California) weatherman believes that global warming is a hoax got up by scientists to benefit their own careers. When Europe suffered through a cold winter in 2008–2009 he took this as "proof" that the earth was not warming. He also believes, correctly, that solar flares can cause warming—but concludes, incorrectly, that any warming is due only to solar flares, not a build up of carbon dioxide in the atmosphere. He believes that all reports of global warming are due to measurement errors. He found, correctly, that many weather stations are in urban areas, where the reflected heat of roadways and parking lots drives the temperature up. Presumably, we would get the "truth" if weather stations reported from national parks. We know it is 110°F in downtown Phoenix, because the thermometer says so. He concludes, incorrectly, that there is no global warming, because he does not know that measures of global warming are not based on weather reports; they are based on readings at the earth's poles. A local citizen and letter writer denies the globe is rapidly warming and says that, even if it is, the benefits outweigh the disadvantages. The "evidence" he found is on a website that celebrates the possibility that one will be able to raise sheep in Greenland. Others go to the website for the Oregon Institute for Science and Medicine (www.oism.org/project), where they not only can have their beliefs confirmed but are invited to sign a petition that denies rapid climate change is a reality, based on "scientific" evidence provided by the institute. This petition has in turn been used, regardless of who signed it, to "prove" that "thousands of scientists believe global warming is a hoax."

Efforts to discredit science cannot be discounted as a factor in shaping public beliefs (Freudenberg et al. 2008; McCright, et al. 2008). P. J. Jacques and his colleagues (2008) have demonstrated through an analysis of conservative think tanks that "Skepticism is a tactic of an elite driven countermovement designed to combat environmentalism, and that the successful use of this tactic has contributed to the weakening of U.S. commitment to environmental protection." For many, the idea of rapid climate change is just another version of environmentalism, which is an ideology invented to challenge the foundation of our economic system (Antilla 2005).

The U.S. Chamber of Commerce argued in 2009 that the UN Climate conference that was to take place in Copenhagen, Denmark, in 2009 should be used, not to reduce carbon emissions, but to put the science of global warming on "trial." As a result of the chamber's position, two large energy providers (PG&E and the Public Service Co. of New

Mexico) cancelled their chamber memberships, noting that the chamber's position is inconsistent with that of the utilities, which spent billions on developing renewable sources of energy (Bryan and Williams 2009). Paul Krugman (2009c), in a *New York Times* Op-Ed piece wrote, "Cold, calculated lies await debate on global warming" and added that the attack on global warming is likely to take the form of arguing that it will cost consumers millions and countless jobs to lower carbon emissions.

The implications here are many. When we organize events to discuss the causes and consequences of rapid climate change, we are often speaking to the like-minded. How do we reach other audiences who do not share our understanding about the nature of the problem? Overcoming people's well-structured system of beliefs requires us to understand the reasoning and motivation behind their ideas. People do not believe in rapid climate change because

- They do not trust those delivering the message.
- They see rapid climate change as a form of radical environmentalism.
- They believe action will increase government power.
- They believe action will cost them more to heat and cool their homes.
- They believe others will benefit while they suffer.

For the West Coast logger without a job, the problem is just that: no job. These loggers blame environmentalists, "activist" judges, other countries (Canada in particular, since it ships milled wood to the United States), "rich" people who moved into the woods and do not want logging operations, the economy, and the spotted owl. Any discussion about rapid climate change has to take shape as a discussion about sustainability and how to strengthen rural communities and provide jobs through stewardship of the forests. Opportunities to increase the number of jobs by planting more trees, sequestering carbon, and managing our forests to reduce fire danger are all ways to connect to those who might share different views about the kind of problems we face.

An important and useful exercise for those leading sustainability efforts, whether in the community, businesses, colleges, or universities, is to identify the key issue or problem with which the target group or audience is concerned. For example, if one approaches community leaders and presses the topic of sustainability and climate change, a frequent response will be this: How does this help my town? What a business leader can hear and respond to is not a discussion about sustainability but a call to find out how to lower the cost of doing business. The information provided, whether in a workshop or some other forum, can then address the concrete business benefits of reducing energy use, going solar, or retrofitting buildings. The consequence, of course, is to reduce the carbon footprint.

3.4 PEOPLE CAN WORRY ABOUT ONLY A LIMITED SET OF FACTORS AT ANY GIVEN TIME

When people are asked to list the things that worry them, they focus on the basics: food, shelter, economic welfare, and health (Jacoby et al. 1974; Perkins et al. 1983). Sometimes, when we craft programs to reduce energy consumption, we ground them in the assumption that what people should really care about is the concentration of parts per million (cppm) of CO_2 in the atmosphere. In the face of a person's confusion or lack of attention, our response has been to redouble our efforts, by providing more information. *Providing more information is not going to be effective unless people have already made a decision to change their behavior.* So, the construction of "rich" websites offering details about global warming, energy regulations, destruction of ecosystems, or systems of incentives are usually useful only to those who are looking for a new hot water tank and have thought about "going solar." When people experience an economic environment that may cause them to lose their jobs, homes, and health insurance (if they have any), they are not ready to hear messages about global warming or the need to install a new energy efficient cooling system to reduce greenhouse gas emissions. What they might be prepared to hear about are issues that relate to food prices, economic welfare, and health. The sociologist C. Wright Mills (1959) noted that being unemployed is a *private trouble*, but that 15 million people being unemployed is a *public trouble*. In the case of climate change, it would mean that we need to demonstrate to people that we are in public trouble and the immediate quality of our lives is affected by climate change. Strong arguments can be made that, if we are to have a robust economy, we must shift to one in which "green" jobs are predominant, because the current economic models are not only causing people to lose their homes and resources, they will not work in the near future; rather they will beggar them, their children, and their children's children.

3.5 PEOPLE FOCUS ON SHORT-TERM, NOT LONG-TERM, THREATS

How do you get people to focus on something that might not happen until a century later (Bosettil et al. 2006; Dietz et al. 2007; Dietz 2008)?

Jane Goodall (2009) noted that we wrecked the planet because we focus on what benefits us now or tomorrow. The solution she says is to "think about future generations." One suggestion offered by others is that we should learn to think like the Iroquois, who said that every action needed to be considered in terms of what its effect would be on the seventh generation. Many in the sustainability movement explain their involvement by noting that they are helping to save the planet for their children or grandchildren. They are thinking in the long term.

Getting college students to think seven generations down the line is more difficult. What I personally find useful in a classroom is to ask students to characterize the world of their grandparents (or some other elderly person with whom they are or were close) and, having done that, to characterize the kind of world in which they wish to live in 25 or 50 years from now. I find it perfectly interesting that the world of their grandparents and its values (a good day's pay for a hard day's work, honesty, community, family) is almost the same as the world in which they want to live. The next questions I ask are, "And how are you going to create that world? What will you do as an individual and what will you do as a member of society to make that happen?" In short, we need to personalize discussions about climate change so that the individual can connect short-term behaviors to long-term goals. And, we need to connect to the way in which people are actually making decisions.

Most Consumption Behavior Results Not From Rational Deliberation but From Automatic Responses or Emotional Responses

Most of you reading this will have shopped at Costco, Wal-Mart, or an equivalent. As you enter Costco you are faced with literal mounds of bounty—boxed televisions stacked in large numbers, plasma screen TVs of various sizes tuned to people's favorite programs or sports channels, mounds of clothing, bedding, shoes, toothpaste, pots and pans, chairs, sofas, mattresses—things you never knew you needed. You might decide you need a new laptop computer and the price is right so you take one home. The production of that laptop computer created almost 5000 pounds of waste. That was not what was on your mind when you bought it (Hoffman et al. 2007).

Because so many of our decisions are based on similar impulses, we need to tap into our feelings in designing programs designed to reduce energy consumption (Wilson and Dowlatabadi). Focusing on family comfort is a better strategy than focusing on energy savings (Dietz 2008). Television and print ads from virtually all the major manufacturers of heating and cooling systems focus on "instant" comfort. The manufacturer of a 110 W portable heater disguises it as a portable fireplace and pictures the happy family gathered around its glow. (It also saves energy if you heat only one room in your home and leave the others without heat though that is not mentioned as a selling point.)

As Thaler and Sunstein (2008) note in their work, *Nudge*, we are creatures of habit as well as emotion. We tend to do something one way because we have always done it that way. One of the many ideas they have for getting us to be better planetary stewards is to change the default setting on thermostats, washing machines, and other electrical appliances so that when we turn them on we automatically use less power. Technology, if we choose, can help us make right decisions. As the new "smart" power strips become available, they will lower power use, because we will not have to crawl under our desks to turn off the power strip; it will go off automatically when we are not using our computers and printers. Imagine buying a new programmable thermostat for your home that automatically sets your heating at 68°F, your cooling at 76°F, and when no motion was detected in your home, sets it to prevent freezing, 55°F, or overheating, 85°F. The fewer decisions we have to make the more we are likely to save.

People Receive Conflicting Messages About Which Actions They Should Take

We are urged to buy new cell phones, updated computers, desktops, laptops, Blu-ray players, high-definition television sets, all of which increase power consumption in the home. At the same time, we receive messages that we need to lower our carbon footprints. We are told to respect Mother Nature and at the same time provided with zero-interest financing and dealer discounts to buy SUV's with large carbon footprints to escape to the "wilderness," with all of our toys.

Conflicting messages about consumption and saving energy are ubiquitous (Amting et al. 2009). Yet, we are no more clear about which technologies are actually the most efficient and cost effective when it comes to energy use. It is difficult for the average homeowner to find out what the cheapest way is to heat water for his or her home—electric, gas, on-demand, solar. Of course, many variables—number of family members, number of teenage children—need to be taken into account in making an informed decision. It is hard to find well-informed vendors who can install what we need. The problem of heating and cooling a home is even more complicated, considering the available technologies from which to choose—high efficiency gas or electric furnaces, passive, solar, or geothermal. Nevertheless, entire villages in China use the sun to heat water for their homes, and Germany demands that all new homes be zero-energy homes by 2012. The technologies to reduce greenhouse gases are here. We need to be clear and concise about the costs and benefits of the alternatives, and we need clear guideposts on the path to action.

People Need to have Pathways to Efficiency Pointed Out to Them and They Need to Be Encouraged to Take Small Steps to Reduce Their Carbon Footprints

Let us assume that people know the polar icecaps are melting, ocean levels are rising, and this will cause millions of people in low-lying areas to be driven inland from their homes; let us also assume that people know the price of all natural resources will increase, and it will get more and more expensive to commute from a residential suburb. If we would expect them to be stirred to action on the basis of this knowledge, we would be wrong (Mulder 2005; Wilson and Dowlatabadi 2007).

Decision making has several steps (Newman 1980). The first requires recognition that there is an actual problem (and remember not everyone believes rapid climate change *is* a problem) then definition of a goal or goals. Another step is to identify *feasible* and meaningful alternatives, weigh the differences among the alternatives, and choose the best one available. That is informed decision making but, as you will have already intuited, it is not always the path chosen, because emotions and habits can get in the way. Public information campaigns, then, must do more than raise awareness; they need to define alternatives and explain why the alternative chosen is the best one. In the debates about how we should lower our carbon footprint in this country, we have two broadly defined alternatives—cap carbon emissions and allow trading of credits or a simple tax on carbon consumption. One is the better alternative (see later) but public debates have not yet clearly laid out the differences between the two alternatives.

As people move toward action and an agreed-upon goal (saving on your energy bill), clear benefits need to be defined at each stage. If I get rid of my old electric hot-water heater and replace it with an on-demand water heater (assuming this is a viable alternative), I need to see real and meaningful savings on my energy bill. I need to know something as basic as this: I have more money to spend because of my decision. To get people to take the first step, some municipalities are providing the funding for solar heating systems, efficient heating and cooling systems, and adding the payback to the property tax, which gets paid off much like a home loan (Downing 2009). This has the benefit of keeping the savings with the house, and it also has the benefit of getting rid of a major barrier—cost.

Also clear feedback loops need to be created to help reinforce the behavior desired. It is a basic premise of learning theory that every parent and K–12 teacher knows: *Positive feedback is essential to learning.* One way in which we can encourage people is to let them know how other people in their neighborhood are doing. It has proven effective to let people know how others in their neighborhood are doing in terms of consuming energy, as are reminders from your utility company about ways to reduce energy consumption. If you are in a large organization, it is important to receive feedback on what it means to turn off power strips at the end of the day, to install new software on computers that put them into a "sleep" mode when not in use, what the savings are now that you have gone to "cloud" computing or the use of virtual servers. In short, people need to have *information that is meaningful to them.*

Barriers to adoption, as well as steps to success, must be identified (Bontempo 2008). A barrier to putting up solar panels to provide power for one's home is the cost. Few homeowners can assume the up-front costs of installation without taking out a loan or getting a second mortgage. Recognizing this barrier and overcoming it through new means of financing will advance the solar industry. Likewise, it will be important to identify other barriers: lack of knowledge, lack of tradespeople to install energy-efficient systems, or a community environment in which one's friends and neighbors do not support change.

People Follow Leaders

This is not an argument for the "great man" theory of history; it is to underscore an important finding from the social sciences (Flaum 2003; Smith 2009; Tyler 2002; Valikangas and Okumura 1997). Social feedback is critical in shaping our behavior. Drawing on the work of classical sociologists (Durkheim, Mead), Perkins and Berkowitz (1986) introduced *social norms theory* as a way to help understand drinking behavior on college campuses. Social norms are what the majority of people in a group, whether a college, club, community, or town, do and think is okay. Norms guide our behavior; we want to be like other people. In the case of alcohol abuse, Perkins and Berkowitz developed campaigns organized around what students actually do—most students, most of the time, are not binge drinkers. The real norm was responsible drinking behavior not binge drinking.

Using the same kind of logic, campaigns to reduce greenhouse gas emissions need to provide information to members of a community about what the norms really are. It is important to add that norms do not just refer to what people do but to what people are expected to do. Here is where early adopters and key community leaders are critical.

We want to know that *leaders* in the community who are committed to reducing greenhouse gas emissions are driving hybrid cars, putting up solar panels, building energy-efficient homes, and retrofitting older ones to be energy efficient. We need *real identifiable* people, who are well known and respected, doing real things from which others can learn. Demonstration projects by community colleges, universities, municipalities, and "green" builders are key components in changing behavior.

Social Networks are Key to Change

In his famous 1960s small-world experiment, the psychologist Stanley Milgram (1967) asked some people in Omaha, Nebraska, to send a package to a stranger in Boston via somebody they knew. Without using the mail service, on average, it took six people to get the letter to its destination. The experiment was repeated in 2002, using email with the same result; we are a country of dense networks with only *six degrees of separation* between us. More recently, Christakis and Folwer (2009) found that, for some behaviors, there are only *three degrees of separation* between us. Our moods, whether we are thin or fat, happy or depressed, even whether or not we vote, are contagious up to three degrees of separation. Social networks do have "minds" of their own; they are the result of the collective, not just based on the individual. In terms of health, Christakis and Folwer note that campaigns to reduce obesity should focus on the hub of the network and not the individual. So, for campaigns to reduce energy, we would need to use the same approach. Identify social networks, then focus on the key actors in those networks to effect change (Busken and Yamaguchi 1999; Haythornthwaite 1996; Jung 2009; Wasserman and Faust 2004). For instance, the California Farm Bureau is a dense network of major agriculturalists in the state. If a goal were getting farmers to switch from diesel powered generators used to pump irrigation water to solar pumps, we would focus on the entire network. But we would also want to have key members of the organization involved in demonstration projects. (This is precisely the strategy being employed and is a familiar one to agricultural extension agents across the country—use dense networks and seek out the support of key members of the networks to help initiate change.)

Information is More Readily Diffused if the Source is Regarded as Trustworthy and Reliable and is Factually Correct

Local campaigns for the elimination of plastic bags have often argued that the plastic bags blowing in the wind represent a dependence on foreign oil, and if we stop using them, we will not only move to energy independence but lower greenhouse gas emissions. Most plastic bags are made of natural gas, so if we stop making them, that will have little impact on greenhouse gas emissions. They are also mostly recyclable. Paper is not a good trade-off either, because a paper bag takes about four times as much energy to produce as a plastic bag. The point is that campaigns can founder if the information on which the campaign is built is weak or wrong.

So, who do we trust? Unfortunately, many organizations are involved in "greenwashing," which refers to the fact that a company is spending more time advertising and marketing "greenness" than in engaging in business practices that would minimize environmental impact (EnviroMedia 2009). Most of us have checked into a "green" hotel that allows us to reuse our towels and keep the sheets on the bed for another day. Perhaps, it even provides a blue bin for our newspaper. Does the hotel ask us to save water, or on the social side of the sustainability equation, does it pay a living wage to the employees who clean their rooms or bus their tables? Virtually all major oil companies tout the fact that they are "green," because they invest in renewable energy sources. A closer look at the portion of resources devoted to the development of renewable energy reveals that it is a small percentage of their total investment effort.

Universities and colleges are uniquely positioned to provide "neutral" information about sustainability and rapid climate change, and they need to provide this information without prejudice. Detailed factual information about rapid climate change is not, however, as effective as information that is simple, understandable, and personally relevant. Making climate change personal is a key to the development of successful programs.

The Theory of Moral Hazards Derives from the Work of Behavioral Economists.

It suggests that people behave differently if no risk is associated with their actions. The recent financial crisis is frequently offered as an example. People behaved recklessly and maximized short-term gains because there were no negative consequences. This is like the gambler who goes to the casino with somebody else's money and an unlimited line of credit. With the odds on the side of the house, he or she will eventually lose all of somebody else's money. Many of our common resources are being destroyed in the name of market rationality.

Garrett Hardin's (1968) famous essay on the *tragedy of the commons* explains how multiple individuals, each behaving rationally and consulting his or her self-interest, will ultimately destroy a common resource. The herdsman who grazes cattle on the village green will keep cows there as long as possible, and so will his or her neighbor. The quick result is that nobody has any pasturage left. Fishing fleets decimated ocean fisheries because they acted on the basis of individual self-interest. Such behaviors have given rise to government programs to manage goods owned collectively, so that fisheries can rebound and forests can regrow. Rapid climate change truly presents a problem about how to manage the commons, and the commons includes the entire planet: its water, air, forests, and plains.

There needs to be a direct relationship between people's actions and the consequences. A challenge is that energy is invisible. We flip a switch and the lights go on or the house cools down. We do not normally think about the oil pumped in the Niger delta or the social, economic, and environmental consequences of turning on the switch. This is one reason why it is argued that a direct tax on carbon is needed, instead of a cap and trade system, which disguises the consequences of our actions. A direct tax on carbon would translate into a price increase at the pump, which in turn would, among other things, spur the development of fuel-efficient cars. Gasoline at $10 a gallon is normal in the European Union, not because oil costs more there than here but because EU governments have chosen to use the revenue from that tax to reduce their dependence on oil and instead encourage the development of renewable energy, the development of energy-efficient homes, and the infrastructure for public transportation. The political will and history does not exist at this time for such a tax in the United States. We did not experience, for example, the postwar rationing of goods nor the economic hardships that befell the Western European powers.

Our Use of Energy is Socially Constructed

How much energy we use depends on social and cultural norms (Hannigan 2006; McKibben 1989, 2007; Rosa et al. 2004; Stehr and Storch 2009; York et al. 2003). Energy use is embedded in cultural practices, style-of-life choices, and the social context. Consider our homes. In the 19th century, large homes were built, but most people lived in houses of between 600 and 800 ft^2. In 1950, the average home had grown to 1000 ft^2, and by 2000 to 2000 ft^2 (Ward 1999). Since 2000, they crept up another 500 ft^2, and now total close to 2500 ft^2. What happened? First, people wanted more room, and they wanted to separate functions, like visiting, sleeping, cooking, and bathing. For centuries, humans did all these things in the same space and included the animals for warmth during the winter. Many people in the world still live this way. Today, in the United States, people define a house as a place where there are separate sleeping rooms, master suites, walk-in closets, a separate entertainment space, separate places to bathe and cook, and so forth. People also rent storage sheds for the overflow.

Technology helped make such behemoths possible. Just heating and cooling our houses accounts for 50% of all home-energy use. Instead of passive heating and cooling, we have heating and cooling systems that use fossil fuels. Instead of overhanging roofs and porches, we have walls of glass that look out on a backyard "vista." Instead of low ceilings, we build high ceilings, and we do so because large open spaces fit with our cultural definitions of what a living space should be. What will get people to want less? Interestingly enough, price does not seem to be a driver of behavior. (It might be if the price shocks are sufficient.) The California utility, PG&E, will go to real-time pricing in 2012, with the assumption that this will drive down energy use. Studies have shown, though, that, even when the cost of peak pricing is eight times higher, price accounts for only 11% of the shift in consumption behavior. We want to wash our clothes when it is convenient, and we want to cool down our houses when we get home at the end of the day.

We demand energy based on our concepts of cleanliness and comfort (Liu et al. 2003). We do not wear the same garments, unless we have to, more than one day. We bathe once a day or oftener. We wash, vacuum, clean, heat, and cool based on our cultural norms. We drive our children to school so they will be safe; we buy a large vehicle to transport them in for the same reason. We buy all-terrain vehicles, ski mobiles, jet skis, jet boats, and campers because that is our idea of fun. The reality is that we U.S. citizens use up 20% of the entire world's energy every year, because of our cultural norms.

Much of our energy use is embedded this way; that is, it is locked up in our habits relating to child care, cooking, cooling, and lighting. We do things this way, because we always have. We do not think about it, *until we get the bill.* (And, we do not see the bill for the damage done to the planet.) A simple solution at the household level is a real-time meter that shows how much energy is being used at any given moment and what the cost is; also, new meters are coming to market that will allow us to determine which of the appliances in our home are drawing power and what they cost. And there are the simple globes, connected to our meters, that glow green or red depending on our energy consumption.

There is No Silver Bullet

There are no one-shot solutions to problems of rapid climate change and energy use. Kermit, of *Sesame Street*, used to say, "It's not easy being green." But, based on what we know about how human beings behave and how we are motivated, there *are* solutions (Tennant 2009). We should be mindful of the fact that campaigns to reduce energy use and limit our impact on the planet need to be multidimensional and should include a focus on those issues that are

- **Meaningful to people**. People connect to issues that affect the health and well-being of their friends, families, children, and grandchildren.
- **Immediate and real**. Rising energy costs are a real and immediate issue.
- **Cost effective**. Benefits (savings, rebates, and tax incentives) need to be associated with actions.
- **Actionable**. People need to be able to act and need to receive positive feedback for doing the right thing. Acting together helps.
- **Relevant to the social networks to which people belong**. The language used to describe problems and the solutions needs to be grounded in the norms of the groups.
- **Positive**. A focus on solutions, as opposed to a focus on problems, is more effective; people need to be able to act on the information provided.
- **Simple**. The complexity of sustainability is a given but people need to act on those things they understand well and that have an immediate payoff.

All the things we want people to do to reduce energy consumption or the consumption of any natural resource, relate to quality-of-life issues, how people define what a community is, what a civilization is, what it means to be a responsible citizen. Answers to these questions have been driven in the last decades by mass marketing campaigns that encourage us to consume more, right now, and without regard for others. Changing will take concerted and focused effort. As Pogo said, "We are the enemy." We are also the solution.

REFERENCES

Amting, J.M., et al., 2009. Getting mixed messages: the impact of conflicting social signals on the brain's target emotional response. NeuroImage 47 (4), 1950–1959.

Antilla, L., 2005. Climate of skepticism: U.S. newspaper coverage of the science of climate change. Glob. Environ. Change 15, 338–352.

Ariely, D., 2008. Predictably Irrational: The Hidden Forces That Shape Our Decisions. Harper Collins, New York.

Becker, G., 1993. Human Capital, third ed. University of Chicago Press, Chicago.

Bontempo, B.D., 2008. Public perception of climate change voluntary mitigation and barriers to behavior change. Am. J. Prev. Med. 35 (5), 479–487.

Bosettil, V., Galeotti, M., Lanza, A., 2006. How consistent are alternative short-term climate policies with long-term goals? Climate Policy (Earthscan) 6 (3), 295–312.

Bryan, S.M., Williams, M., September 28, 2009. Second utility breaks with chamber over warming. Sacramento Bee.

Buskens, V., Yamaguchi, K., 1999. A new model for information diffusion in heterogeneous social networks. Sociol. Methodol. 29 (1), 281–326.

Chandler, M.A., 2009. Mid-Pilocene warming. In: Gornitz, V. (Ed.), Encyclopedia of Paleoclimatology and Ancient Environments. Springer, New York, pp. 566–568.

Christakis, N., Fowler, J., 2009. Connected: The Surprising Power of Our Social Networks and How They Shape Our Lives. Little, Brown, Boston.

Dietz, T., 2008. Environmentally efficient well-being: rethinking sustainability as the relationship between human well-being and environmental impacts. Human Ecol. Rev. 16, 113–122.

Dietz, T., Rosa, E.A., York, R., 2007. Driving the human ecological footprint. Frontiers Ecol. Environ. 5, 13–18.

Downing, J., September 29, 2009. Energy-saving incentives gain. Sacramento Bee B8.

Dukas, R., 2004. Causes and consequences of limited attention. Brain Evol. 63 (4), 197–210.

Dumanoski, D., 2009. The End of the Long Summer: Why We Must Remake Our Civilization to Survive on a Volatile Earth. Crown, New York, pp. 31-32.

Dunlap, R.E., Xiao, C., McCright, A.M., 2001. Politics and environment in America: partisan and ideological cleavages in public support for environmentalism. Env. Polit. 10, 23–48.

EnviroMedia Greenwashing Index, by EnviroMedia Social Marketing and the University of Oregon, 2009, www.greenwashingindex.com/what.php/.

Evans, J.St.B.T., 1997. Are people rational? Yes, no, and sometimes. Psychologist 10 (9), 403–406.

ExxonMobil. Saving Energy and Reducing Greenhouse Gas Emissions. www.exxonmobil.com/.

Flaum, S.A., 2003. When ideas lead, people follow. Leader to Leader 30, 7–12.

Fleming, J., 2007. The climate engineers. Wilson Q. 31 (2), 46–60.

Frank, R.H., September 13, 2009. Flaw in free markets: humans. NY Times B4.

Freudenburg, W.R., Gramling, R., Davidson, D.J., 2008. Scientific certainty argumentation methods (SCAMS): science and the politics of doubt. Sociol. Inq. 78, 2–38.

Goodall, J., September 19, 2009. Big thinkers, big ideas. New Sci. 35.

Greenfield, D., 2009. Energy management: first steps toward greater efficiency. Control Eng. 56 (1), 32–35.

Hannigan, J., 2006. Environmental Sociology: A Constructivist Perspective, second ed. Routledge, New York.

Hardin, G., December 1968. The tragedy of the commons. Science 162 (3859), 1243–1248.

Haythornthwaite, C., 1996. Social network analysis: an approach to the technique for the study of information exchange. Libr. Inf. Sci. Res. 18 (4), 323–343.

Hoffman, W., Rauch, W., Gawronski, B., 2007. And deplete us not into temptation: automatic attitudes, dietary restraint, and self-regulatory resources as determinants of eating behavior. J. Exp. Soc. Psychol. 43 (3), 497–504.

Holdren, J.P., 2008. Meeting the Climate Change Challenge. National Council for Science and the Environment, Washington, DC, p.15.

Jacoby, J., Speller, D.E., Kohn, C.A., 1974. Brand choice behavior as a function of information load. J. Mark. Res. 11 (1), 63–69.

Jacques, P.J., Dunlap, R.E., Freeman, M., June 2008. The organization of denial: conservative think tanks and environmental skepticism. Env. Polit. 17 (3), 349–385.

Jung, J.J., 2009. Trustworthy knowledge diffusion model based on risk discovery in peer-to-peer networks. Expert Syst. Appl. 36 (3), 7123–7128. Part II.

Koopmans, C.C., Velde, D.W., 2001. Bridging the energy efficient gap: using bottom-up information in a top-down energy demand model. Energy Econ. 23 (1), 57–74.

Krugman, P., June 28, 2009a. Betraying the planet. NY Times. Op-Ed Page.

Krugman, P., September 6, 2009b. How did economists get it so wrong? NY Times Mag. 36–43.

Krugman, P., September 27, 2009c. Cold, calculated lies await debate on global warming. NY Times. Op-Ed Page.

Kunsler, J.H., 2008. A World Made by Hand. Atlantic Monthly Press, New York.

LePage, M., September 19, 2009. Get real. New Sci. 31.

Levitt, S.D., Dubner, S.J., 2005. Freakonomics: A Rogue Economist Explores the Hidden Side of Everything. HarperCollins, New York.

Liu, J., Daily, G.C., Ehrlich, P.R., Luck, G.W., 2003. Effects of household dynamics on resource consumption and biodiversity. Nature 421, 530–533.

Lucas, R.E., 2002. Lectures on Economic Growth. Harvard University Press, Cambridge, MA.

MacCoun, R.J., 2002. Comparing micro and macro rationality. In: Gowda, R., Fox, J.C. (Eds.), Judgments, Decisions, and Public Policy. Cambridge University Press, New York, pp. 116–137.

McCright, A.M., Shwom, R.L., 2008. Defeating kyoto: the conservative movement's impact on U.S. climate change policy. Soc. Probl. 50 (3), 348–373.

McKibben, B., 1989. The End of Nature. Random House, New York.

McKibben, B., 2007. Deep Econony: The Wealth of Communities and the Durable Future. Henry Holt, New York.

McMakin, A.H., Malone, E.L., Lundgren, R.E., 2002. Motivating residents to conserve energy without financial incentives. Environ. Behav. 34 (6), 848–864.

Milgram, S., 1967. The small world problem. Psychol. Today 2, 60–67.

Mills, C.W., 1959. The Sociological Imagination. Oxford University Press, New York, pp. 8–9.

Mlodinow, L., 2008. The Drunkard's Walk: How Randomness Rules Our Lives. Random House, New York.

Mouwad, J., Galbraith, K., September 20, 2009. Plugged-in age feeds a hunger for electricity. NY Times. A 1, 26.

Mulder, P., 2005. The Economics of Technology Diffusion and Energy Efficiency. Edward Elgar Publishing, Cheltenham, UK.

National Intelligence Council, 2008. Global Trends 2025: A Transformed World. U.S. Government Printing Office, Washington, DC.

Newman, D.G., 1980. Engineering Economic Analysis, second ed. Engineering Press, San Jose, CA.

Office of Technology Assessment, U.S. Congress, 1993. Energy Efficiency: Challenges and Opportunities for Electric Utilities. Government Printing Office, Washington, DC.

Pearce, F., 2007. With, Speed and Violence: Why Scientists Fear Tipping Points in Climate Change. Beacon Press, Boston, p.242.

Perkins, D.N., Allen, R., Hafner, J., 1983. Difficulties in everyday reasoning. In: Maxwell, W. (Ed.), Thinking: The Expanding Frontier. Franklin Institute Press, Philadelphia.

Perkins, H.W., Berkowitz, A.D., 1986. Perceiving the community norms of alcohol use among students: some research implications for campus alcohol education programming. Int. J. Addict. 21, 961–976.

Polski, M.M., 2009. Wired for Survival: The Rational (and the Irrational) Choices We Make, from the Gas Pump to the Terrorism. FT Press, Upper Saddle River, NJ.

Rosa, E., York, R., Dietz, T., 2004. Tracking the drivers of ecological impacts. AMBIO: A J. Hum. Environ. 33, 509–512.

Shogren, J., Toman, M., 2000. Climate Change Policy. Resources for the Future, Washington, DC, p.4.

Smith, K. The wisdom of crowds. Nature Reports Climate Change, online July 30, 2009.

Speth, J.G., 2008. The Bridge at the Edge of the World: Capitalism, the Environment, and Crossing from Crisis to Sustainability. Yale University Press, New Haven, CT.

Stehr, N., von Storch, H., 2009. Climate and Society: Climate as Resources, Climate as Risk. Scientific Publishing Co., Hackensack, New Jersey.

Tennant, D., 2009. Changing behavior. Computerworld 43 (14), 4.

Thaler, R.H., Sunstein, C.R., 2008. Nudge: Improving Decisions about Health, Wealth, and Happiness. Yale University Press, New Haven, CT.

Thompson, P.B., 2002. Consumer theory, home production, and energy efficiency. Contemp. Econ. Policy 20 (1), 50–60.

Tyler, T.R., 2002. Leadership and cooperation in groups. Am. Behav. Sci. 45 (5), 7769–7782.

U.S. Department of Energy, 2009. Residential Energy Use. www.energy.gov

Valikangas, L., Okumura, A., 1997. Why do people follow leaders? A study of a U.S. and Japanese change program. Leadersh. Q. 8 (3), 313–338.

Ward, P.W., 1999. A History of Domestic Space. University of Vancouver Press, Vancouver, B.C.

Wasserman, S., Faust, K., 2004. Social Network Analysis: Methods and Applications. Cambridge University Press, Cambridge, UK.

Weber, M., 1978. The sociology of religion (E. Fischoff, Trans.). In: Roth, G., Wittich, C. (Eds.), Economy and Society. University of California Press, Berkeley.

Wilson, C., Dowlatabadi, H., 2007. Models of decision making and residential energy use. Annual Review of Environmental Resources 32, 169–203.

York, R., Rosa, E.A., Dietz, T., 2003. Footprints on the earth: the environmental consequences of modernity. Am. Sociol. Rev. 68, 279–300.

CHAPTER 4

Achieving More with Less: The State of Energy Conservation and Energy Efficiency

Michael F. Hoexter, Ph.D

Contents

4.1	Introduction	45
4.2	Electrical Energy Efficiency: Generating "Negawatts"	49
4.3	Utility Revenue Decoupling and Energy Efficiency	50
4.4	Green Design: Guiding Natural Energy Flows	51
4.5	Near-zero, Net-Zero, and Plus-Energy Buildings	53
4.6	Electricity and Energy Efficiency Retrofits of Existing Buildings	55
4.7	Efficient Lighting: Efficient Fluorescents, Induction Lighting, and LEDs	56
4.8	Heat Pumps: Ground Source and BTES Linked	56
4.9	Key Technologies for More Energy Efficient, Eventually Carbon-Neutral Living	58
4.10	Quality Assurance and Certification in Energy Efficiency	59
4.11	Energy Efficiency in Transport: Short-Term and Long-Term Solutions	59
	Short-Term Solutions	60
	Longer-Term Measure: Shifting to Electric Drive	61
4.12	Price Signals and Energy Efficiency	62
4.13	Conclusion: Energy Efficiency in the United States Today	62
References		62

4.1 INTRODUCTION

Increased energy conservation and energy efficiency are critical elements of any future sustainable society. Energy conservation is both an ethic that inspires the wise use of energy in general and a description of a specific group of strategies to save energy. Conservation, as a strategy, means reducing energy use by voluntary human effort to save energy (Figure 4.1 shows Gifford Pinchot, an early conservationist); by contrast, energy efficiency is a strategy of saving energy by means of installing devices that use less energy to do the same amount of useful work. With historically inexpensive energy and a cultural and political attraction to excess, the United States has lagged in its efforts to conserve energy, with the notable exception of the state of California. Raising the price of energy, mandating energy efficiency, and incentivizing conservation

Sustainable Communities Design Handbook
ISBN: 978-1-85617-804-4, DOI: 10.1016/B978-1-85617-804-4.00004-5
45

Figure 4.1 Gifford Pinchot (1865–1946) was one of the founders of the movement toward conservation of natural resources. Observing the destruction that the family lumber business had wrought on America's forests, Pinchot's father encouraged him to study forestry in Europe. Coining the term *the conservation ethic*, Pinchot was the first leader of the U.S. Forest Service, appointed by Theodore Roosevelt.

and energy efficiency are all effective strategies to increase energy efficiency. The economics of energy efficiency in the electrical sector is further improved by the sharing of costs between beneficiaries of more efficient energy end use devices through rebate programs and other incentives. Multiple technologies and strategies are now available that can deliver more utility for less energy expenditure. In the area of building design, two main methods save the most energy: utilizing naturally occurring energy to heat, cool, ventilate, and light buildings and routing and controlling those flows using technologies like insulation, heat pumps, and heat exchangers. Near-zero energy buildings can be achieved by applying the tactics of, among others, passive house technology, but to achieve net-zero or "plus-energy" houses requires the addition of local electricity generation like photovoltaic (PV) panels or wind turbines. Electric heat pumps offer a means to cool, heat, and provide hot water to more conventional houses efficiently without fossil fuels on site. A list of recommended building efficiency measures is provided. In transportation, the largest efficiency gains are achieved through the transition to electric drive technology. With or without electric drive, reducing vehicle weight through the use of advanced materials is also a promising method for saving energy.

A key challenge along the path to sustainability is using energy more wisely. Over the past 20 years, we have discovered that the resource that may be in shortest supply is the ability of the atmosphere and the oceans to absorb our carbon emissions, many of which originate in the use of fossil energy, largely in the developed and now the rapidly developing world. The United States and Canada are the two most energy inefficient large nations in the world, with relatively cheap energy prices, political cultures that are heavily influenced by the fossil fuel industries, and a lax attitude toward wasting energy resources.

Most analysts acknowledge that the least expensive and most rapid way to supply the first several "tranches" of carbon emissions reduction and take the first steps toward energy independence is to avoid having to generate as much electricity, use as much natural gas, or drill for as much fossil fuel for transportation in the first place (McKinsey & Company 2009). Furthermore, if our economy becomes more efficient and conserving of energy resources, far less new, nonpolluting electricity generation and other clean energy sources will need to be brought online, reducing the overall expense of energy and its contribution to the overall costs associated with building a sustainable economy.

Energy efficiency and energy conservation are different but related concepts, although they are often confused. Energy efficiency means that users of powered devices can get the same mechanical or useful work out of a more efficient device that uses less energy as an input. Energy conservation is an intentional pattern of human action by which energy use is avoided; one strategy under the umbrella of energy conservation is to install or use more energy efficient devices, while other strategies include avoiding the use of energy altogether.

Energy efficiency and energy conservation can be more or less linked together. As a concrete day-to-day measure, energy efficiency is considered to be more effective than energy conservation, because once a device is installed, it takes the choice to waste energy out of the hands of people, while conservation as an on-the-ground energy-saving tactic requires human effort and choice. On the other hand, the value of energy efficiency is enhanced and its implementation facilitated by a preexisting ethic of energy conservation that may permeate a society as a whole; investors and governments are more likely to give high priority to energy efficiency investments if they believe that resources are valuable, limited, and ought to be conserved. Furthermore, without an overarching ethic of energy conservation, efforts to save energy by energy efficiency measures can become victim to the economic phenomenon called "Jevon's Paradox": When a resource is used more efficiently, paradoxically more of that resource ends up being used because its use becomes cheaper to society.

An example may highlight the subtle difference between pure energy efficiency and energy efficiency combined with conservation. Using an occupancy sensor in lighting has now become standard practice in many offices and public spaces, turning lights on and off conditional on the presence of people (communicated via sounds or body heat). A combination approach is a lighting control that has a manual on switch but an occupancy sensor that shuts off the light when the room is once again unoccupied. The conservation portion involves a human being deciding whether he or she needs artificial light. It becomes a matter of preference, use requirements, and cultural patterns to choose between these two options.

The importance of an ethic of conservation in promoting both energy efficiency and renewable energy has in part been underplayed in the United States over the past three decades because of the political defeat of Jimmy Carter in 1980, who was until recently the most powerful public figure to actively promote energy efficiency and resource conservation. Historian and political commentator Andrew Bacevich contrasts the unpopularity of Carter's image as a prudent conservator of resources versus the then more attractive image of the swashbuckling Ronald Reagan, who painted the picture of an America of infinite resources and prosperity. Bacevich sees Reagan as the "prophet of profligacy," an attitude that, because of Reagan's political influence, to this day has colored the American view of the ethic of conservation (Bacevich 2008). At the moment, we seem to be at a turning point against this decades long stereotyping of the pursuit of conservation, where green is fashionable and oil companies are declaring in expensive TV commercials that conservation is an imperative. While there are still significant cultural and infrastructural hurdles in North America, a consciousness of resource limits is key in inspiring leadership on clean energy.

An equally important factor in promoting energy conservation and efficiency is the price of energy. Conservation measures and energy efficiency are far more attractive when wasteful use of energy has immediate and noticeable impacts on familial, business, and institutional budgets. Nations without substantial fossil resources but with high energy demand, most of Europe and Japan, generally pay more for energy on per unit basis than we do in the United States and Canada.[1] In those countries, higher energy taxes make up one portion of the difference especially in the price of petroleum. In the United States, those states with higher energy costs, like California, generally use energy more efficiently than states with lower

---

[1] Relatively current European energy prices can be found at the European Union's energy portal: www.energy.eu/#prices. In 2009, as an example, large industrial users were paying approximately $0.10 to $0.18/kWh in Western Europe, while in the United States excluding Hawaii and Alaska, industrial users were paying from $0.04 to $0.15/kWh, with many states averaging $0.06/kWh.

energy costs, although additional factors are involved, like state energy efficiency programs. Except for energy intensive industries, like electricity generation, cement manufacture, air transport, trucking, and metals processing, energy is still a relatively small portion of the overall budget for individual or corporate economic actors.

While, on a national level, support for energy efficiency has been inconsistent, California's state government, since the initial oil shocks of the 1970s, has developed a set of energy efficiency oriented regulations for utilities and building standards that remain the state of the art within the United States. California's electrical energy use per capita has remained steady since the 1970s due to a successful energy regulatory environment and despite the rising population in hotter areas of the state, away from the temperate Pacific coast (Roland-Holst 2008). Some of the early, fairly easily achieved national measures for energy efficiency can be obtained by adopting wholesale or revised versions of California's regulatory culture.

4.2 ELECTRICAL ENERGY EFFICIENCY: GENERATING "NEGAWATTS"

Energy efficiency is a measurable quantity, the percentage of energy or work that results from energy that is put into a process. Efficiency is expressed as a percentage between 0 and 100; for example, a (very efficient) process with 95% efficiency converts 95% of the energy input into useful work:

$$\text{Efficiency} = \frac{E(\text{out})}{E(\text{in})} \times 100$$

The energy guru Amory Lovins coined the term *negawatts*, meaning "avoided megawatts," to describe how gains in energy efficiency can avoid the production of large quantities of energy (Lovins 1990). Lovins likes to call energy efficiency and negawatts "not a free lunch: it's a lunch that you're paid to eat." Highly influential, Lovins is relentlessly upbeat about how energy efficiency is a sound business and product design practice, although his enthusiasm downplays the challenges facing energy efficiency in the American context, where energy is still relatively cheap. While, in Europe and Japan, the higher cost of energy facilitates investment in energy efficiency without incentives, in the United States, systems of incentives have been necessary, most notably successful in California, to encourage significant adoption of energy efficiency measures.

One can compare the price of negawatts to megawatts as a decision making tool. A modern combined cycle natural gas power plant can cost somewhere around $800–1000/kW of power to build without fuel costs. The cost of natural gas contributes currently about 73% to the cost of power from this plant and rises as the price of the fuel (inevitably) goes up (California Energy Commission 2006). On the other hand, an efficient lighting project, where there is a substantial step downward in wattage between old and new fixtures, can cost $500–1500/kW, fuel "included," which in effect becomes more competitive as the price of power rises. In addition, if the environmental externalities and risks are appropriately priced into the costs of the new power plant, the efficiency project wins hands down. Furthermore, if the generator and the user split the cost of the efficiency project, as it can benefit both economically under a number of regulatory conditions, the cost of the project is reduced. Not all energy efficiency projects are as inexpensive, but the same principle applies that, as the price of power goes up, the return on investment on an installed energy efficiency project gets more favorable; if pricing of environmental and climate effects are considered inevitable, the favorability of negawatts increases.

If energy efficiency and new clean generation are not played off as an either/or proposition, the extra expense of new clean generation will spur energy efficiency investment, as the higher per-kilowatt-hour costs of a new technology make investment in energy efficiency all the more attractive. More efficient use of energy in turn lowers the overall costs of building a new clean infrastructure, as less generation capacity must be built. The interplay between new clean generation and energy efficiency then functions as a "virtuous circle."

4.3 UTILITY REVENUE DECOUPLING AND ENERGY EFFICIENCY

In 1982, to align the interests of the investor-owned utilities with the state of California's goal to increase energy efficiency, the California Public Utilities Commission (Figure 4.2) created an innovative system by which utilities would not suffer decreases in revenue by reducing power sales (Canine 2007). The decoupling of utility revenues mandated that utilities invest a certain amount in energy efficiency programs, usually through rebates for energy-efficient devices and device installation, yet allowed the utilities to recover lost revenues from these reductions in power sales by increases in

Figure 4.2 California energy commissioner Art Rosenfeld is sometimes called the "father of energy efficiency" in California. A trained physicist, Rosenfeld in the 1970s realized that many of the energy challenges facing the United States could be met by increasing the efficiency of devices and processes. Many of the efficiency programs in California were devised or influenced by Rosenfeld, whose current interests include "cool-colored" materials and designing HVAC systems with local climatic conditions in mind.

power rates the subsequent years. These increases, in turn, facilitated further investments in energy efficiency as higher power costs spurred power end-users to put more money into more efficient end-use devices. California has higher power costs than surrounding states but power use has remained around 7500 kWh per year per person since 1977 as power use has risen throughout the United States to an average of 12,000 kWh per year.

Utilities under decoupling regulation have found that investment in energy efficiency is a way for them to avoid or postpone large-scale capital investments in new power contracts or transmission and distribution infrastructure. Northern California's large investor-owned utility PG&E, for instance, invested three times as much in energy efficiency as is mandated by the state for just these reasons. In addition, investment in energy efficiency is good public relations in an era in which being green is considered a public virtue.

4.4 GREEN DESIGN: GUIDING NATURAL ENERGY FLOWS

Energy supply in a sustainable energy economy means tapping into natural energy flows or gradients and using them to generate electricity or provide heat to power useful devices. But, what if those currents of natural energy and material flow had desirable uses in their stronger, unconverted natural forms?

As we already established, renewable generators are, at least with current technology, not inexpensive and, like most electric generators, convert only a fraction (10–40%) of the primary energy they receive into electricity.

One way to think of green design and building principles is that they are able to route natural energy flows to serve a desired human end, avoiding the losses and expense associated with converting the energy into a new form, like electricity. For instance the heat from sunlight or from the bodily warmth of people and animals can be used to keep the interior of buildings warm during the winter with the proper materials and construction (Figure 4.3). Or natural light can be used to light the interior of buildings through windows and skylights or through new fiber-optic daylighting systems and solar tubes. Wind can be used to cool a building through wind towers in hot dry climates. An awareness of these natural flows and gradients is one of the most important tools of the green architect or urban planner.

Specifying the right materials, design, and mechanicals also allows green buildings to simultaneously gather in natural energy from the environment and work against the natural tendency of energy to dissipate by keeping a space warm or cold or reducing the need for artificial light. Superinsulation

Figure 4.3 Making a statement about green design, the Alberici construction company of Missouri built its new headquarters as one of the highest scoring LEED™ platinum buildings. (LEED™, Leadership in Energy and Environmental Design, is the green building certification program of the U.S. Green Building Council, a nonprofit trade association in the building industry. LEED™ attempts to encompass all environmental impacts of building. LEED™ platinum is the most demanding of four levels of certification.) The architects reused the shell of a 50-year-old manufacturing and office facility, orienting the new facades of the rebuilt structure to the south to capture more winter sun and optimized natural ventilation flows to increase energy efficiency and improve indoor air quality.

and advanced window technologies allow buildings to use almost no energy to maintain comfortable interior temperatures with minimal heating or cooling energy required. Older technologies like straw-bale design and adobe walls can have a similar effect in realizing our intentions to keep a space warm or cool, fighting against the entropic tendency for heat and air to evenly disperse across natural barriers. Prefabricated building and building parts allow for more precise design tolerances and tighter buildings, as factory construction can benefit from computer aided design and manufacture.

4.5 NEAR-ZERO, NET-ZERO, AND PLUS-ENERGY BUILDINGS

While green building encompasses more than a focus on energy usage, reducing the energy use and attributable greenhouse gas emissions of buildings is a key concern of green builders today, contributing, for instance, approximately one third of the potential points to the LEED™ green building rating systems. Near-zero energy buildings are achieved with the application of efficient building technologies, green building principles, and some on-site renewable energy generators, most often photovoltaic (PV) solar panels. However, a near-zero energy residential building can also be achieved exclusively through the application of hyperefficient building technologies without on-site renewable energy capture and generation (Figure 4.4).

One building system that can produce near-zero energy buildings (Figure 4.5) are "passive" buildings, houses that use ambient energy from the sun for heat in the winter and cool from the upper layers of the ground in the summer. Passive houses or buildings are superinsulated and use an air-to-air heat exchanger (driven by small electric motors) to preheat or precool incoming air with exhaust air, thereby keeping interior air fresh while preserving the desired interior temperature. A passive house can use 15% of the energy of a nonpassive house or less for space conditioning; furthermore, the heat given off by lighting can contribute significantly to the warmth of the house in the winter leading to a two-for-one effect (Feist, Wolfgang et al. 2005).

Building closer to the ground or using thick earthen or naturally insulated walls can reduce the need for space conditioning in almost all climates, as the temperatures of the ground and groundwater remain fairly constant relative to the air temperature. Also the introduction of walls or floors as thermal masses gives architects another tool to reduce building energy usage by storing heat or "cool" in these masses for slow release later on. The "earthships," by New Mexico architect Mike Reynolds, use

Figure 4.4 Superinsulation is a characteristic of many near-, net-, and plus-energy buildings. In these infrared thermograms, the passive building on the right emits much less heat than the ordinary building on the left, as it is more tightly constructed and has walls with a much higher insulation value; this allows the passive building to use 15% of the energy of ordinary buildings to heat, cool, and ventilate. (Source: Juergen Schneiders and Andreas Hermelink, "CEPHEUS Results: Measurements and Occupants' Satisfaction Provide Evidence for Passive Houses Being an Option for Sustainable Building," Energy Policy 34 (2) (2006): 151–171.)

Figure 4.5 This near-zero energy building in Los Angeles, the Audubon Center at Debs Park, has an innovative system of rooftop solar thermal collectors and absorption cooling, which use solar heated water to both heat and cool the well-insulated interior space (also a LEED™ platinum building).

the thermal mass of thick walls and thoughtful design in relationship to their environment to reduce or eliminate the need for space conditioning. A new technology, borehole thermal energy storage (BTES) is a means to heat buildings using installations of thermal masses in the ground to store the heat of the sun during the summer, which remarkably is still available during the winter, 6 to 9 months later.

To push beyond the near-zero energy threshold, net-zero and plus-energy buildings require the application, sometimes liberally, of solar PV or wind turbine technologies to cover the internal uses of energy in the building, as modern building users draw energy with the local utility via the grid. The building efficiency vs. on-site power generation technologies are influenced by the relative cost of these technologies; the uses of the building (residential, office, industrial); the local climate; the intentions, commitments, and budget of the builders and owners; and the renewable energy resources available. It may be more inexpensive at one point in time or place to apply efficient building technologies; but at a point of diminishing returns, the purchase of PV panels or an on-site wind turbine may become the more feasible option. With more power usage per square foot, to achieve net zero or plus energy requires, of necessity, more on-site generation. Compared to the building techniques of the last two centuries, which depend on energy subsidy from coal, gas, oil, or wood for comfort and functionality, using current and emerging building technologies in new buildings makes it easier to approach the net-zero energy ideal (Torcellini et al. 2006).

4.6 ELECTRICITY AND ENERGY EFFICIENCY RETROFITS OF EXISTING BUILDINGS

Reaching the extremes of energy efficiency is easier in new construction using the latest or revived ancient energy efficient techniques. One key policy measure for enhancing the future energy efficiency of buildings is national building standards that may be based on California's Title 24, a system by which new construction is pushed to become more efficient with every successive generation of buildings (California Energy Commission 2008). Just as in its utility laws, California now has three decades of experience in designing effective building laws from which most other states and the national government can draw in designing a broader system. More aggressive policies are possible, for instance, basing future building standards on near-zero benchmarks.

However, for the next half a century or so, wherever we live, we will be living with many buildings that were built without much regard for their energy use. Many of these buildings can be made tighter and better insulated but only in some cases will achieve the standards of hyperefficient new construction without major renovations.

Buildings, now, typically draw their energy from a combination of wholesale generated electricity from the grid, piped-in natural gas, propane from tanks, and occasionally wood and wood pellets. It is unfortunate that fossil fuels predominate in this mix. As it turns out, if more buildings used electricity for more of their daily operations, building energy use could be halved for most energy intensive tasks. Electrical energy, which once came from fossil sources, can be generated by renewable electric generators, thereby giving all-electric buildings the potential to become carbon neutral in their operations now or at some point in the future.

Furthermore, as we do not have the luxury of building an entire new building stock of near-zero and net-zero buildings from the ground up, high efficiency electric appliances and systems are fairly easy retrofits for existing buildings, although to implement these on a large scale sometimes requires incentives to facilitate the move.

4.7 EFFICIENT LIGHTING: EFFICIENT FLUORESCENTS, INDUCTION LIGHTING, AND LEDs

For most applications, the still common incandescent lightbulb is now obsolete, because it converts less than 5% of the electricity that it receives into light. For many ambient light applications, modern linear fluorescents are six times more efficient, while compact fluorescents are three to four times more efficient. Fluorescents are still the most efficient producers of white and white-yellow light, although in some applications, newer LEDs use less wattage and are therefore more appropriate than fluorescents. In terms of the economics of lighting energy efficiency, there are now three tools for different applications: high-efficiency modern linear and compact fluorescents, electrodeless fluorescent induction lamps, and LED lighting. LED lighting is now the subject of a good deal of hype and already a viable alternative in some directional and sign lighting. LED and induction lighting are economically viable for some inaccessible fixtures because of their long life. As there is a mass-market for both linear and compact fluorescent lamps and they are very efficient, they still command an economic advantage in most applications.

4.8 HEAT PUMPS: GROUND SOURCE AND BTES LINKED

About 45% of the 40% of total U.S. energy consumption (meaning 18% of total U.S. energy use) attributable to buildings is used by heating, ventilation, and cooling systems, also known as space conditioning or HVAC, and water heating (US Department of Energy 2009). Even in severe climates, this amount can be cut to half or less of current usage by the use of more efficient HVAC technologies, most of which require only electricity as its energy input. Daily combustion of fossil fuels for space conditioning can be eliminated in most climates by the use of (electrically driven) heat pumps, which can pull heat out of or put heat into spaces as desired by building users (Figure 4.6). Heat pumps in combination with fans and water pumps distribute heat or cool either using an air duct or a fluid-based radiant heat or cool distribution system in a building, thus they can substitute for both an air conditioning and a heating system. Heat pumps operate using the same principle as a refrigerator but, unlike a refrigerator, can work in reverse. Not only can energy use be cut by using properly designed heat pumps, dependence on fossil fuels can be eliminated for space conditioning, allowing, at some point in the future, all energy for a building to come from renewable electric generators.

The most efficient, though highest price, heat pumps used for space conditioning are ground source and groundwater source heat pumps (GSHPs), which use the substantial thermal mass, conductivity, and consistent year-round temperature of the ground or groundwater as either the heat source or the heat sink. The expense of GSHPs comes from the requirement to build a

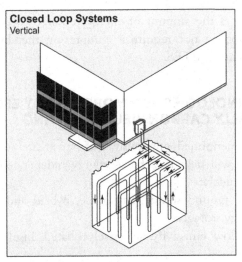

Figure 4.6 A ground source heat pump is a refrigerator-sized appliance inside a building that either extracts heat from or pushes heat into the ground through a heat exchange fluid. The pictured configuration shows vertical boreholes, through which is threaded a precisely engineered length of flexible pipe for the heat exchange fluid for that building's cooling and heating load.

ground loop by trenching at a 10-ft depth laterally or, with limited yard space, drilling boreholes several hundred feet deep, through which a tube with a heat transfer fluid is drawn. The size of the GSHP's ground loop has to do with the heating and cooling load and soil characteristics. Sometimes called *geothermal* or *geoexchange heat pumps*, they can also use the excess heat extracted from the building or the ground to heat some of the hot water used in the building, although one would need to build a dedicated heat pump for domestic hot water for year-round hot water or install a supplementary water heating system.

GSHPs can reduce the energy needed to cool by half and to heat a house by as much as two thirds with the energy requirements purely electric: the fan, compressor, and pump energy required to circulate the heat exchange fluid, extract the heat, and distribute the heat or cool throughout the building. However, to reduce the size of the ground loop and therefore initial expense, it makes sense to tighten up and insulate the house. In super-insulated passive houses in colder or hotter climates, a miniaturized ground source heat or simple "earth tubes" can supplement passive heating and cooling or precondition air. These miniaturized GSHP's are quite affordable and represent only a minor addition to the capital cost of the building.

With the advent of BTES, electric heat pumps can be used to deposit or extract heat from the seasonal thermal energy store, which, in some applications, reduces the amount of energy required to condition buildings. These pumps do not require a compressor, thereby reducing the energy requirement for BTES.

4.9 KEY TECHNOLOGIES FOR MORE ENERGY EFFICIENT, EVENTUALLY CARBON-NEUTRAL LIVING

Including those mentioned previously, listed next are some of the key technologies that will help us achieve energy independence and carbon neutrality more quickly:

1. Heat pumps: ground source, air source, hybrid and with borehole thermal energy storage.
2. Super glass (low emissivity, selectively coated, insulated) and super windows.
3. High-R insulation and structural insulated panels.
4. Efficient fluorescent, induction, and LED lighting.
5. Fiber-optic solar lighting and advanced skylights for day lighting.
6. Intelligent building, lighting, and appliance controls.

7. Light-colored and "cool-colored" building and paving materials (which reduce the heat island effect of the built environment and building heat loads).

8. Solar thermal water and space heating.

9. Variable frequency drives (electronically adjusting pump and fan speeds to energy demand).

10. Weatherproofing and tighter building envelope standards (with testing).

11. Radiant heating (using water rather than air as the heat transfer medium in a building).

12. Induction cooktops, convection ovens, and electric infrared grilling.

4.10 QUALITY ASSURANCE AND CERTIFICATION IN ENERGY EFFICIENCY

More so than in the generation of electricity or extraction of energy, the implementation of energy efficient technologies, through either the private market or government programs, requires extensive testing by government or trusted third party agencies to make sure that promised energy savings are realized by a new technology. The potential for fraud in promising "more for less" or for improper installation of a technology requires oversight by both private and public regulators. Paired with the decoupling of utility revenues and combined with a mandate to invest in energy efficiency, power utilities have an interest in monitoring the effectiveness of energy efficiency measures. Furthermore, if energy efficiency is to preempt future capital spending by the utility, the incentive to invest in and monitor energy efficiency is even greater.

4.11 ENERGY EFFICIENCY IN TRANSPORT: SHORT-TERM AND LONG-TERM SOLUTIONS

A key feature of a sustainable energy system is the replacement of petroleum with electricity as the energy carrier for transportation. However, this transfer will take place at varying speeds, depending on the future cost and availability of petroleum, battery technologies, as well as political support for electrification of transportation. Petroleum and natural gas will be around for at least a decade or two in force and in vestiges in the following decades. Increasing the efficiency of internal combustion drive vehicles will play a role even as we transition to vastly more efficient electric transport.

One of the motivations to transfer transport energy to electricity is the threefold increase in efficiency that electric motors represent over petroleum and natural gas fueled internal combustion engines: The 90%–95% efficiency of electric motors contrasts favorably with the 25–30% efficiency of the modern internal combustion engine.

Short-Term Solutions

In the near future, some measures can continue to increase the efficiency of vehicles short of full electrification. The following are some specific measures that can be applied to vehicles themselves.

Lightweight Vehicles

While the internal combustion engine is near the end of its development trajectory, a number of innovators in the area of vehicle materials are attempting to show that the use of lightweight body materials, such as carbon fiber, can reduce conventional vehicle mass substantially without endangering vehicle safety. Amory Lovins has long championed the use of carbon fiber to double vehicle efficiency, claiming that bulky vehicles with advanced lightweight materials could have superior mileage and comfort compared to today's economy cars. The German company Loremo and the American company Aptera (Figure 4.7) have also suggested radical, lightweight vehicle designs as ways to create hyperefficient vehicles that would either have a small internal combustion engine or an electric motor.

Vehicle Efficiency Standards and Automaker Penalties versus Gas Taxes

Mandating vehicle efficiency standards has been an uphill battle in the United States, requiring American automakers to work against their own design

Figure 4.7 Aptera, with its revolutionary Typ-1, is radically restyling passenger vehicles to save weight and energy. Although classified as a motorcycle, Aptera has targeted exceeding passenger car safety standards in its design.

culture and the tendencies of American auto buyers to prefer large, powerful vehicles in an environment of cheap, abundant petroleum. While vehicle efficiency standards are, in the culture of environmental reform and public virtue, viewed to be a necessity to impress on on both automakers and the public the optimality of fuel efficiency, higher gas taxes in Japan and European countries have been a far more effective means of compelling automakers and auto buyers to conserve energy and choose more efficient vehicles.

If U.S. legislators and environmental pressure groups are at all serious about encouraging gasoline-powered vehicles to use gasoline more wisely, they need to challenge the taboo against being seen politically to raise the price of fuel, by instituting substantial increases in fuel taxes. This will take more courage on the part of politicians than simply asking for higher fuel efficiency standards, which puts the onus on automakers to lead the market. While the shortsightedness of U.S. automakers is truly lamentable, legislators so far have not succeeded in transforming that culture through vehicle efficiency mandates. Those who cite the success of foreign carmakers vis-à-vis U.S. carmakers forget that, among other things, the headquarters of these companies are in countries with fuel that costs at least twice as much as in the United States. Fuel efficiency standards require U.S. automakers to lead the efficiency charge, which requires them to occupy a position of moral and environmental leadership without the aid of high fuel prices.

A compromise that avoids some of the negative political fallout of an across-the-board gas tax hike is a varying tax surcharge that keeps the price of fuel above a certain level, blocking efforts by oil producers to artificially lower prices or smooth over the effects of temporary drops in demand. This fuel "price floor" would be explainable to constituents who should at some point understand that the movement to higher fuel prices is inevitable and energy efficiency in transport socially desirable.

Longer-Term Measure: Shifting to Electric Drive

As discussed previously, the shift to electric drive is by far the most effective means of conserving energy resources. The current generation of hybrids use electric motors to provide assistance for relatively inefficient gasoline internal combustion engines. Plug-in hybrids and electric vehicles have the potential to double or treble the efficiency of automobile drivetrains. The serial hybrid design, as with GM's Chevrolet Volt, enables electric drive vehicles to use a small diesel or gasoline engine as a generator to extend the range of a vehicle while sizing the battery for everyday commutes.

4.12 PRICE SIGNALS AND ENERGY EFFICIENCY

Just as with the finance of new clean energy generation technologies, the price of energy is key in spurring energy efficiency investment and energy conservation. As indicated previously, price signals are some of the most effective ways to spur private parties to cut their energy use; the implementation of those price signals through policy instruments needs to proceed at an urgent pace yet not so rapidly as to encourage backlash against the necessary efforts that we all must undertake to help preserve a favorable climate. Carbon taxes, fees, and, with less certainty and efficacy, cap and trade systems in all likelihood will spur investment in energy efficiency, although the degree to which they do will depend on the level of the resulting carbon price as well as the ultimate efficiency of the chosen mechanism. These instruments in their early stages, in all probability, will be more effective in spurring energy efficiency investments than they will in stimulating the building of new clean electricity generation, as the relative cost of the latter is in many cases higher.

4.13 CONCLUSION: ENERGY EFFICIENCY IN THE UNITED STATES TODAY

There are signs that Americans are paying more attention to energy efficiency than they have in the last decade or so. Concern about global warming has spurred a revival in green consciousness, and additionally, businesses have started to realize that they can benefit from investment in energy efficiency. Nevertheless, especially in our buildings and transportation sectors, we have a long way to go to catch up with most other developed economies, where energy has been more expensive and concern about climate change is more widespread and intense. Among the many products and services vying for the attention of the American consumer or professional buyer, energy efficiency does not yet have the priority it needs to make serious headway in reducing our carbon emissions nor to rival many other developed nations in implementing the latest energy efficient technology.

REFERENCES

Bacevich, A., 2008. The Limits of Power: The End of American Exceptionalism. Henry Holt, New York.
California Energy Commission, "Comparative Costs of California Central Station Electricity Generation Technologies," draft Staff Report June 2007.
California Energy Commission, 2008. Building Energy Efficiency Standards for Residential and Non-residential Buildings. Document # CEC-400-2008-001-CMF.

California Public Utilities Commission and California Energy Commission, 2006. Energy Efficiency: California's Highest Priority Resource. accessed 20.2.2010 from ftp://ftp. cpuc.ca.gov/Egy_Efficiency/CalCleanEng-English-Aug2006.pdf

Canine, C., 2007. California Illuminates the World, On Earth, Spring 2006. Natural Resources Defense Council; Jonas Ketterle, " Decoupling: Removing Market Barriers to Energy Efficiency ". Review of Policy Research 24 (5), 467–533.

D&R International for US Department of Energy, 2009. 2009 Buildings Energy Data Book. US DOE October 2009.

Feist, W., Peper, S., Görg, M., (2001). CEPHEUS Final Technical Report [Electronic version] Darmstadt: Passivhaus Institut Accessed February 20, 2010 from http://www.passiv. de/07_eng/news/CEPHEUS_final_long.pdf

Feist, W., et al., 2005. Re-inventing air heating: Convenient and comfortable within the frame of the Passive House concept. Energy and Buildings 37 (11), 1186–1203.

Ketterle, J., 2007. Decoupling: Removing market barriers to energy efficiency. The Review of Policy Research 24 (5), 467–533.

Lovins, A., 1990. The Negawatt Revolution. Across the Board 27 (9), 21–22.

Lovins, A., 2004. Winning the Oil Endgame. Rocky Mountain Institute, Boulder, CO.

McKinsey & Company., 2009. Pathways to a Low Carbon Economy (Version 2) Accessed from https://solutions.mckinsey.com/ClimateDesk/default.aspx February 16, 2010.

Miller, C., 2001. Gifford Pinchot and the Making of Modern Environmentalism. Island Press, Washington.

Parker D., et al., 1998. Field Evaluation of Efficient Building Technology with Photo-voltaic Power Production in New Florida Residential Housing. Report No. FSEC-CR-1044-98. Florida Solar Energy Center, Cocoa, FL. Accessible from http://www.fsec.ucf.edu/en/publications/html/FSEC-CR-1044-98/index.htm.

Roland-Holst, D., 2008. Energy Efficiency, Innovation, and Job Creation in California. Center for Energy, Resources and Economic Sustainability, University of California, Berkeley, California.

Simon, S., 2010. Even Boulder Finds that It Isn't Easy Being Green. Wall Street Journal February 13.

Torcellini, P., et al., 2006. "Zero Energy Buildings: A Critical Look at the Definition" National Renewable Energy Laboratory. NREL/CP-550-39833.

U.S. Department of Energy., 2010. Energy Star Program Website accessed 20.2.2010 from <http://www.energystar.gov>

U.S. Department of Energy, October 2009. 2009 Buildings Energy Data Book. Washington DC.

CHAPTER 5

Renewable Energy

Joe Kantenbacher

Contents

5.1	Introduction: Energy Use in the United States	65
5.2	Wind	67
5.3	Solar	70
	Solar Photovoltaics	70
	Solar Thermal	72
5.4	Geothermal	73
5.5	Biopower	76
5.6	Marine	78
5.7	Advanced Renewables Deployment	78
	Renewables and Buildings	79
	Vehicle-to-Grid	79
	Hybrid Systems	80
5.8	Summary	80

5.1 INTRODUCTION: ENERGY USE IN THE UNITED STATES

The United States is the largest consumer of energy in the world. In 2007, the U.S. consumed nearly 100 quadrillion Btus (quads) of primary energy, approximately 20% of the global total for that year. A full 84% of that energy was supplied by fossil fuels (coal, petroleum, and natural gas; see Figure 5.1). Nuclear and renewable power sources provide the balance of the U.S. energy mix.

The Unites States consumed 4100 TWh of electricity in 2007, an average of 13,600 kWh per capita. Coal dominates the American electricity market, providing the fuel for more than half of the total electricity consumed in large part because of abundant and cheap domestic resources. For several years, the capacity additions of natural gas plants have been greater than those of any other fuel type and now natural gas provides 21% of U.S. electricity. Nuclear power provides a further 20% of U.S. electricity and has maintained its market share for years despite a freeze on the construction of new nuclear power plants, in large part because of substantially improved capacity factors in existing plants. Large-scale hydroelectric facilities form the other major, noncarbon electricity source, contributing about 6% of the

Sustainable Communities Design Handbook
ISBN: 978-1-85617-804-4, DOI: 10.1016/B978-1-85617-804-4.00005-7

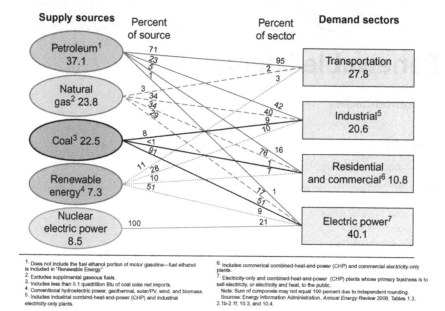

Figure 5.1
(Source: U.S. Energy Information Agency, *Annual Energy Review 2008*).

electricity consumed. Conventional renewable energy sources like wind or photovoltaics contribute roughly 1%.

Buildings are large energy consumers in the United States, accounting for 39% of the primary energy consumption. Natural gas and electricity are the principal energy sources in the built environment, with temperature regulation, lighting, and water heating being the chief energy-consuming activities. Renewable energy sources such as solar, geothermal, and biomass contribute 8% of the on-site energy consumption in buildings.

The transportation sector, which comprehends personal, public, and commercial ground travel as well as aircraft and marine vessels, is almost exclusively powered by fossil fuels. Petroleum products accounted for 94% of the primary energy input in the transportation sector in 2008, and natural gas contributed an additional 2.4%. Biomass-derived enthanol and biodiesel are the chief sources of renewable transportation fuel, providing 3%. At present, electricity is a negligible component of the transportation energy system, a mere 0.3%. Significant growth in electric vehicle use in the coming years is forecast, however.

This chapter discusses the various ways in which renewable energy sources serve to satisfy U.S. demand for electricity and heating. The chapter is organized by source of renewable energy according to current market penetration, with emerging technologies discussed at the end. Although large hydrokinetic power plants currently constitute the largest source of renewable energy in the United States, they are not discussed in this chapter due to limited opportunities for further domestic development.

5.2 WIND

Wind turbines translate the kinetic energy of moving air into mechanical energy, which in turn powers a generator that produces electrical energy. Conventional turbines have a horizontal-axis design, in which two or three rotor blades are mounted atop a tower and arrayed such that they resemble airplane propellers. When oriented into the wind, the movement of air across the blades generates lift, spinning a shaft connected to an electric generator. Wind power output from this conventional turbine design is a function of two factors: swept area and wind speed. As the rotor area (determined by the blade length) of the turbine doubles, the power output quadruples. A doubling of incoming wind speed translates to an eightfold increase in power output.

Over the past three decades, the size of wind turbines has substantially increased, with rotor diameter increasing eightfold and tower height quadrupling. This led to a 200-fold increase in power output and contributed to the price of wind-generated electricity dropping from about 40 cents/kWh in the early 1980s to around 5–8 cents/kWh today. Modern turbines can reach peak power outputs in the megawatt range, meaning that utility-scale aggregations of turbines (wind farms) can readily scale up to several gigawatts in size. While arraying turbines in wind farms can help achieve economies of scale (particularly with respect to transmission costs), the use of stand-alone turbines can be an economically viable means of providing power to systems in remote locations, such as communications towers or rural irrigation networks.

Also in operation are various vertical-axis designs, which have a vertically oriented main rotor shaft and are usually situated on the ground or rooftops. Because wind speeds tend to be lower closer to the ground and these vertical-axis systems tend to be small in size, the power output from this category of turbines is usually low (in the watt or kilowatt range). However, vertical-axis designs are increasingly popular installations for distributed, renewable energy generation.

Although wind power has been employed in the United States since the mid-19th century, the U.S. modern wind industry did not develop until the 1970s, when it was launched in response to the increasing cost of oil-based

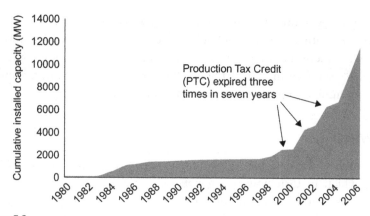

Figure 5.2
(Source: Department of Energy, Office of Energy Efficiency and Renewable Energy, *20% Wind by 2030*).

electricity generation. By the mid-1980s, California had 1.2 GW of installed wind capacity, which accounted for more than 90% of the global total. California's dominance in this area is commonly attributed to the federal and state investment tax credits that were in place, as well as state-mandated utility contracts for wind power. After the expiration of state and federal investment incentives, the U.S. wind industry stagnated until the late 1990s, at which time the first in a series of production tax credits (giving renewable power producers a rebate for each kilowatt hour generated) sparked renewed growth in domestic installations. In recent years, wind power has been a major component of nationwide generation capacity addition, growing from 10% of new capacity additions in 2005 to more than 40% of new additions in 2008.

In 2009, nearly 10,000 MW of new wind generation capacity was brought online in the United States, bringing the cumulative installed capacity to more than 35,000 MW (Figure 5.2). At 9400 MW, Texas leads the country in terms of total wind capacity, trailed by Iowa, California, and Washington. Fourteen states have greater than 1 GW of installed capacity, and 36 states have at least 1 MW.

While the wind power resource in the United States is substantial, it is also, significantly, distributed across the continent with substantial variation (see Figure 5.3). Though several areas, particularly those on the East and West coasts, feature a coincidence of both strong continental wind resources and high-density population centers, the broadest swath of wind resource is located in the Plains states, whose low population levels and densities decrease the economic efficiency of employing those wind resources locally.

Figure 5.3
Source: Department of Energy Office of Energy Efficiency and Renewable Energy "Renewable Resources Maps." http://www1.eere.energy.gov/maps_data/renewable_resources.html
Darker colors represent higher wind speeds.

Capitalizing on the resources of the plains will require the build-out of expensive transmission lines, increasing a given project's expense and challenging its viability. Offshore wind resources, which are generally stronger and steadier than their continental counterparts, have also attracted the attention of wind power developers. However, the inaccessibility of marine locations and the added expense of undersea transmission lines diminish the attractiveness of investment in offshore systems, dimming the prospects that the Unites States will have a sizeable sea-based wind fleet in the near future.

Though utility-scale turbines dominate the U.S. wind market, small wind installations (under 100 kW) are an active and expanding portion of the industry. In 2008, domestic small turbine installations grew by nearly 80%, placing an additional 17 MW in service. Although two-thirds of installations in that year were off-grid and less than 1 kW in peak output, nearly 80% of the new wattage was on account of grid-tied systems. Spurred by the passage of a new, eight-year federal investment tax credit, the wind industry projects that the cumulative installed small wind capacity will grow from under 100 MW today to 1700 MW by 2013. The most common applications of small wind systems include off-grid electricity provision, diesel/wind microgrids, and powering agricultural operations such as irrigation.

5.3 SOLAR

The sun is the primary energy input for virtually all of the processes, biological and otherwise, that occur on the earth's surface. While in the broadest sense several of the energy sources currently employed by contemporary society, including wind, biomass, and fossil fuels, have a solar genesis, there is a class of technologies directly converts electromagnetic energy from the sun into useful energy. The two main categories of such technology are: solar photovoltaics and solar thermal.

Solar Photovoltaics

Solar photovoltaic (PV) technologies harness the photoelectric effect, using sunlight to excite electrons in a semiconductor and generate direct current electricity. Three main categories of PV technologies which currently exist are: silicon crystal, thin film, and third generation.

Today, silicon is the most commonly used (greater than 90% on a wattage basis) semiconductor in the PV industry, on account of both the relatively high efficiencies it can achieve as well as its relative abundance as a raw material, though the process of refining silicon to a sufficient level of purity is expensive. Thin-film technologies, such as cadmium telluride (CdTe) or copper indium gallium selenide (CIGS), promise to address the high materials costs of silicon PV, combining low-cost materials (such as glass or plastic) with thinly spread semiconductors. While thin-film technologies have limited efficiencies, they have the potential to provide solar power at a lower cost per watt than silicon crystals. The upcoming third generation of PV designs currently features a variety of low-cost, low-efficiency materials combinations, including organic dyes, nano-structure silicon, and iron pyrites. These have yet to come to market and may not achieve significant scale in the near future.

In the United States, PV originally had a niche application in the space program, providing power to shuttles and orbiting satellites. With improved efficiencies (from laboratory best figures of 15% in the early 1980s to better than 40% today) and decreasing production costs (by a factor of nearly 100 since the 1950s), PV came to be an economical power source of a variety of earth-bound applications, including off-grid living and remote communication devices (see Figure 5.4). Prior to 2005, the majority of cumulative U.S. PV installations were dedicated to off-grid applications even though utility-scale PV arrays were deployed beginning in the early 1980s. Since 2005, however, 80% of added PV capacity has been grid-tied, bringing the grid-tied total

Figure 5.4
Source: DOE photo.

Figure 5.5
Source: Department of Energy Office of Energy Efficiency and Renewable Energy "Renewable Resources Maps." http://www1.eere.energy.gov/maps_data/renewable_resources.html
Lighter colors indicate higher average insolation.

to more than 1 GW , the sizeable majority of which is decentralized (rooftop). See Figure 5.5 for the distribution of solar resources in the contiguous United States.

Electricity is produced by current solar PV systems at a cost of about 18-25 cents per kWh, which is substantially higher than that of conventional sources. As a result, accelerated deployment of PV in the United States and around the world has been based on government support. In Germany, for example, an aggressive feed-in tariff program has led the country to be the world leader in terms of installed capacity despite relatively modest solar resources. In Japan, a well-coordinated effort to fund research and development of PV technology as well as promote its dissemination, leading to a "technological push" and "demand pull" that has significantly advanced both PV engineering and deployment. The latest innovation in public solar policy was developed in Berkeley, California, in 2008, where the BerkeleyFIRST program allows homeowners to amortize the cost of a system over twenty years and pay for it through property taxes, thereby sidestepping the high upfront cost that is an insuperable barrier to many would-be installers. By November 2009, 16 states had passed legislation to permit the use of this financial mechanism, now referred to as property assessed clean energy.

Solar Thermal

Solar thermal technologies harness sunlight to produce thermal energy. This heat is then used directly or to generate electricity.

In contrast with solar PV, electricity generation by solar thermal technology is almost exclusively the province of large, utility-scale generators. Concentrated solar power (CSP) systems use mirrors or lenses to focus solar energy, heating a working fluid (such as water or oil) that drives an electricity-generating-turbine. Varying by how they collect solar energy, there are four mainstream CSP designs - trough, linear Fresnel reflector, tower, and dish - though one (trough) has dominated the U.S. CSP market to date. These systems employ parabolic mirrors that concentrate solar heat on fluid-filled receiver that runs the length of each trough (see Figure 5.6).

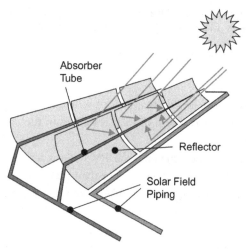

Figure 5.6
Source: National Renewable Energy Laboratory. http://www.nrel.gov/csp/troughnet/solar_field.html

The operating of concentrated solar plants requires both large tracts of contiguous land for siting and substantial volumes of water to provide a cooling reservoir for the steam turbine. Many of the prime locations suitable for CSP siting (those with high insolation and a flat grade, among other criteria) are remote - requiring additional transmission to connect to the grid - and arid - placing a burden on scarce local water resources.

Photovoltaic systems and CSP plants are dependant on the same intermittent power input, sunlight. However, unlike PV, CSP systems can be readily and economically fitted with thermal storage systems such as molten salt. Adding storage allows for operating at night or during cloudy conditions and turns CSP into a dispatchable power resource. CSP systems can also be paired with other thermal energy-based systems, such as natural gas power plants, to increase reliability and efficiency.

Despite producing electricity at a lower levelized cost than PV, CSP systems have not achieved greater penetration in the United States than photovoltaics There is currently 430 MW installed CSP capacity in the United States, 360 MW of which is located in California. However, the CSP industry is on the verge of a rapid expansion, with more than 10,000 MW of new capacity either under construction or development. Most new projects are sited in California, Arizona, and Nevada.

On a megawattage basis, small-scale, distributed applications of solar thermal energy have the biggest market share. In 2008, U.S. consumers installed 900 MW-th (thermal equivalent) of solar thermal systems, including 760 MW-th of pool heaters and 140 MW-th of water heating systems. In 2008 and 2009, the investment tax credit for solar water heaters was extended to 2016 and then uncapped, events projected to promote the continued expansion of the residential solar thermal market.

Smaller-scale applications of solar thermal energy include heating water and cooking.

5.4 GEOTHERMAL

Geothermal energy systems tap into underground heat reservoirs, utilizing the stored thermal energy directly or as a feedstock for electricity production.

Hydrothermal resources exist where magma comes close enough to the surface to transfer heat to groundwater reservoirs, producing steam or high-pressure hot water. When hydrothermal resources are sufficiently hot (several hundred degrees Fahrenheit) and close to the surface (within a few miles), it can be economically sensible to drill a well and use the steam or hot water either as a direct power input into a turbine (as with dry and flash steam plants) or as a heat source to produce steam with a secondary fluid (as with binary-cycle plants) (see Figure 5.7). Shallow-depth hydro-thermal resources of more moderate temperature, which in the United States are located primarily in Alaska, Hawaii, and many western continental states, are commonly used directly to provide heat for buildings, agriculture, and industrial processes (see Figure 5.8).

Figure 5.7
Source: DOE photo
Dry steam geothermal plant at The Geysers.

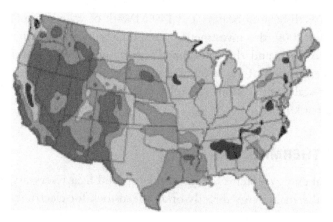

Figure 5.8
Source: Department of Energy Office of Energy Efficiency and Renewable Energy "Renewable Resources Maps." http://www1.eere.energy.gov/maps_data/renewable_resources.html
Warmer colors indicate higher temperatures at a depth of 6 kilometers.

Even in the absence of hydrothermal resources, geothermal energy can be harnessed for use. Geothermal heat pumps for buildings take advantage of the constant-temperature nature of subsurface earth, cooling warm summer air in the underground or, in the winter, drawing up heat from the relatively warm ground. As this technology does not require the presence of hydrothermal resources, its adoption is feasible in all regions of the country.

In the domain of electricity production, next-generation "enhanced geothermal systems" are being designed to access the hot, dry rock at several miles' depth and, through the injection of water, artificially create a hydrothermal resource.

One key feature of geothermal energy is that it is continuously available, as opposed to intermittent resources like sunshine and wind. As such, geothermal power is one of the leading options for renewable resource-based base load electricity generation and provides a technically viable option for supplanting coal power plants, which are arguably the most environmentally pernicious class of generators currently operating.

Though in the late 19th century numerous communities in the United States made use of surface-level hydrothermal resources for residential and commercial heating services, it was not until 1922 that the first geothermal electricity generator was brought online at The Geysers near San Francisco, California. In 1960, the first U.S. utility-scale geothermal plant, a 11 MW facility, was completed. Beginning in the early 1970s, the average growth rate of electrical output from the geothermal industry was nearly 13% per year.

In 2007, geothermal energy resources provided 4% of U.S. renewable electricity generation, and as of 2009, there were more than 3100 MW of geothermal electricity generators online. Owing to its location on a series of tectonic plate conjunctions, California enjoys considerable hydrothermal resources and, at 2600 MW of installed capacity, is the premier geothermal power producer in the country. With 450 MW of installed capacity and more projects under development than any other state, Nevada's fleet is also a significant component of the U.S. geothermal power plant stock. In addition, about 470 MW of power is provided by the direct use of hydrothermal resources, and it is estimated that about three million people use geothermal heat pumps to heat and cool their homes.

Reported projects currently under development are set to triple the installed geothermal power plant capacity in the United States (see Figure 5.9). Geothermal heat pump shipments have surged in recent years, with capacity

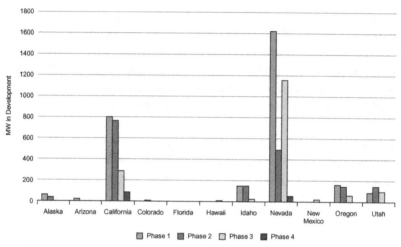

Figure 5.9
Source: Geothermal Energy Association "U.S. Geothermal Power Production and Development Update." 2009.

additions per year more than tripling between 2003 and 2008. It is anticipated that the Emergency Economic Stabilization Act of 2008, which provides investment credits for heat pump system installation, will continue to spur the expansion of small-scale geothermal energy use, though the market currently suffers from order backlogs and a deficit of trained installers.

5.5 BIOPOWER

Biomass resources exist where solar energy is stored in an organic form as plant matter or other biological material. Biopower technologies utilize biomass to generate electricity.

Direct-combustion steam production is the most common mode for generating biopower, though the scale and thermal efficiency of those boilers tend to be smaller and lower than their coal counterparts. Co-firing biomass in conventional power plants can simultaneously scale up the use of biopower and mitigate the environmental impacts of fossil fuel combustion. Gasification systems and organic, anaerobic digestion can also be used to convert biomass into syngas (a mixture of carbon monoxide and hydrogen gas) and methane, respectively (see Figure 5.10). Biomass-dervied gases can be used in high-efficiency combined cycle generators and other modular systems, making them prime distributed generation devices.

Figure 5.10
Source: DOE photo
An NREL biomass gasification plant in Golden, CO.

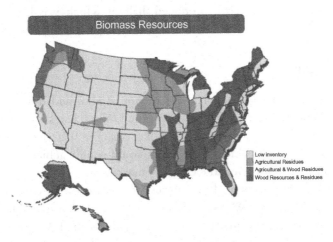

Figure 5.11
Source: Department of Energy Office of Energy Efficiency and Renewable Energy
"Renewable Resources Maps." http://www1.eere.energy.gov/maps_data/renewable_
resources.html

The broad distribution of resources and diversity of methods for utilizing them makes biomass a viable renewable energy source for much of the contiguous United States (see Figure 5.11). It is a particularly important renewable resource in the Southeast, which has a modest endowment of solar, wind, and geothermal resources. A survey of available biomass resources from forest and agricultural lands determined that upwards of 1.3 billion tons of dry biomass, nearly an order of magnitude greater than current production levels, could be eventually harvested for energy-production purposes. Transportation costs and seasonal availabilities diminish the attractiveness of biomass as an energy resource, particularly in electricity markets, where it competes with coal's low costs and all-year availability.

In 2008, Americans consumed 3.9 quads of biomass-based energy, about 4% of the total energy used in the United States that year. Woody biomass and wood byproducts account for about 60% of the primary energy content of biomass feedstocks and is used primarily for electricity production and industrial heating. Biofuels make up 28% of the biomass total, and combusted agricultural and municipal solid waste supply most of the biomass balance. Taken together, these three sources of energy provide for more than half of the renewable energy generated in the United States. Some 12,500 MW of currently installed electricity generating capacity have some form of biomass as its primary feedstock. In 2008, 56 TWh of biopower electricity were produced, half of which was by industrial facilities.

5.6 MARINE

In the oceans exist large, regular, and untapped ebbs and flows of water. Nearly 80% of the electricity provided in the United States is consumed in a state that borders an ocean or Great Lake, meaning that electricity based on these off-shore hydrokinetic resources could readily achieve substantial penetration into high-demand electricity markets.

There are three potential categories of marine water currents: oceanic, tidal, and riverine. Waves add a fourth concentrated hydraulic power source. As a whole, the marine energy industry is in the early stages of development, and while there is a broad array of concepts and designs for capturing marine energy, there has been relatively little in the way of deployment or standardization, particularly at the commercial level of deployment. As of this writing, there are three U.S.-based projects in full or partial deployment, summing to 250 kW of total installed capacity.

5.7 ADVANCED RENEWABLES DEPLOYMENT

Through the experience of deployment and the engineering breakthroughs of the laboratory, renewable energy technologies have over the last decades steadily improved in terms of performance (efficiencies, capacity factors) and economics (capital costs). These iterative advances have brought renewable technologies ever closer to shedding the long-held perception that they are "alternative" energy sources.

Moving forward, in addition to the regular technological advance of renewable energy devices, there are several measures and developments that can further enhance the utility of renewable energy systems. Broadly, these categories include: building integration, vehicle integration, and hybrid systems.

Renewables and Buildings

One of the fastest-growing segments of the solar industry is that of building-integrated photovoltaics (BIPV). BIPV designs seek to replace or enhance certain elements of a building, such as the roof, window overhangs, or walls, with solar panels. This reduces both the materials cost of building construction and the installation cost of the PV panels, and ensures that the PV panels will be optimally situated on the finished structure. Passive solar building design can also take advantage of solar energy, using windows and interior surfaces to regulate indoor air temperatures.

Building-integrated wind designs have also been proposed and implemented, but such arrangements have so far have featured substandard turbine performance and unappealing impacts on building inhabitants.

Vehicle-to-Grid

Plug-in hybrid vehicles (PHEVs), which feature both internal combustion engines and electric motors, have the potential not only to reduce consumption of petroleum products but to facilitate greater penetration of renewable power sources, as well. The electric battery in PHEVs can both be charged by and discharge into the electric grid, turning the car into a mobile, distributed electricity storage device. This storage capability is thought to be of particular benefit to wind turbines, the power output of which is generally greatest at night when demand for electricity is lowest. PHEVs would allow for higher penetrations of wind power than might otherwise be economically viable, storing excess generating at night and dispatching that electricity to meet greater loads during the day. An intermediate step in the vehicle-to-grid system might be a vehicle-to-home approach, wherein electricity is delivered to a household through a direct connection with a PHEV.

Hybrid Systems

Pairing renewable energy systems has the potential to improve economics or performance over what could be achieved by each system working in isolation. For example, siting solar and wind systems together can reduce overall transmission costs as there is a day/night complementarity in peak output that could reduce grid congestion and allow for smaller transmission lines. Similarly, combining off-shore wind turbines with marine power stations has been proposed as a means of reducing the construction and maintenance costs for each. Hydroelectric systems can serve as energy storage units for wind farms, making use during the day of water pumped uphill by the excess generation of turbines at night. Such hybrid systems have yet to achieve substantial deployment but may have promise as a means to overcome the limitations of individual renewable systems.

5.8 SUMMARY

Fossil fuels are the dominant energy source in the United States, providing over four-fifths of all primary energy consumed. However, renewable energy resources are abundant, and when harnessed can contribute significantly to the satisfaction of U.S. heat and power demands. Over the past several decades, renewable energy technologies have advanced significantly, and have achieved ever greater levels of deployment, both by utilities and by individual consumers:

- Wind turbines currently produce electricity at costs competitive with other, fossil fuel-based generators, leading annual capacity additions of wind power to be among the highest in the electric power sector.
- Solar technologies can now efficiently utilize sunlight to provide both heat and electricity. Photovoltaic and thermal systems both have become popular modes of distributed energy provision.
- Geothermal deposits provide consistent, clean energy for both electricity generation and heating. The number of hydrothermal power plants is rapidly increasing, particularly in the western United States, through which heat pumps bring geothermal energy to households in every region.
- Biomass is one of the largest sources of non-fossil power in the United States, especially for industrial applications. The quantity and variety of feedstocks available for combustion is substantial, providing for multiple avenues for expanding the biopower sector.
- Marine energy resources are vast, but the technologies to exploit them are still immature.

Public policy support mechanisms have been and are likely to continue to be instrumental in renewables deployment. A variety of new systems of renewables deployment, including those involving buildings or vehicles as well as mixed renewables use, are under development, and may serve to overcome certain limitations of current renewables technologies.

CHAPTER 6

How Energy Conservation Fits in an Existing Facilities Master Plan: A Case Study

Alison Gangl, CRM and Ben Johnson, PE, LEED AP, CEM

Contents

6.1	Introduction	83
6.2	Getting Started	84
6.3	Investment Grade Audit	86
	Establishing a Baseline	87
	Selecting Energy Conservation Measures	88
	Savings Calculations	92
6.4	Installation	92
6.5	Measurement and Verification	95
6.6	Conclusion	97
References		97

6.1 INTRODUCTION

Facility owners are faced with a number of challenges when it comes to constructing, operating, and maintaining their facilities. With the growing focus on "going green," perhaps the most prominent challenge is identifying the appropriate strategic time to implement sustainability measures. This case study focuses on a California College's efforts to implement an energy conservation program within the context of other key facility initiatives. With an extensive facilities master plan program in place, the college turned to performance contracting to reduce their energy usage across their campus.

Performance contracting is a turnkey contracting method whereby the design, construction, commissioning, and performance measurements are incorporated into one guaranteed, fixed price contract. Performance contracting is implemented through the use of an energy services company (ESCO), that acts as the single provider accountable for all aspects of the

Sustainable Communities Design Handbook
ISBN: 978-1-85617-804-4, DOI: 10.1016/B978-1-85617-804-4.00019-7

project. The energy savings from the performance contract are guaranteed by the ESCO, meaning that if the savings do not materialize, the ESCO agrees to pay the difference. Committed to being a good steward of tax-payer dollars, the California College chose performance contracting as an ideal procurement method because of its focus on identifying the most cost effective, yet comprehensive, measures for energy conservation.

6.2 GETTING STARTED

Understanding the goals, motivations, and requirements is a key first step to any energy conservation project. The California College had two priorities that motivated their energy conservation efforts:

1. **Moving toward grid neutrality**. Energy conservation was one com-ponent of the college's effort to become grid-neutral. According to the California Division of the State Architect, the steps to grid neutrality are to (a) set energy performance goals, (b) implement and maintain appro-priate energy efficiency and conservation measures, (c) install renewable energy systems to meet remaining needs, (d) maintain energy systems. By implementing energy conservation efforts prior to renewable energy generation, the college will reduce the amount of on-site electricity needed, thereby lowering the cost of the system.
2. **Saving operational dollars**. The U.S. Department of Energy esti-mates that at least 25% of the $6 billion colleges and universities spend annually on energy could be saved through energy conservation. In a time of severe budget cuts across the nation, many public insti-tutions are looking for ways to free up the general fund. To accom-plish this, the college utilized restricted funds (i.e., bond dollars) to implement the energy conservation upgrades and reduce operat-ing expenditures. This enabled the college to lessen the strain on the general fund—money that can be used across multiple operational functions.

Once the priorities and objectives of the project were determined, the college needed to select an ESCO partner. As previously mentioned, under a performance contract, the ESCO is the single point of accountability for all aspects of the process, including the initial analysis, the development and design, the project installation, and the project guarantee lasting up to 15-years. Clearly, it is vital to find an ESCO that is trustworthy, reputable, and able to deliver on its promises long term.

There are several methods used to select an ESCO partner, depending on the legislation of the particular state. The college elected to issue a request for proposals/qualifications (RFP/Q) followed by an interview process. As part of the RFP/Q, responding firms were asked to provide a sample project to the college. The proposals were to include examples of recommended energy conservation measures, the savings and estimated cost associated with those types of measures, and the qualifications of the firm to complete the proposed work.

Because the sample scope of work proposed varies firm by firm, the selection is not based upon the lowest bidder, as is typically the case in construction projects. Rather, the selection is based upon the qualifications of the responding firms and their ability to meet the needs of the college. This long-term partnership aspect of performance contracting makes the interview process an important component of the ESCO selection. Identifying a trusted partner that is easy to work with is vital to the success of an energy conservation project. Some key questions to consider when selecting an ESCO are:

1. **In the company's performance contracting project history, what portion of the total guaranteed savings is energy savings versus nonenergy savings (i.e., maintenance, operational, capital cost avoidance, or the like)? If maintenance and operational savings are used as part of the guarantee, how are they verifiable?** Nonenergy savings are considered "soft savings," meaning they are difficult to measure and verify. Operational and maintenance savings often are based on assumptions that may not be true. For example, claiming "labor savings" assumes that reducing the man-hours necessary for maintenance or operation of a particular system will result in the institution realizing cost savings. Most good performance contracts will reduce the amount of time spent maintaining buildings; however, are these savings really seen on the bottom line? Generally, no. Even if maintenance personnel spend less time on energy-related tasks, there are many other tasks to attend to. Therefore, while maintenance time may be more productive, they are still working the same number of hours and receiving the same pay. If the increased productivity from the energy project enables the institution to pay fewer overtime hours or reduce the operations staff, there may be hard dollar savings justified; however, that is a decision that should be made and quantified by the institution not the ESCO.

2. **Describe a typical method of calculating dollar savings and how it protects the college from risk.** Understanding how the dollar savings for a project are calculated is important to ensure the guarantee provided is real. Is the guarantee based on actual rates from the utility bills or a blended rate? How are differences in weather adjusted for? Some calculations can falsely inflate savings numbers, so it is important to check the logic behind a contractor's equations to make sure they are accurately determining the project savings. One of the most common forms of savings inflation occurs in the escalation of utility rates. Acceptable escalation varies depending on the utility provider, but it is important to confirm that any utility rate escalation is realistic.

3. **Does the contractor provide payment and performance bonds?** A payment and performance bond is the insurance that a performance contract will hold until the end of the contract term. It guarantees that if the ESCO should go out of business or fail to meet its promises, their obligation to the institution will be honored. The bonding company will assume the ESCO's "debt" if the ESCO shuts down. Even if a company seems unlikely to go out of business, a payment and performance bond is security against unforeseen circumstances. Be wary of any ESCO that cannot (or will not) provide a bond upon request.

4. **Has the company ever been involved in a lawsuit or litigation regarding a performance contract?** Understand if and why the company was involved in any lawsuits to avoid potential problems before you select an ESCO. Always check references before selecting an ESCO. Ensure they can provide contact names and phone numbers for several projects of different size and scope. Ask the ESCO to include references of projects that fell short of the savings guarantee. These contacts are the best resource for finding out the ESCO's reliability. Did they promptly reimburse the customer for the shortfall? The reputation of the ESCO is extremely important, because they are becoming a partner to the institution throughout the life of the guarantee.

6.3 INVESTMENT GRADE AUDIT

Upon completion of the selection process, the college chose Schneider Electric as the preferred ESCO for their performance contract. They entered into an investment grade audit (IGA) contract with Schneider Electric to proceed with the development and design of an energy conservation project for the campus.

The comprehensive nature of a performance contract requires a high level of communication and coordination between various stakeholders in the project. For the college, this included administration, maintenance personnel, the construction program manager (CPM), energy team personnel, and consultants to the college for the energy program. With so many vested parties, Schneider Electric focused early on identifying the specific interests and opinions within the group, as well as the process to obtain project approval.

With a current bond program in place, the campus was undergoing many changes at the time of the IGA. The relationship between Schneider Electric and the college CPM was crucial in identifying how the performance contract would interact with other campus projects. Because many buildings were undergoing major renovations, being completely demolished, or newly constructed, the campus master plan significantly impacted how an energy conservation project could be implemented and tracked.

Establishing a Baseline

The first step as part of the IGA was to establish a baseline of the energy usage on campus. To do this, Schneider Electric collected the most recent 26 months of utility data (electricity, natural gas, water), entered it into a utility analysis program, and compared usage over a two-year period to ensure the most accurate baseline period was selected. The data was prorated to determine electricity, natural gas, and water usage for each month over a two-year period. The electricity and natural gas data were then compared to actual weather data during that same two-year period. A correlation between daily energy usage and cooling and heating degree days was determined. This correlation was then applied to typical meteorological year weather data to get weather-normalized utility data. Weather-normalizing utility data removes any anomalies from utility data due to extreme weather conditions. No weather adjustments were made to the water data as there is no strong correlation between water consumption and weather.

Schneider Electric chose to use utility data from June 2007 through May 2008 as the baseline year data. This range of dates was selected as it represented the most recent utility data prior to major changes at the campus under the bond program. In the summer of 2008, the campus added the first of a new group of buildings, began the renovation of a classroom building, and began modifications to the central plant.

Actual tariffs from the utility company were used to establish a baseline and to estimate savings. The most recent two months of bills were analyzed along with the tariff structure from the utility company to create a tariff model. The weather-normalized baseline data was run through the tariff model to create a new utility cost baseline. This new baseline represents the cost of utilities during a typical meteorological year using current utility rates.

One challenge for Schneider Electric was to determine the amount of energy each campus building was using. This was challenging because the majority of the campus is tied to one central meter. To address the challenge, Schneider Electric built computer energy models for each of the buildings on campus. Inputs for these models were determined from information gathered during site visits, interviews of campus operations personnel, previous load studies, Schneider Electric's load studies, and as-built mechanical, electrical, plumbing, and architectural drawings.

Once the models were complete, it was necessary to create an adjusted energy baseline to account for the recent bond projects completed. This would ensure that when savings were measured, only the savings attributable to the performance contract were counted, not those of the other projects. To account for the changes, Schneider Electric developed energy models for the new buildings and made changes to the model for the central plant. The inputs for these models were determined from interviews with operations personnel, interviews with design architects and engineers, and as-built mechanical, electrical, plumbing and architectural drawings. Certain operating parameters had to be assumed, because some of these buildings had not been in operation long enough to determine a baseline.

Selecting Energy Conservation Measures

Schneider Electric investigated over 65 energy conservation measures for the college as part of the performance contract. The campus buildings were split into five categories according to the master plan:

1. Existing buildings with no planned major renovations in the next five to seven years.
2. Existing buildings with planned major renovations or demolition in the next five to seven years.
3. Renovated or new buildings with complete design drawings.
4. New buildings not yet in design.
5. Buildings to be demolished in the next two years.

The various categories dictated the payback requirements for the energy conservation measures investigated for each building. Figure 6.1 is a campus map showing the various buildings color-coded based on which category they fell into. Table 6.1 is an example of a chart created for a building in the first category.

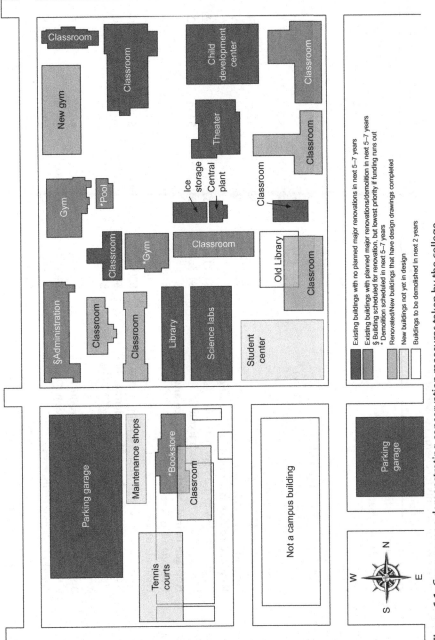

Figure 6.1 Campus map demonstrating conservation measures taken by the college.

Table 6.1 Payback Requirements for Energy Conservation Measures

ID #	EWCM Description	Savings	Cost	Rebates	Payback
1.1	**4'T8/T12 lighting upgrade**				
1.1b	Upgrade to Super T8	$2,170	$42,694	$1,350	19.1
1.1c	Upgrade to LED T8	$3,798	$95,906	$2,363	24.6
1.1a	Upgrade to T5	$1,492	$77,344	$928	51.2
1.1d	Upgrade to LED Panel	−$271	$281,531	$0	No savings
1.3	**Lighting controls upgrade**				
1.3a	Dual tech occupancy sensors (stand alone)	$27,641	$154,688	$17,719	5.0
1.3b	Advanced lighting controls package	$32,248	$253,125	$20,672	7.2
1.6	**Upgrade exit signs**				
1.6a	Upgrade to LED	$33	$900	$21	26.4
1.6b	Upgrade to photoluminescent	$44	$600	$28	12.9
2.1	**Replace toilets with new fixture and valves**				
2.1a	Upgrade toilets to 1.28 gpf fixtures	$1,214	$7,875	$404	6.2
2.1b	Upgrade toilets to dual-flush fixtures	$1,368	$10,275	$455	7.2
2.2	**Replace urinals with new fixture and valves**				
2.2a	Upgrade urinals to waterless	$328	$10,200	$3,931	19.1
2.2b	Upgrade urinals to 1/8 gpf	$301	$11,400	$3,604	25.9
2.3	Replace sink aerators with 1.0gpm	$202	$105	$134	−0.1
3.1	Replace existing boilers with condensing boilers	$8,098	$105,000	$0	13.0
6.1	Upgrade/replace/integrate existing EMS (scheduling, setpoint control)	$5,655	$84,525	$3,609	14.3
6.3	Install CO_2 monitoring in all classrooms	$1,845	$23,625	$622	12.5
6.4	On/off interlock on dampers/windows/doors	$4,453	$72,000	$3,082	15.5
7.3	Window glazing treatments	$1,812	$29,709	$2,321	15.1
8.2	Premium efficient motors on fans	$391	$2,625	$393	5.7

8.6	Phantom load control	$936	$18,675	$960	18.9
2.5	Install motion sensors on sinks	$191	$4,725	$127	24.0
5.1	Convert from local constant volume AHU to central VAV AHU	$19,101	$750,000	$13,080	38.6
7.1	Upgrade façade	$2,715	$108,933	$2,321	39.3
3.2	Connecting building to HW system	$1,884	$78,750	$0	41.8
2.4	Upgrade sinks from dual knob to single knob	$17	$900	$11	51.4
7.2	Install operable windows	$2,226	$148,545	$1,541	66.0
3.3	Convert constant flow HW system to variable flow HW system	$382	$37,500	$392	97.2
8.5	Two phase shift transformers/eliminate harmonics	$133	$45,000	$134	337.5
2.6	Install motion sensors on toilets	$0	$3,750	$0	No savings
2.10	Water leakage rate detection	$0	$51,750	$0	No savings
9.1	Pulsed output meters/HW/CHW meters	$0	$8,256	$0	No savings
9.2	Air quality annunciators	$0	$18,900	$0	No savings
9.3	Predictive maintenance	$0	$18,900	$0	No savings
9.4	Install kW meters on all buildings	$0	$4,725	$0	No savings
EWCMs already in place					
4.4	Connecting building to CHW system				
5.9	Air filtration				

Savings Calculations

The savings component of a performance contract is extremely important. Because performance contracting provides a comprehensive building solution, savings need to be calculated and measured in the same manner. When multiple energy conservation measures are implemented in a building together, they interact and impact the savings for each one. Furthermore, it is important that the baseline established is accurate, since that is what savings measurements will be based on.

Table 6.2 is an example of a common savings calculation that was provided to the college for LED lamps. The column on the left was completed by the lamp manufacturer while the column on the right was adjusted by Schneider Electric to incorporate proper savings calculation methodologies. As demonstrated, savings projections were skewed to reflect a payback of almost seven years faster due to inaccurate savings calculations, expressing the need for verification of savings calculations.

The college evaluated the data provided by Schneider Electric and selected energy conservation measures based on the payback requirements and needs of the campus. Table 6.3 shows the final list of measures implemented under the performance contract Figure 6.2 shows pictures and details of a few selected measures.

6.4 INSTALLATION

Once the IGA was complete, the college signed a construction contract with Schneider Electric to implement the selected energy conservation measures. A detailed schedule was made and weekly construction coordination meetings were held to ensure the project installation did not interfere with ongoing master plan projects.

Performance contracting brings many benefits to the construction phase of a project that differ from a traditional design-bid-build project.

1. **Selection of equipment and subcontractors**. The college has complete control over the type of equipment installed and the subcontractors utilized, rather than being required to use the low bid.
2. **Project management**. The ESCO acts as the general contractor overseeing all aspects of the project. This includes management of all subcontractors.
3. **Comprehensive project**. The project is implemented in one installation period, optimizing system interaction and maximizing savings over time.

Table 6.2 Savings Calculations for LED Lamps

Cost/Benefit Analysis	Manufacturer Methodology		Schneider Electric Corrections	
	Current Lighting	LED	Current Lighting	LED
Bulb cost	$2.00	$61.80	$2.00	$61.80
Ballast cost	$20.00	—	$20.00	—
Life of bulb	10,000	—	10,000	—
Life of ballast	20,000	—	20,000	—
Labor cost to replace bulb	$2.00	$0.00	$2.00	$0.00
Labor cost to replace ballast (prevailing wage)	$24.00	$0.00	$24.00	$0.00
Watts per bulb	32	15	29[1]	15
Number of fixtures/bulbs	402	402	402	402
Number of ballasts	201	0	201	0
Hours per day	24.00	24.00	17.04[2]	17.04[2]
kWh per day	340	144.72	199	102.778
Number of working days	30	30	30	30
Total kWh/month	10,188	4,342	5,961	3,083
kWh cost	$ 0.1400	$ 0.1400	$0.1100[3]	$0.1100[3]
Maintenance costs/bulb/month	$2.15	—	—[4]	—
HVAC load	10%	—	0%[5]	—
Cost per month to burn	$2,433.30	$607.82	$655.72	$339.17
Total annual cost to operate	$29,200.76	$7,293.89	$7,868.68	$4,070.01
Cost of waiting to retrofit (monthly)		$1,825.47		$316.56

(*Continued*)

Table 6.2 Savings Calculations for LED Lamps (Continued)

Cost/Benefit Analysis	Manufacturer Methodology		Schneider Electric Corrections	
	Current Lighting	LED	Current Lighting	LED
Total cost of bulbs		$24,844		$24,844
Total cost of labor		$6,030		$6,030
Total cost of retrofit		$30,874		$30,874
Total retrofit cost (benefit value)		$30,874		$30,874
Total savings monthly with LED		$1,825.47		$316.56
Total savings annually with LED		$21,905.67		$3,798.67
Total savings over 10 years		$219,056.68		$37,986.75
ROI in months		16.9		97.5
ROI in years		1.4		8.1

[1] A three-lamp T8 fixture draws 87 W including ballast, which equates to 29 W/lamp.

[2] Logging data at the college indicated a building average of 118 hours/week; at 30 days/month, this equates to 17 hours/day.

[3] The rate used by the manufacturer did not accurately reflect the college's utility data. Savings would need to be run through the actual utility tariff for accuracy. Schneider Electric determined a blended rate based on the college's baseline profile and current utility rates for comparison purposes only.

[4] The college instructed Schneider Electric to not include maintenance savings.

[5] HVAC load would save cooling energy, but increase heating energy. Building modeling is required for accurate analysis.

Table 6.3 Final List of Measures Implemented

ID #	EWCM Description	Utility Savings	Rebates	Estimated Cost	Simple Payback
1.0	Interior lighting upgrades	$72,896	$67,590	$1,222,722	15.8
1.1	Lighting control upgrades	$14,347	$7,340	$167,028	11.1
2.0	Plumbing fixture upgrades	$12,777	$19,350	$201,001	14.2
6.0	Upgrade and integrate EMS	$250,551	$27,141	$1,996,901	7.9
7.0	Window glazing upgrades	$10,535	$6,723	$156,482	14.2
8.0	Fan/pump motor upgrades	$2,275	$1,759	$35,308	14.7
9.0	Install utility submeters	$0	$0	$498,416	N/A
	Totals	$363,380	$129,903	$4,277,858	11.4

4. **Project savings guarantee**. Because there is a guarantee tied to the performance of the systems, the ESCO has a vested interest in making the systems operate optimally.
5. **Project cost**. There are no contractor-initiated change orders in performance contracting. Change orders occur only if a change is requested by the building owner. The price given is a guaranteed, fixed price.
6. **Rebates and incentives**. The ESCO will assist with applying for and obtaining the rebates and incentives associated with the project. Because the ESCO leads the development and design process, they have the necessary knowledge of the project to take this process on.
7. **Ongoing training and support**. The college receives postinstallation training and support to ensure the systems are being operated at maximum capability, generating the maximum savings.

6.5 MEASUREMENT AND VERIFICATION

The guarantee in a performance contract is meaningless if the measurement and verification is not done properly. Schneider Electric's scope for the California College project was limited to certain buildings on campus that were not planned to undergo major renovation as part of the bond program. Because of this, Schneider Electric needed a method to measure and verify the impact of the project apart from the impact of other concurrent projects. Schneider Electric and the college agreed on the solution of installing submeters at each building. These submeters will measure the amount of energy each building uses by tracking the electricity, hot water, chilled water, and natural gas.

EWCM-1.1 Lighting Control Upgrades

Lighting controls throughout the campus were retrofitted with dual-technology occupancy sensors. Occupancy sensors are designed to turn off lights when spaces are unoccupied. The dual-technology sensors use passive infrared and ultrasonic technologies to ensure accuracy in detecting occupancy.

EWCM-2.0 Plumbing Fixture Upgrades

Plumbing fixtures throughout the campus were retrofitted to replace 1.5 gallon per flush urinals with waterless urinals, 3.5 gallon per flush toilets with 1.28 gallon per flush toilets, and 2.2 gallon per minute sink aerators with 1.0 gallon per minute sink aerators.

EWCM-6.0 Upgrade & Integrate EMS

As part of the performance contract, the existing energy management system (EMS) was replaced with a direct digital control (DDC) system, allowing the maintenance team to easily monitor HVAC equipment from a central location. It also provides a much faster network for the EMS to communicate on, helping to minimize the amount of time spent viewing and troubleshooting the system.

The college had some buildings that had already been upgraded to a DDC system and those will be integrated into the new EMS making it possible to view a single front-end rather than multiple systems from various vendors.

EWCM-9.0 Install Utility Sub-Meters

Sub-meters for electricity, hot water, chilled water, and natural gas were installed on buildings as required for measurement and verification of the savings from the performance contract. In addition, electric meters were installed on every building on campus to assist the college with tracking utility usage on a building-by-building basis. Sub-meters were integrated with the campus Energy Management System for ease of use.

Figure 6.2 Examples of how the college will implement conservation data.

Performance contracting measurement and verification methodologies are described under the International Performance Measurement and Verification Protocol. There are four options under this protocol:

Option A: Partially measured retrofit isolation.

Option B: Retrofit isolation.

Option C: Whole facility.

Option D: Calibrated simulation.

Because the college did not have usable baseline data for individual buildings, making it impossible to compare pre- and postretrofit usage, Schneider Electric selected option D for the measurement and verification methodology. This method uses, a computer simulation to determine the baseline energy usage. The submeters on each building provide the postproject energy use, that calibrates a computer simulation of all energy conservation measures installed on campus. This calibrated computer model will be modified by removing the effect of the energy conservation measures to generate a simulation of the baseline energy use. The energy savings from the project at each facility will be the difference between the baseline and postretrofit computer models' energy use. The utility tariff will then be applied to the energy savings to determine the dollar savings from the project.

6.6 CONCLUSION

Performance contracting is an excellent option for facility owners to meet the challenge of going green. Beginning with conservation tied to a guarantee of performance, facility owners reap the benefits of sustainability without risk. An energy services company brings the expertise and flexibility necessary to accommodate any situation from tight budgets to ongoing master plan improvements.

REFERENCES

California Department of General Services – Division of the State Architect, 2009. Grid Neutral: Electrical Independence for California Schools and Community Colleges. Sacramento, CA.
U.S. Department of Energy, 2002. Myths About Energy in Schools. Washington, DC.

CHAPTER 7

Life-Cycle Analysis

The Economic Analysis of Demand-Side Programs and Projects in California

Don Schultz, Woodrow W. Clark, II, Ph.D, and Arnie Sowell*

Contents

7.1	Preface	100
7.2	Introduction	101
7.3	The Basic Methodology	103
	Demand-Side Management Categories and Program Definitions	104
	Basic Methods	106
	Balancing the Tests	109
	Limitations: Externality Values and Policy Rules	109
	Externality Values	109
	Policy Rules	110
7.4	Participant Test	110
	Definition	110
	Benefits and Costs	110
	How the Results Can Be Expressed	111
	Strengths of the Participant Test	112
	Weaknesses of the Participant Test	112
	Formulas	112
7.5	Ratepayer Impact Measure Test	115
	Definition	115
	Benefits and Costs	115
	How the Results Can Be Expressed	116
	Strengths of the Ratepayer Impact Measure Test	117
	Weaknesses of the Ratepayer Impact Measure Test	117
	Formulas	118
7.6	Total Resource Cost Test	121
	Definition	121
	How the Results Can Be Expressed	122
	Strengths of the Total Resource Cost Test	124
	Weakness of the Total Resource Cost Test	125
	Formulas	125

*With the California State Government Green Accounting Team (2001–2003) of forty active members.

Sustainable Communities Design Handbook
ISBN: 978-1-85617-804-4, DOI: 10.1016/B978-1-85617-804-4.00007-0

	7.7	Program Administrator Cost Test	126
		Definition	126
		Benefits and Costs	127
		How the Results Can Be Expressed	127
		Strengths of the Program Administrator Cost Test	128
		Weaknesses of the Program Administrator Cost Test	128
		Formulas	128
References			129
Appendix a		Inputs to Equations and Documentation	129
Appendix b		Summary of Equations and Glossary of Symbols	131
		Basic Equations	131
		Benefits and Costs	133
		Glossary of Symbols	134
Appendix c		Derivation of Rim Life-Cycle Revenue Impact Formula	136
		Rate Impact Measure	136

7.1 PREFACE

This chapter was completed in 2001 and made public policy in California initially by the California Public Utility Commission (CPUC) and later by the State of California General Services Agency for all government building in California. Now, ten years later, the Standard Practices Manual (SPM) has been implemented and enforced state-wide in government and now non-government organization sectors. Other states have followed too. However, the private sector in most communities has resisted the "life-cycle analysis" approach to economics and accounting. Today, in the second decade of the 21st Century, there is a growing understanding and need to see the costs for environmentally sound technologies and practices to be long term costs rather than short term (cost-benefit analysis). The result is that more changes and climate control processes can be installed and used in buildings and infrastructures ranging from individual sites to complexes of buildings, from homes to offices, and the infrastructures of energy, water, waste, and transportation needed to serve every community. They will see more significant technological changes (in fact the normal) that are less costly due to the longer term pay-back periods. These advances will most likely be reflected in standard accounting mechanisms such as leases, mortgages, and long term financial contracts. As of 2010, these standard accounting and economic practices are becoming more and more applicable in all sectors of the building and construction markets. Consider the question: how often in developed versus developing nations do people have outside toilets and no air conditioning or heating? The costs for these "standard" elements of any building that are now part of the mortgage or long term financing are the model in which renewable energy systems for buildings and communities will be established and common place.

7.2 INTRODUCTION

In 2001, the California governor's office organized and cochaired an Interagency Green Accounting Working Group (IGAWG), whose goal was to revise the *California Standard Practice Manual* (CSPM). The cochairs of that effort were Clark, Sowell, and Schultz.[1] The CSPM revision process was badly needed, since the manual had not been revised since 1987 under a Republican governor. It was necessary at this time, in particular, due to the California energy crisis (2000–2002 but continues into 2010) to do so and under a Democrat governor in order to:

- Ensure that all California state agencies, departments, and commissions apply a standard and common set of procedures when accounting for the life-cycle costs and associated benefits for projects.
- Provide guidelines for the billions of dollars that have been appropriated from a variety of sources (including ratepayers, the general fund, and bonds) to meet current and future energy and environmental project needs.
- Provide financial assistance to customers choosing to reduce their electricity loads on the central electric grid.
- Give financial guidelines to both public and private sectors.
- Create a "life-cycle analysis" accounting system that works to replace or offset the conventional (for-profit businesses) "cost/benefit" analysis system.
- Take into account the need to calculate the externalities, which include health costs and environmental impact.

The CSPM was revised, finalized, and published in a journal (2002), yet subject to periodic updates and revisions. State agencies, departments, and commissions are applying the methodology described in the CSPM

---

[1] Over 50 state analysts and experts participated in the *Standard Practices Manual* revision. They are listed in Appendix A at the end of this book, which reproduces the manual. The revision of the manual was truly a collaborative effort that lasted over eight months but resulted in significant accounting and analytical changes for projects and programs in California.

Purpose

These worksheets, developed by the California Interagency Green Accounting Working Group, are designed to calculate the various cost-benefit tests as prescribed in the **Standard Practice Manual: Economic Analysis of Demand-Side Programs and Projects** (October, 2001). Currently in place on the sheets are examples of the application of the cost effectiveness tests to various self-generation and energy efficiency programs. Future versions of this workbook may include sample calculations for load management programs.

Using the Spreadsheet

The input values in the worksheets can be modified for those who wish to use these worksheets to evaluate the cost-effectiveness of an actual energy efficiency or self-generation program or project. All values that should be modified are in blue. Changes in the input values on these worksheets to conduct analyses of actual programs will produce cost-benefit results that conform with the *SPM*.

All other values in black or red; any changes to the formulas or values in the black cells may produce cost-effectiveness results that do not conform with the *SPM*.

Avoided Costs

The avoided cost values used in the analysis in this spreadsheet---the primary parameter for establishing the benefits of reduced purchases of electricity from the central grid--are based on long term forecasts developed in the year 2000, and are currently used to estimate the life-cycle costs and benefits of energy efficiency and self generation programs under the regulatory oversight of the California Public Utilities Commission. When these avoided cost forecasts are updated, the updated forecasts will be incorporated into this spreadsheet by replacing the values shown in the Avoided Cost worksheet.

Box 7.1 CSPM Spreadsheet.

when planning and assessing the costs and benefits of various energy efficiency and self-generation programs being administered by various state agencies. Such use, however, does not assure approval of funding for individual projects financed by these programs, rather it informs the decision-making process by using a standardized methodology.

The IGAWG also contributed to the development of an economic spreadsheet (Box 7.1) that provides:

- Case example calculations of cost-effectiveness for portfolios of energy efficiency and self-generation programs and projects that conform with the CSPM methodology.
- Forecast of long-term avoided costs that are currently being used at the California Public Utilities Commission to forecast the life-cycle benefits of reductions in demand for energy provided by the major investor-owned utility distribution companies.

The CSPM spreadsheet (should be a useful tool for all state agencies to institutionalize the capability to conduct life-cycle benefit/cost analyses on programs and projects that reduce the demand for energy.

7.3 THE BASIC METHODOLOGY

Efficiency, conservation, and load management programs have been promoted since the 1970s by the California Public Utilities Commission (CPUC) and the California Energy Commission (CEC) as alternatives to power plant construction and gas supply options. Conservation and load management (C&LM) programs have been implemented in California by the major utilities through the use of ratepayer money and by the CEC pursuant to the CEC legislative mandate to establish energy efficiency standards for new buildings and appliances. The result is that California has been ranked consistently as one of the most energy efficient states.

While cost-effectiveness procedures for the CEC standards are outlined in the Public Resources Code, no such official guidelines existed for utility-sponsored programs. With the publication of the *Standard Practice for Cost-Benefit Analysis of Conservation and Load Management Programs* in February 1983, this void was substantially filled. With the informal "adoption" one year later of an appendix that identified cost-effectiveness procedures for an "all ratepayers" test, C&LM program cost-effectiveness consisted of the application of a series of tests representing a variety of perspectives—participants, nonparticipants, all ratepayers, society, and the utility. The *Standard Practices Manual* was revised again in 1987–1988. The primary changes (relative to the 1983 version) were:

1. Renaming the "nonparticipant test" the *ratepayer impact test*.
2. Renaming the "all-ratepayer test" the *total resource cost test*.
3. Treating the "societal test" as a variant of the "total resource cost test."
4. An expanded explanation of "demand-side" activities that should be subjected to standard procedures of benefit/cost analysis.

Further changes to the manual captured in the 2002 version (Clark et al 2002) were prompted by the cumulative effects of changes in the electric and natural gas industries and a variety of changes in California statutes related to these changes. As part of the major electric industry restructuring legislation of 1996 (AB 1890), for example, a public goods charge was established that ensured minimum funding levels for "cost-effective conservation and energy efficiency" for the 1998–2002 period, then (in 2000) this was extended until the year 2011. Additional legislation in 2000 (AB 1002) established a natural gas surcharge for similar purposes. Later in that year, the Energy Security and Reliability Act of 2000 (AB 970) directed the California Public Utilities Commission to establish, by spring 2001, a distribution charge to provide revenues for a self-generation program and a directive to consider changes to cost-effectiveness methods to better account for reliability concerns.

In spring 2001, a new state agency, the Consumer Power and Conservation Financing Authority, was created. This agency was expected to provide additional revenues, in the form of state revenue bonds, that could supplement the amount and type of public financial resources to finance energy efficiency and self-generation activities. By 2003, the agency closed due to lack of demand for funds.

The modifications to the *Standard Practices Manual* reflect these more recent developments in several ways. First, the utility cost test has been renamed the *program administrator test* and includes the assessment of programs managed by other agencies. Second, a definition of self-generation as a type of demand-side activity is included. Third, the description of the various potential elements of externalities in the societal version of the total resource cost test is expanded. Finally, the limitations section outlines the scope of the manual and elaborates on the processes traditionally instituted by implementing agencies to adopt values for these externalities and adopt the policy rules that accompany the manual.

Demand-Side Management Categories and Program Definitions

An important aspect of establishing standardized procedures for cost-effectiveness evaluations is the development and use of consistent definitions of categories, programs, and program elements.

This chapter employs the use of general program categories that distinguish among different types of demand-side management programs—conservation, load management, fuel substitution, load building, and self-generation. Conservation programs reduce electricity or natural gas consumption during all or significant portions of the year. *Conservation* in this context includes all "energy efficiency improvements." An energy efficiency improvement can be defined as reduced energy use for a comparable level of service, resulting from the installation of an energy efficiency measure or the adoption of an energy efficiency practice. The level of service may be expressed in such ways as the volume of a refrigerator, temperature levels, production output of a manufacturing facility, or lighting level per square foot. Load management programs may either reduce electricity peak demand or shift demand from peak to non-peak periods.

Fuel substitution and load building programs share the common feature of increasing annual consumption of either electricity or natural gas relative to what would have happened in the absence of the program. This effect is accomplished in significantly different ways, by inducing the choice of one fuel over another (fuel substitution) or by increasing sales of electricity, gas, or electricity and gas (load building). *Self-generation* refers to distributed generation (DG) installed on the customer's side of the electric utility meter, which serves some or all of the customer's electric load, that otherwise would have been provided by the central electric grid.

In some cases, self-generation products are applied in a combined heat and power manner, in which case the heat produced by the self-generation product is used on-site to provide some or all of the customer's thermal needs. Self-generation technologies include, but are not limited to, photovoltaics, wind turbines, fuel cells, microturbines, small gas-fired turbines, and gas-fired internal combustion engines.

Fuel substitution and load building programs were relatively new to demand-side management in California in the late 1980s, born out of the convergence of several factors that translated into average rates that substantially exceeded marginal costs. Proposals by utilities to implement programs that increase sales had prompted the need for additional procedures for estimating program cost-effectiveness. These procedures may be applicable in a new context. AB 970 amended the Public Utilities Code and provided the motivation to develop a cost-effectiveness method that can be used on a common basis to evaluate all programs that will remove electric load from the centralized grid, including energy efficiency, load control/demand responsiveness programs, and self-generation. Hence, self-generation was also added to the list of demand-side management programs for cost-effectiveness evaluation. In some cases, self-generation programs installed with incremental loads are also included, since the definition of self-generation is not necessarily confined to projects that reduce the electric load on the grid. For example, suppose an industrial customer installs a new facility with a peak consumption of 1.5 MW, with an integrated on-site 1.0 MW gas-fired DG unit. The combined impact of the new facility is *load building*, since the new facility can draw up to 0.5 MW from the grid, even when the DG unit is running. The proper characterization of each type of demand-side management program is essential to ensure the proper treatment of inputs and the appropriate interpretation of cost-effectiveness results.

Categorizing programs is important because in many cases the same device can be and should be evaluated in more than one category. For example, the promotion of an electric heat pump can and should be treated as part of a conservation program if the device is installed in lieu of a less efficient electric resistance heater. If the incentive induces the installation of an electric heat pump instead of gas space heating, however, the program needs to be considered and evaluated as a fuel substitution program. Similarly, natural-gas-fired self-generation units, as well as self-generation units using other nonrenewable fossil fuels, must be treated as fuel substitution. In common with other types of fuel substitution, any costs of gas transmission and distribution and environmental externalities must be accounted for. In addition, cost-effectiveness analyses of self-generation should account for utility interconnection costs. Similarly, a thermal energy storage device should be treated as a load management program when the predominant effect is to shift the load. If the acceptance of a utility incentive by the customer to install the energy storage device is a decisive aspect of the customer's decision to remain an electric utility customer (i.e., to reject or defer the option of installing a gas-fired cogeneration system), then the predominant effect of the thermal energy storage device has been to substitute electricity service for the natural gas service that would have occurred in the absence of the program.

In addition to fuel substitution and load building programs, recent utility program proposals have included reference to load retention, sales retention, market retention, and customer retention programs. In most cases, the effect of such programs is identical to either a fuel substitution or a load building program—sales of one fuel are increased relative to sales without the program. A case may be made, however, for defining a separate category of program called *load retention*. One unambiguous example of a load retention program is the situation where a program keeps a customer from relocating to another utility service area. However, computationally, the equations and guidelines included in the manual to accommodate fuel substitution and load building programs can handle this special situation as well.

Basic Methods

The chapter identifies the cost and benefit components and cost-effectiveness calculation procedures from four major perspectives: participant, ratepayer impact measure (RIM), program administrator cost (PAC), and total resource cost (TRC). A fifth perspective, the societal, is treated as a variation on the total resource cost test. The results of each perspective can be expressed in a variety of ways, but in all cases, it is necessary to calculate the net present value of program impacts over the life-cycle of those impacts.

Table 7.1 Cost-Effectiveness tests

Primary	Secondary
Participant	
Net present value (all participants)	Discounted payback (years)
	Benefit/cost ratio (BCR)
	Net present value (average participant)
Ratepayer impact measure	
Life-cycle revenue impact per unit of energy (kilowatt or therm) or demand customer (kilowatt)	Life-cycle revenue impact per unit Annual revenue impact (by year, per kilowatt hour, kilowatt, therm, or customer) First-year revenue impact (per kilowatt hour, kilowatt, therm, or customer)
Net present value	Benefit/cost ratio
Total resource cost	
Net present value (NPV)	Benefit/cost ratio
	Levelized cost (cents or dollars per unit of energy or demand)
	Societal (NPV, BCR)
Program administrator cost	
Net present value	Benefit/cost ratio
	Levelized cost (cents or dollars per unit of energy or demand)

Table 7.1 summarizes the cost-effectiveness tests addressed in the manual. For each of the perspectives, the table shows the appropriate means of expressing test results. The primary unit of measurement refers to the way of expressing test results considered by the staff of the two commissions as the most useful for summarizing and comparing demand-side management (DSM) program cost-effectiveness. Secondary indicators of cost-effectiveness represent *supplemental* means of expressing test results that are likely to be of particular value for certain types of proceedings, reports, or programs.

This chapter does not specify how the cost-effectiveness test results are to be displayed or the level at which cost-effectiveness is to be calculated (e.g., groups of programs, individual programs, and program elements for all or some programs). It is reasonable to expect different levels and types of results for different regulatory proceedings or different phases of the process used to establish proposed program funding levels. For example, for summary tables in general rate case proceedings at the CPUC, the most appropriate tests may be the RIM life-cycle revenue impact, total resource cost, and program administrator cost test results for programs or groups of programs. The analysis and review of program proposals for the same proceeding may include participant test results and various additional indicators of cost-effectiveness from all tests for each program element. In the case of cost-effectiveness evaluations conducted in the context of integrated long-term resource planning activities, such detailed examination of multiple indications of costs and benefits may be impractical.

Rather than identifying the precise requirements for reporting cost-effectiveness results for all types of proceedings or reports, the approach taken in the manual is to:

- Specify the components of benefits and costs for each of the major tests.
- Identify the equations to be used to express the results in acceptable ways.
- Indicate the relative value of the different units of measurement by designating primary and secondary test results for each test.

It should be noted that, for some types of demand-side management programs, meaningful cost-effectiveness analyses cannot be performed using the tests in the CSPM. The following guidelines are offered to clarify the appropriated "match" of different types of programs and tests:

1. For generalized information programs (e.g., when customers are provided generic information on means of reducing utility bills without the benefit of on-site evaluations or customer billing data), cost-effectiveness tests are not expected because of the extreme difficulty in establishing meaningful estimates of load impacts.

2. For any program where more than one fuel is affected, the preferred unit of measurement for the RIM test is the life-cycle revenue impacts per customer, with gas and electric components reported separately for each fuel type and for combined fuels.

3. For load building programs, only the RIM tests are expected to be applied. The total resource cost and program administrator cost tests are intended to identify cost-effectiveness relative to other resource options. It is inappropriate to consider increased load as an alternative to other supply options.

4. Levelized costs may be appropriate as a supplementary indicator of cost per unit for electric conservation and load management programs relative to generation options and gas conservation programs relative to gas supply options, but the levelized cost test is not applicable to fuel substitution programs (since they combine gas and electric effects) or load building programs (which increase sales).

The delineation of the various means of expressing test results in Table 7.1 is not meant to discourage the continued development of additional variations for expressing cost-effectiveness. Of particular interest is the development of indicators of program cost-effectiveness that can be used to assess the appropriateness of the program scope (i.e., level of funding) for general rate case proceedings. Additional tests, if constructed from the net present worth in conformance with the equations designated in the manual, could prove useful as a means of developing methodologies that will address issues such as the optimal timing and scope of demand-side management programs in the context of overall resource planning.

Balancing the Tests

The tests set forth in the manual are not intended to be used individually or in isolation. The results of tests that measure efficiency, such as the total resource cost test, the societal test, and the program administrator cost test, must be compared not only to each other but also to the ratepayer impact measure test. This multiperspective approach requires program administrators and state agencies to consider trade-offs among the various tests. Issues related to the precise weighting of each test relative to other tests and to developing formulas for the definitive balancing of perspectives are outside the scope of the manual. The manual, however, does provide a brief description of the strengths and weaknesses of each test (Sections 7.3, 7.4, 7.5, and 7.6) to assist users in qualitatively weighing test results.

Limitations: Externality Values and Policy Rules

The list of externalities identified in Section 7.5, in the discussion on the societal version of the total resource cost test is broad, illustrative, and by no means exhaustive. Traditionally, implementing agencies have independently determined the details, such as the components of the externalities, the externality values, and the policy rules that specify the contexts in which the externalities and the tests are used.

Externality Values

The values for the externalities have not been provided in the manual. Separate studies and methodologies are used to arrive at these values. Also separate processes must be instituted by implementing agencies before such values can be adopted formally.

Policy Rules

The appropriate choice of inputs and input components vary by program area and project. For instance, programs for low-income groups are evaluated using a broader set of nonenergy benefits, which have not been provided in detail in the manual. Implementing agencies traditionally have had the discretion to use or to not use these inputs and benefits on a project- or program-specific basis. The policy rules that specify the contexts in which it is appropriate to use the externalities, their components, and tests mentioned in the manual are an integral part of any cost-effectiveness evaluation. These policy rules are not a part of the manual.

To summarize, the manual provides the methodology and the cost/benefit calculations only. The implementing agencies (such as the CPUC and the CEC) have traditionally utilized open public processes to incorporate the diverse views of stakeholders before adopting externality values and policy rules that are an integral part of the cost-effectiveness evaluation.

7.4 PARTICIPANT TEST

Definition

The participant test is the measure of the quantifiable benefits and costs to the customer due to participation in a program. Since many customers do not base their decision to participate in a program entirely on quantifiable variables, this test cannot be a complete measure of the benefits and costs of a program to a customer.

Benefits and Costs

The benefits of participation in a demand-side program include the reduction in the customer's utility bill(s), any incentive paid by the utility or other third party, and any federal, state, or local tax credit received. The reductions to the utility bill(s) should be calculated using the actual retail rates that would have been charged for the energy service provided (electric demand or energy or gas). Savings estimates should be based on gross savings, as opposed to net energy savings.[2]

---

[2] Gross energy savings are considered to be the savings in energy and demand seen by the participant at the meter. These are the appropriate program impacts to calculate bill reductions for the participant test. Net savings are assumed to be the savings attributable to the program. That is, net savings are gross savings minus those changes in energy use and demand that would have happened even in the absence of the program. For fuel substitution and load building programs, gross-to-net considerations account for the impacts that would have occurred in the absence of the program.

In the case of fuel substitution programs, benefits to the participant also include the avoided capital and operating costs of the equipment or appliance not chosen. For load building programs, participant benefits include an increase in productivity or service, which is presumably equal to or greater than the productivity or service without participating. The inclusion of these benefits is not required for this test, but if they are included, then the societal test should also be performed.

The costs to a customer of program participation are all out-of-pocket expenses incurred as a result of participating in a program, plus any increases in the customer's utility bill(s). The out-of-pocket expenses include the cost of any equipment or materials purchased, including sales tax and installation; any ongoing operation and maintenance costs; any removal costs (less salvage value); and the value of the customer's time in arranging for the installation of the measure, if significant.

How the Results Can Be Expressed

The results of this test can be expressed in four ways: through a net present value per average participant, a net present value for the total program, a benefit/cost ratio, or discounted payback. The primary means of expressing test results is the net present value for the total program; discounted payback, benefit/cost ratio, and per participant net present value are secondary tests.

The discounted payback is the number of years it takes until the cumulative discounted benefits equal or exceed the cumulative discounted costs. The shorter the discounted payback, the more attractive or beneficial the program is to the participants. Although *payback period* is often defined as undiscounted in textbooks, a discounted payback period is used here to approximate more closely the consumer's perception of future benefits and costs.[3]

The net present value (NPV_p) gives the net dollar benefit of the program to an average participant or to all participants discounted over some specified time period. A net present value above 0 indicates that the program is beneficial to the participants under this test.

The benefit/cost ratio (BCR_p) is the ratio of the total benefits of a program to the total costs discounted over some specified time period. The benefit/cost ratio gives a measure of a rough rate of return for the program to the participants and is also an indication of risk. A benefit/cost ratio above 1 indicates a beneficial program.

---

[3] It should be noted that, if a demand-side program is beneficial to its participants ($NPV_p \geq 0$ and $BCR_p \geq 1.0$) using a particular discount rate, the program has an internal rate of return (IRR) of at least the value of the discount rate.

Strengths of the Participant Test

The participant test gives a good "first cut" of the benefit or desirability of the program to customers. This information is especially useful for voluntary programs, as an indication of potential participation rates.

For programs that involve a utility incentive, the participant test can be used for program design considerations, such as the minimum incentive level, whether incentives are really needed to induce participation, and whether changes in incentive levels will induce the desired amount of participation.

These test results can be useful for program penetration analyses and developing program participation goals, which minimize adverse ratepayer impacts and maximize benefits.

For fuel substitution programs, the participant test can be used to determine whether program participation (i.e., choosing one fuel over another) will be in the best interest of the customer in the long run. The primary means of establishing such assurances is the net present value, which looks at the costs and benefits of the fuel choice over the life of the equipment.

Weaknesses of the Participant Test

None of the participant test results (discounted payback, net present value, or benefit/cost ratio) accurately captures the complexities and diversity of customer decision-making processes for demand-side management investments. Until or unless more is known about customer attitudes and behavior, interpretations of participant test results continue to require considerable judgment. Participant test results play only a supportive role in any assessment of conservation and load management programs as alternatives to supply projects.

Formulas

The following are the formulas for discounted payback, the net present value (NPV_p) and the benefit-cost ratio (BCR_p) for the participant test:

$$NPV_p = B_p - C_p$$

$$NPV_{avp} = \frac{B_p - C_p}{P}$$

$$BCR_p = \frac{B_p}{C_p}$$

$$DP_p = \text{Min } j \text{ such that } B_j \geq C_j$$

where:

NPV_p = Net present value to all participants

NPV_{avp} = Net present value to the average participant

BCR_p = Benefit/cost ratio to participants

DP_p = Discounted payback in years

B_p = NPV of benefit to participants

C_p = NPV of costs to participants

B_j = Cumulative benefits to participants in year j

C_j = Cumulative costs to participants in year j

P = Number of program participants

j = First year in which cumulative benefits equal cumulative costs

d = Interest rate (discount)

The benefit (B_p) and cost (C_p) terms are further defined as follows:

$$B_p = \sum_{t=1}^{N} \frac{BR_t + TC_t + INC_t}{(1+d)^{t-1}} + \sum_{t=1}^{N} \frac{AB_{at} + PAC_{at}}{(1+d)^{t-1}}$$

$$C_p = \sum_{t=1}^{N} \frac{PC_t + BI_t}{(1+d)^{t-1}}$$

where:

BR_t = Bill reductions in year t

BI_t = Bill increases in year t

TC_t = Tax credits in year t

INC_t = Incentives paid to the participant by the sponsoring utility in year t[4]

PC_t = Participant costs in year t, including:

- Initial capital costs, including sales tax[5]
- Ongoing operation and maintenance costs including fuel cost
- Removal costs, less salvage value
- Value of the customer's time in arranging for installation, if significant

[4] Some difference of opinion exists as to what should be called an *incentive*. The term can be interpreted broadly to include almost anything. Direct rebates, interest payment subsidies, and even energy audits can be called *incentives*. Operationally, it is necessary to restrict the term to include only dollar benefits, such as rebates or rate incentives (monthly bill credits). Information and services such as audits are not considered incentives for the purposes of these tests. If the incentive is to offset a specific participant cost, as in a rebate-type incentive, the full customer cost (before the rebate must be included in the PC_t term.

[5] If money is borrowed by the customer to cover this cost, it may not be necessary to calculate the annual mortgage and discount this amount if the present worth of the mortgage payments equals the initial cost. This occurs when the discount rate used is equal to the interest rate of the mortgage. If the two rates differ (e.g., a loan offered by the utility), then the stream of mortgage payments should be discounted by the discount rate chosen.

PAC_{at} = Participant avoided costs in year t for alternative fuel devices (costs of devices not chosen)

Ab_{at} = Avoided bill from alternative fuel in year t

The first summation in the B_p equation should be used for conservation and load management programs. For fuel substitution programs, both the first and second summations should be used for B_p.

Note that, in most cases, the customer bill impact terms (BR_t, BI_t, and AB_{at}) are further determined by the costing period to reflect load impacts or rate schedules, which vary substantially by time of day and season. The formulas for these variables are as follows:

$$BR_t = \sum_{i=1}^{I}(\Delta EG_{it} \times AE : E_{it} \times K_{it})$$
$$+\sum_{i=1}^{I}(\Delta DG_{it} \times AC : D_{it} \times K_{it}) + OBR_t$$

For AB_{at}, use the BR_t formula, but with rates and costing periods appropriate for the alternate fuel utility.

$$BI_t = \sum_{i=1}^{I}[\Delta EG_{it} \times AC : E_{it} \times (K_{it} - 1)]$$
$$+\sum_{i=1}^{I}[\Delta DG_{it} \times AC : D_{it} \times (K_{it} - 1)] + OBI_t$$

where:

ΔEG_{it} = Reduction in gross energy use in costing period i in year t

ΔDG_{it} = Reduction in gross billing demand in costing period i in year t

$AC{:}E_{it}$ = Rate charged for energy in costing period i in year t

$AC{:}D_{it}$ = Rate charged for demand in costing period i in year t

K_{it} = 1 when ΔEG_{it} or ΔDG_{it} is positive (a reduction) in costing period i in year t, and 0 otherwise

OBR_t = Other bill reductions or avoided bill payments (e.g., customer charges, standby rates).

OBI_t = Other bill increases (i.e., customer charges, standby rates)

I = Number of periods of participant's participation

In load management programs such as time of use (TOU) rates and air-conditioning cycling, there are often no direct customer hardware costs. However, attempts should be made to quantify indirect costs customers may incur that enable them to take advantage of TOU rates and similar programs.

If no customer hardware costs are expected, or estimates of indirect costs and value of service are unavailable, it may not be possible to calculate the benefit/cost ratio and discounted payback period.

7.5 RATEPAYER IMPACT MEASURE TEST

Definition

The ratepayer impact measure test[6] measures what happens to customer bills or rates due to changes in utility revenues and operating costs caused by the program. Rates go down if the change in revenues from the program is greater than the change in utility costs. Conversely, rates or bills go up if revenues collected after program implementation are less than the total costs incurred by the utility in implementing the program. This test indicates the direction and magnitude of the expected change in customer bills or rate levels.

Benefits and Costs

The benefits calculated in the RIM test are the savings from avoided supply costs. These avoided costs include the reduction in transmission, distribution, generation, and capacity costs for periods when the load has been reduced and the increase in revenues for any periods in which the load has been increased. The avoided supply costs are a reduction in total costs or revenue requirements and are included for both fuels for a fuel substitution program. The increase in revenues are also included for both fuels for fuel substitution programs. Both the reductions in supply costs and the revenue increases should be calculated using net energy savings.

The costs for this test are the program costs incurred by the utility or other entities incurring costs and creating or administering the program, the incentives paid to the participant, decreased revenues for any periods in which the load has been decreased, and increased supply costs for any periods when the load has been increased. The utility program costs include initial and annual costs, such as the cost of equipment, operation and maintenance, installation, program administration, and customer dropout, and removal of equipment (less salvage value). The decreases in

---

[6]The ratepayer impact measure test was previously described under what was called the *nonparticipant test*. The nonparticipant test has also been called the *impact on rate levels test*.

revenues and the increases in the supply costs should be calculated for both fuels for fuel substitution programs, using net savings.

How the Results Can Be Expressed

The results of this test can be presented in several forms: the life-cycle revenue impact (cents or dollars) per kilowatt hour, kilowatt, therm, or customer; annual or first-year revenue impacts (cents or dollars per kilowatt hour, kilowatt, therms, or customer); benefit/cost ratio; and net present value. The primary units of measurement are the life-cycle revenue impact, expressed as the change in rates (cents per kilowatt hour for electric energy, dollars per kilowatt for electric capacity, cents per therm for natural gas) and the net present value. Secondary test results are the life-cycle revenue impact (LRI) per customer, first-year and annual revenue impacts, and the benefit/cost ratio. LRI_{RIM} values for programs affecting electricity and gas should be calculated for each fuel individually (cents per kilowatt hour or dollars per kilowatt and cents per therm) and on a combined gas and electric basis (cents per customer).

The life-cycle revenue impact is the one-time change in rates or the bill change over the life of the program needed to bring total revenues in line with revenue requirements over the life of the program. The rate increase or decrease is expected to be put into effect in the first year of the program. Any successive rate changes, such as for cost escalation, are made from there. The first-year revenue impact (FRI) is the change in rates in the first year of the program or the bill change needed to get total revenues to match revenue requirements for only that year. The annual revenue impact (ARI) is the series of differences between revenues and revenue requirements in each year of the program. This series shows the cumulative rate change or bill change in a year needed to match revenues to revenue requirements. Thus, the ARI_{RIM} for year 6 per kilowatt hour is the estimate of the difference between present rates and the rate that would be in effect in year 6 due to the program. For results expressed as life-cycle, annual, or first-year revenue impacts, negative results indicate favorable effects on the bills of ratepayers or reductions in rates. Positive test result values indicate adverse bill impacts or rate increases.

Net present value (NPV_{RIM}) gives the discounted dollar net benefit of the program from the perspective of rate levels or bills over some specified time period. A net present value above 0 indicates that the program benefits (lowers) rates and bills.

The benefit/cost ratio (BCR_{RIM}) is the ratio of the total benefits of a program to the total costs discounted over some specified time period. A benefit/cost ratio above 1 indicates that the program lowers rates and bills.

Strengths of the Ratepayer Impact Measure Test

In contrast to most supply options, demand-side management programs cause a direct shift in revenues. Under many conditions, revenues lost from DSM programs have to be made up by ratepayers. The RIM test is the only test that reflects this revenue shift, along with the other costs and benefits associated with the program.

An additional strength of the RIM test is that the test can be used for all demand-side management programs (conservation, load management, fuel substitution, and load building). This makes the RIM test particularly useful for comparing impacts among demand-side management options.

Some of the units of measurement for the RIM test are of greater value than others, depending on the purpose or type of evaluation. The life-cycle revenue impact per customer is the most useful unit of measurement when comparing the merits of programs with highly variable scopes (e.g., funding levels) and when analyzing a wide range of programs that include both electric and natural gas impacts. Benefit/cost ratios can also be very useful for program design evaluations to identify the most attractive programs or program elements.

If comparisons are being made between a program or group of conservation or load management programs and a specific resource project, life-cycle cost per unit of energy and annual and first-year net costs per unit of energy are the most useful ways to express test results. Of course, this requires developing life-cycle, annual, and first-year revenue impact estimates for the supply-side project.

Weaknesses of the Ratepayer Impact Measure Test

The results of the RIM test are probably less certain than those of other tests, because the test is sensitive to the differences between long-term projections of marginal costs and long-term projections of rates, two cost streams that are difficult to quantify with certainty.

RIM test results are also sensitive to assumptions regarding the financing of program costs. Sensitivity analyses and interactive analyses that capture feedback effects between system changes, rate design options, and alternative means of financing generation and nongeneration options can help overcome these limitations. However, these types of analyses may be difficult to implement.

An additional caution must be exercised in using the RIM test to evaluate a fuel substitution program with multiple end use efficiency options.

For example, under conditions where marginal costs are less than average costs, a program that promotes an inefficient appliance may give a more favorable test result than a program that promotes an efficient appliance. Although the results of the RIM test accurately reflect rate impacts, the implications for long-term conservation efforts need to be considered.

Formulas

The formulas for the life-cycle revenue impact (LRI_{RIM}) on net present value (NPV_{RIM}), benefit/cost ratio (BCR_{RIM}), the first-year revenue impacts (FRI_{RIM}), and annual revenue impacts (ARI_{RIM}) follow:

$$LRI_{RIM} = \frac{C_{RIM} - B_{RIM}}{E}$$

$$FRI_{RIM} = \frac{C_{RIM} - B_{RIM}}{E}, \quad \text{for } t = I$$

$$ARI_{RIM,} = FRI_{RIM}, \quad \text{for } t = I$$
$$= \frac{C_{RIM,} - B_{RIM,}}{E_t}, \quad \text{for } t = 2, ..., N$$

$$NPV_{RIM} = B_{RIM} - C_{RIM}$$

$$BCR_{RIM'} = \frac{B_{RIM}}{C_{RIM}}$$

where:

LRI_{RIM} = Life-cycle revenue impact of the program per unit of energy (kilowatt hour or therm) or demand (kilowatt) (the one-time change in rates) or per customer (the change in customer bills over the life of the program). (Note: An appropriate choice of kilowatt hour, therm, kilowatt, and customer should be made.)

FRI_{RIM} = First-year revenue impact of the program per unit of energy, demand, or per customer

ARI_{RIM} = Stream of cumulative annual revenue impacts of the program per unit of energy, demand, or per customer (Note: The terms in the ARI formula are not discounted; therefore, they are the nominal cumulative revenue impacts. Discounted cumulative revenue impacts may be calculated and submitted if they are indicated as such. Note also that the sum of the discounted stream of cumulative revenue impacts *does not* equal the LRI_{RIM}.)

NPV_{RIM} = Net present value levels

BCR_{RIM} = Benefit/cost ratio for rate levels

B_{RIM} = Benefits to rate levels or customer bills

C_{RIM} = Costs to rate levels or customer bills

E = Discounted stream of system energy sales (kilowatt hours or therms), demand sales (kilowatts), or first-year customers (see Appendix c of this chapter for a description of the derivation and use of this term in the LRI_{RIM} test.)

The B_{RIM} and C_{RIM} terms are further defined as follows:

$$B_{RIM} = \sum_{t=1}^{N} \frac{UAC_t + RG_t}{(1 + d)^{t-1}} + \sum_{t=1}^{N} \frac{UAC_{at}}{(1 + d)^{t-1}}$$

$$C_{RIM} = \sum_{t=1}^{N} \frac{UIC_t + RL_t + PRC_t + INC_t}{(1 + d)^{t-1}} + \sum_{t=1}^{N} \frac{RL_{at}}{(1 + d)^{t-1}}$$

$$E = \sum_{t=1}^{N} \frac{E_t}{(1 + d)^{t-1}}$$

where:

UAC_t = Utility avoided supply costs in year t

UIC_t = Utility increased supply costs in year t

RG_t = Revenue gain from increased sales in year t

RL_t = Revenue loss from reduced sales in year t

PRC_t = Program administrator program costs in year t

E_t = System sales in kilowatt hours, kilowatts, or therms in year t or first-year customers

UAC_{at} = Utility avoided supply costs for the alternative fuel in year t

RL_{at} = Revenue loss from avoided bill payments for alternate fuel in year t (i.e., device not chosen in a fuel substitution program)

For fuel substitution programs, the first term in the B_{RIM} and C_{RIM} equations represents the sponsoring utility (electric or gas), and the second term represents the alternative utility. The RIM test should be calculated separately for electric and gas and combined electric and gas.

The utility avoided cost terms (UAC_t, UIC_t, and UAC_{at}) are further determined by costing period to reflect time-variant costs of supply:

$$UAC_t = \sum_{i=1}^{I}(\Delta EN_{it} \times MC : E_{it} \times K_{it})$$
$$+\sum_{i=1}^{I}(\Delta DN_{it} \times MC : D_{it} \times K_{it})$$

For UAC_{at} use the UAC_t formula, but with marginal costs and costing periods appropriate for the alternate fuel utility:

$$UIC_t = \sum_{i=1}^{I}[\Delta EN_{it} \times MC : E_{it} \times (K_{it} - 1)]$$
$$+\sum_{i=1}^{I}[\Delta DN_{it} \times MC : D_{it} \times (K_{it} - 1)]$$

where (only terms not previously defined are included here):
ΔEN_{it} = Reduction in net energy use in costing period i in year t
ΔDN_{it} = Reduction in net demand in costing period i in year t
$MC{:}E_{it}$ = Marginal cost of energy in costing period i in year t
$MC{:}D_{it}$ = Marginal cost of demand in costing period i in year t

The revenue impact terms (RG_t, RL_t, and RL_{at}) are parallel to the bill impact terms in the participant test. The terms are calculated in exactly the same way, with the exception that the net impacts are used rather than gross impacts. If a net-to-gross ratio is used to differentiate gross savings from net savings, the revenue terms and the participant's bill terms will be related as follows:

$$RG_t = BI_t \times (\text{net-to-gross ratio})$$

$$RL_t = BR_t \times (\text{net-to-gross ratio})$$

$$RL_{at} = AB_{at} \times (\text{net-to-gross ratio})$$

7.6 TOTAL RESOURCE COST TEST

Definition

The total resource cost test[7] measures the net costs of a demand-side management program as a resource option based on the total costs of the program, including both the participants' and the utility's costs.

The test is applicable to conservation, load management, and fuel substitution programs. For fuel substitution programs, the test measures the net effect of the impacts from the fuel not chosen versus the impacts from the fuel chosen as a result of the program. TRC test results for fuel substitution programs should be viewed as a measure of the economic efficiency implications of the total energy supply system (gas and electric).

A variant on the TRC test is the societal test. The societal test differs from the TRC test in that it includes the effects of externalities (e.g., environment, national security), excludes tax credit benefits, and uses a different (societal) discount rate.

This test represents the combination of benefits and costs of a program on both the customers participating and those not participating in a program. In a sense, it is the summation of the benefit and cost terms in the participant and the ratepayer impact measure tests, where the revenue (bill) change and the incentive terms intuitively cancel (except for the differences in net and gross savings).

The *benefits* calculated in the total resource cost test are the avoided supply costs—the reduction in transmission, distribution, generation, and capacity costs valued at the marginal cost—for the periods when there is a load reduction. The avoided supply costs should be calculated using net program savings, savings net of changes in energy use that would have happened in the absence of the program. For fuel substitution programs, benefits include the avoided device costs and avoided supply costs for the energy-using equipment not chosen by the program participant.

The *costs* in this test are the program costs paid by both the utility and the participants plus the increase in supply costs for the periods in which load is increased. Thus, *all* equipment costs, installation, operation and maintenance, cost of removal (less salvage value), and administration costs, no matter who pays for them, are included in this test. Any tax credits are considered a reduction to the costs in this test. For fuel substitution programs, the costs also include the increase in supply costs for the utility providing the fuel that is chosen as a result of the program.

---

[7]This test was previously called the *all ratepayers test*.

How the Results Can Be Expressed

The results of the total resource cost test can be expressed in several forms: as a net present value, a benefit/cost ratio, or as a levelized cost. The net present value is the primary unit of measurement for this test. Secondary means of expressing TRC test results are a benefit/cost ratio and levelized costs. The societal test—expressed in terms of net present value, a benefit/cost ratio, or levelized costs—is also considered a secondary means of expressing results. Levelized costs as a unit of measurement are inapplicable for fuel substitution programs, since these programs represent the net change of alternative fuels, which are measured in different physical units (e.g., kilowatt hours or therms). Levelized costs are also not applicable for load building programs.

The net present value (NPV_{TRC}) is the discounted value of the net benefits to this test over a specified period of time. NPV_{TRC} is a measure of the change in the total resource costs due to the program. A net present value above 0 indicates that the program is a less expensive resource than the supply option on which the marginal costs are based.

The benefit/cost ratio (BCR_{TRC}) is the ratio of the discounted total benefits of the program to the discounted total costs over some specified time period. It gives an indication of the rate of return on this program to the utility and its ratepayers. A benefit/cost ratio above 1 indicates that the program is beneficial to the utility and its ratepayers on a total resource cost basis.

The levelized cost is a measure of the total costs of the program in a form that is sometimes used to estimate costs of utility-owned supply additions. It presents the total costs of the program to the utility and its ratepayers on a per kilowatt, per kilowatt hour, or per therm basis levelized over the life of the program.

The societal test is structurally similar to the total resource cost test. It goes beyond the TRC test in that it attempts to quantify the change in the total resource costs to society as a whole rather than to only the service territory (the utility and its ratepayers). In taking society's perspective, the societal test utilizes essentially the same input variables as the TRC test, but they are defined with a broader societal point of view. More specifically, the societal test differs from the TRC test in at least one of five ways. First, the societal test may use higher marginal costs than the TRC test if a utility faces marginal costs that are lower than other utilities in the state or its out-of-state suppliers. Marginal costs used in the societal test would reflect the cost to *society* of the more expensive alternative resources. Second, tax credits are treated as a transfer payment in the societal test and therefore are left out. Third, in the case of capital expenditures, interest payments are considered a transfer payment, since society actually expends the resources in the first year. Therefore, capital costs enter the calculations in the year in which they occur. Fourth, a societal discount rate should be used.[8] Finally, marginal costs used in the societal test would also contain externality costs of power generation not captured by the market system. An illustrative and by no means exhaustive list of externalities and their components follows. (Refer to the "Weakness of the total resource cost test" section for elaboration.) These values are also referred to as *adders*, designed to capture or internalize such externalities. The list of potential adders would include, for example,

1. **The benefit of avoided environmental damage.** The CPUC policy specifies two "adders" to internalize environmental externalities, one for electricity use and one for natural gas use. Both are statewide average values. These adders are intended to help distinguish between cost-effective and non–cost-effective energy efficiency programs. They apply to an average supply mix and would not be useful in distinguishing among competing supply options. The CPUC electricity environmental adder is intended to account for the environmental damage from air pollutant emissions from power plants. The CPUC-adopted adder is intended to cover the human and material damage from sulphur oxides (SO_X), nitrogen oxides (NO_X), volatile organic compounds (VOCs, sometimes called *reactive organic gases*, ROGs), particulate matter at or below 10 micron diameter (PM_{10}), and carbon. The adder for natural gas is intended to account for air pollutant emissions from the direct combustion of the gas. In the CPUC policy guidance, the adders are included in the tabulation of the benefits of energy efficiency programs. They represent reduced environmental damage from displaced electricity generation and avoided gas combustion. The environmental damage is the result of the net change in pollutant emissions in the air basins, or regions, in which there is an impact. This change is the result of direct changes in power plant or natural gas combustion emission resulting from the efficiency measures and changes in emissions from other sources that result from those direct changes in emissions.

---

[8] Many economists have pointed out that use of a market discount rate in social cost/benefit analysis undervalues the interests of future generations. Yet, if a market discount rate is not used, comparisons with alternative investments are difficult to make.

2. **Avoided transmission and distribution costs.** The benefit of avoided transmission and distribution costs, energy efficiency measures that reduce the growth in peak demand, decrease the required rate of expansion to the transmission and distribution network, eliminating costs of constructing and maintaining new or upgraded lines.

3. **Avoided generation costs.** The benefit of avoided generation costs, energy efficiency measures to reduce consumption and hence avoid the need for generation, include avoided energy costs, capacity costs, and a T&D line.

4. **Increased system reliability.** The reductions in demand and peak loads from customers opting for self-generation provide reliability benefits to the distribution system in the forms of:
   - Avoided costs of supply disruptions.
   - Benefits to the economy of damage and control costs avoided by customers and industries that need greater than a 99.9 level of reliable electricity service from the central grid, since these industries depend on the electronics delivered from electrical systems.
   - Marginally decreased system operator's costs to maintain a percentage reserve of electricity supply above the instantaneous demand.
   - Benefits to customers and the public of avoiding blackouts.

5. **Nonenergy benefits.** Nonenergy benefits might include a range of program-specific benefits, such as saved water in energy efficient washing machines or self-generation units, reduced waste streams from an energy efficient industrial process.

6. **Nonenergy benefits for low-income programs.** The low-income programs are social programs, which have a separate list of benefits included in what is known as the *low-income public purpose test.* This test and the specific benefits associated with it are outside the scope of this chapter.

7. **Fuel diversity.** Benefits of fuel diversity include considerations of the risks of supply disruption, the effects of price volatility, and the avoided costs of risk exposure and risk management.

Strengths of the Total Resource Cost Test

The primary strength of the total resource cost test is its scope. The test includes total costs (participant plus program administrator) and also has the potential for capturing total benefits (avoided supply costs plus, in the case of the societal test variation, externalities). To the extent supply–side project evaluations also include total costs of generation and transmission, the TRC test provides a useful basis for comparing demand- and supply-side options.

Since this test treats incentives paid to participants and revenue shifts as transfer payments (from all ratepayers to participants through increased revenue requirements), the test results are unaffected by the uncertainties of projected average rates, thus reducing the uncertainty of the test results. Average rates and assumptions associated with how other options are financed (analogous to the issue of incentives for DSM programs) are also excluded from most supply-side cost determinations, again making the TRC test useful for comparing demand-side and supply-side options.

Weakness of the Total Resource Cost Test

The treatment of revenue shifts and incentive payments as transfer payments—identified previously as a strength—can also be considered a weakness of the TRC test. While it is true that most supply-side cost analyses do not include such financial issues, it can be argued that DSM programs *should* include these effects, since in contrast to most supply options, DSM programs do result in lost revenues.

In addition, the costs of the DSM "resource" in the TRC test are based on the total costs of the program, including costs incurred by the participant. Supply-side resource options are typically based only on the costs incurred by the power suppliers.

Finally, the TRC test cannot be applied meaningfully to load building programs, thereby limiting the ability to use this test to compare the full range of demand-side management options.

Formulas

The formulas for the net present value (NPV_{TRC}), the benefit/cost ratio (BCR_{TRC}), and levelized costs follow:

$$NPV_{TRC} = B_{TRC} - C_{TRC}$$

$$BCR_{TRC} = \frac{B_{TRC}}{C_{TRC}}$$

$$LC_{TRC} = \frac{LCRC}{IMP}$$

where:

NPV_{TRC} = Net present value of total costs of the resource

BCR_{TRC} = Benefit/cost ratio of total costs of the resource

LC_{TRC} = Levelized cost per unit of the total cost of the resource (cents per kilowatt hour for conservation programs; dollars per kilowatt for load management programs)

B_{TRC} = Benefits of the program

C_{TRC} = Costs of the program

LCRC = Total resource costs used for levelizing

IMP = Total discounted load impacts of the program

PCN = Net participant costs

The B_{TRC}, C_{TRC}, LCRC, and IMP terms are further defined as follows:

$$B_{TRC} = \sum_{t=1}^{N} \frac{UAC_t + TC_t}{(1+d)^{t-1}} + \sum_{t=1}^{N} \frac{UAC_{at} + PAC_{at}}{(1+d)^{t-1}}$$

$$C_{TRC} = \sum_{t=1}^{N} \frac{PRC_t + PCN_t + UIC_t}{(1+d)^{t-1}}$$

$$LCRC = \sum_{t=1}^{N} \frac{PRC_t + PCN_t + TC_t}{(1+d)^{t-1}}$$

$$IMP = \sum_{t-1}^{N} \frac{\left[\left(\sum_{i=1}^{I} \Delta EN_{it} \right), \text{ or } (\Delta DN_{it}, \text{ where } i = \text{peak period}) \right]}{(1+d)^{t-1}}$$

where all terms have been defined in previous sections.

The first summation in the B_{TRC} equation should be used for conservation and load management programs. For fuel substitution programs, both the first and second summations should be used.

7.7 PROGRAM ADMINISTRATOR COST TEST

Definition

The program administrator cost test measures the net costs of a demand-side management program as a resource option based on the costs incurred by the program administrator (including incentive costs) and excluding any net costs incurred by the participant. The benefits are similar to the TRC benefits. Costs are defined more narrowly.

Benefits and Costs

The benefits for the program administrator cost test are the avoided supply costs of energy and demand—the reduction in transmission, distribution, generation, and capacity valued at their marginal costs—for the periods when there is a load reduction. The avoided supply costs should be calculated using net program savings, savings net of changes in energy use that would have happened in the absence of the program. For fuel substitution programs, benefits include the avoided supply costs for the energy-using equipment not chosen by the program participant only in the case of a combination utility where the utility provides both fuels.

The costs for the program administrator cost test are the program costs incurred by the administrator, the incentives paid to the customers, and the increased supply costs for the periods in which the load is increased. Administrator program costs include initial and annual costs, such as the cost of utility equipment, operation and maintenance, installation, program administration, and customer dropout and removal of equipment (less salvage value). For fuel substitution programs, costs include the increased supply costs for the energy-using equipment chosen by the program participant only in the case of a combination utility, as previously.

In this test, revenue shifts are viewed as a transfer payment between participants and all ratepayers. Although a shift in revenue affects rates, it does not affect revenue requirements, which are defined as the difference between the net marginal energy and capacity costs avoided and program costs. Thus, if $NPV_{pa} > 0$ and $NPV_{RIM} < 0$, the administrator's overall total costs decrease, although rates may increase because the sales base over which revenue requirements are spread has decreased.

How the Results Can Be Expressed

The results of this test can be expressed either as a net present value, benefit/cost ratio, or levelized costs. The net present value is the primary test, and the benefit/cost ratio and levelized cost are the secondary tests.

Net present value (NPV_{pa}) is the benefit of the program minus the administrator's costs, discounted over some specified period of time. A net present value above 0 indicates that this demand-side program decreases costs to the administrator and the utility.

The benefit/cost ratio (BCR_{pa}) is the ratio of the total discounted benefits of a program to the total discounted costs for a specified time period. A benefit/cost ratio above 1 indicates that the program benefits the combined administrator and utility's total cost situation.

The levelized cost is a measure of the costs of the program to the administrator in a form that is sometimes used to estimate costs of utility-owned supply additions. It represents the costs of the program to the administrator and the utility on a per kilowatt, per kilowatt hour or per therm basis levelized over the life of the program.

Strengths of the Program Administrator Cost Test

As with the total resource cost test, the program administrator cost test treats revenue shifts as transfer payments, meaning that test results are not complicated by the uncertainties associated with long-term rate projections and associated rate design assumptions. In contrast to the total resource cost test, the program administrator test includes only the portion of the participant's equipment costs paid for by the administrator in the form of an incentive. Therefore, for purposes of comparison, costs in the program administrator cost test are defined similarly to those supply-side projects, which also do not include direct customer costs.

Weaknesses of the Program Administrator Cost Test

By defining device costs exclusively in terms of costs incurred by the administrator, the program administrator cost test results reflect only a portion of the full costs of the resource.

The program administrator cost test shares two limitations noted previously for the total resource cost test:
1. By treating revenue shifts as transfer payments, the rate impacts are not captured.
2. The test cannot be used to evaluate load building programs.

Formulas

The formulas for the net present value, the benefit/cost ratio, and levelized cost follow:

$$NPV_{pa} = B_{pa} - C_{pa}$$

$$BCR_{pa} = \frac{B_{pa}}{C_{pa}}$$

$$LC_{pc} = \frac{LC_{pc}}{IMP}$$

where:

NPV_{pa} = Net present value of program administrator costs

BCR_{pa} = Benefit/cost ratio of program administrator costs

LC_{pa} = Levelized cost per unit of program administrator cost of the resource

B_{pa} = Benefits of the program

C_{pa} = Costs of the program

LC_{pc} = Total program administrator costs used for levelizing

The formulas for the last three are:

$$B_{pa} = \sum_{t=1}^{N} \frac{UAC_t}{(1+d)^{t-1}} + \sum_{t+1}^{N} \frac{UAC_{at}}{(1+d)^{t-1}}$$

$$C_{pa} = \sum_{t=1}^{N} \frac{PRC_t + INC_t + UIC_t}{(1+d)^{t-1}}$$

$$LC_{pc} = \sum_{t=1}^{N} \frac{PRC_t + INC_t}{(1+d)^{t-1}}$$

where all the variables are defined in previous sections.

The first summation in the B_{pa} equation should be used for conservation and load management programs. For fuel substitution programs, both the first and second summations should be used.

REFERENCE

Clark II, W.W., Sowell, A., Schultz, D., 2002. Standard practice manual: the economic analysis of demand − side programs and projects in California. Int. J. Revenue Manage. 10.

APPENDIX a: INPUTS TO EQUATIONS AND DOCUMENTATION

A comprehensive review of procedures and sources for developing inputs is beyond the scope of this chapter. It would also be inappropriate to attempt a complete standardization of the techniques and procedures for developing inputs for such parameters as load impacts, marginal costs, or average rates.

Nevertheless, a series of guidelines can help establish acceptable procedures and improve the chances of obtaining reasonable levels of consistent and meaningful cost-effectiveness results. The following "rules" should be viewed as appropriate guidelines for developing the primary inputs for the cost-effectiveness equations contained in the chapter:

1. In the past, marginal costs for electricity were based on production cost model simulations that clearly identify the key assumptions and characteristics of the existing generation system as well as the timing and nature of any generation additions or power purchase agreements in the future. With a deregulated market for wholesale electricity, marginal costs for electric generation energy should be based on forecast market prices, which are derived from recent transactions in California energy markets. Such transactions could include spot market purchases as well as longer-term bilateral contracts and the marginal costs should be estimated based on components for energy as well as demand and capacity costs, as is typical for these contracts.

2. In the case of submittals in conjunction with a utility rate proceeding, average rates used in DSM program cost-effectiveness evaluations should be based on proposed rates. Otherwise, average rates should be based on current rate schedules. Evaluations based on alternative rate designs are encouraged.

3. Time-differentiated inputs for electric marginal energy and capacity costs, average energy rates, and demand charges, and electric load impacts should be used for:
   • Load management programs.
   • Any conservation program that involves a financial incentive to the customer.
   • Any fuel substitution or load building program.
   Costing periods used should include, at a minimum, summer and winter, on-, and off-peak periods; further disaggregation is encouraged.

4. When program participation includes customers with different rate schedules, the average rate inputs should represent an average weighted by the estimated mix of participation or impacts. For general rate case proceedings, it is likely that each major rate class within each program will be considered as a program element, requiring separate cost-effectiveness analyses for each measure and each rate class within each program.

5. Program administration cost estimates used in program cost-effectiveness analyses should exclude costs associated with the measurement and

evaluation of program impacts unless the costs are a necessary component to administer the program.

6. For DSM programs or program elements that reduce electricity and natural gas consumption, costs and benefits from both fuels should be included.

7. The development and treatment of load impact estimates should distinguish between gross (i.e., impacts expected from the installation of a particular device, measure, or appliance) and net (impacts adjusted to account for what would have happened anyway and therefore not attributable to the program). Load impacts for the participants test should be based on the gross amount, whereas for all other tests the use of the net amount is appropriate. Gross and net program impact considerations should be applied to all types of demand-side management programs, although in some instances there may be no difference between the gross and net amounts.

8. The use of a sensitivity analysis, that is, the calculation of cost-effectiveness test results using alternative input assumptions, is encouraged, particularly for the following programs: new programs, programs for which authorization to substantially change direction is being sought (e.g., termination or significant expansion), major programs that show marginal cost-effectiveness or particular sensitivity to highly uncertain inputs).

The use of many of these guidelines is illustrated with examples of program cost-effectiveness contained in Appendix b.

APPENDIX b: SUMMARY OF EQUATIONS AND GLOSSARY OF SYMBOLS

Basic Equations

Participant Test

$$\text{NPV}_p = B_p - C_p$$

$$\text{NPV}_{avp} = \frac{B_p - C_p}{P}$$

$$\text{BCR}_p = \frac{B_p}{C_p}$$

$$\text{DP}_p = \text{Min } j \text{ such that } B_j \geq C_j$$

Ratepayer Impact Measure Test

$$\mathrm{LRI}_{\mathrm{RIM}} = \frac{C_{\mathrm{RIM}} - B_{\mathrm{RIM}}}{E}$$

$$\mathrm{FRI}_{\mathrm{RIM}} = \frac{C_{\mathrm{RIM}} - B_{\mathrm{RIM}}}{E}, \quad \text{for } t = 1$$

$$\mathrm{ARI}_{\mathrm{RIM}_t} = \mathrm{FRI}_{\mathrm{RIM}}, \quad \text{for } t = 1$$
$$= \frac{C_{\mathrm{RIM}_t} - B_{\mathrm{RIM}_t}}{E_t}, \quad \text{for } t = 2,\ldots, N$$

$$\mathrm{NPV}_{\mathrm{RIM}} = B_{\mathrm{RIM}} - C_{\mathrm{RIM}}$$

$$\mathrm{BCR}_{\mathrm{RIM}} = \frac{B_{\mathrm{RIM}}}{C_{\mathrm{RIM}}}$$

Total Resource Cost Test

$$\mathrm{NPV}_{\mathrm{TRC}} = B_{\mathrm{TRC}} - C_{\mathrm{TRC}}$$

$$\mathrm{BCR}_{\mathrm{TRC}} = \frac{B_{\mathrm{TRC}}}{C_{\mathrm{TRC}}}$$

$$\mathrm{LC}_{\mathrm{TRC}} = \frac{\mathrm{LCRC}}{\mathrm{IMP}}$$

Program Administrator Cost Test

$$\mathrm{NPV}_{\mathrm{pa}} = B_{\mathrm{pa}} - C_{\mathrm{pa}}$$

$$\mathrm{BCR}_{\mathrm{pa}} = \frac{B_{\mathrm{pa}}}{C_{\mathrm{pa}}}$$

$$\mathrm{LC}_{\mathrm{pc}} = \frac{\mathrm{LC}_{\mathrm{pc}}}{\mathrm{IMP}}$$

Benefits and Costs
Participant Test

$$B_p = \sum_{t=1}^{N} \frac{BR_t + TC_t + INC_t}{(1+d)^{t-1}} + \sum_{t=1}^{N} \frac{AB_{at} + PAC_{at}}{(1+d)^{t-1}}$$

$$C_p = \sum_{t=1}^{N} \frac{PC_t + BI_t}{(1+d)^{t-1}}$$

Ratepayer Impact Measure Test

$$B_{RIM} = \sum_{t=1}^{N} \frac{UAC_t + RG_t}{(1+d)^{t-1}} + \sum_{t=1}^{N} \frac{UAC_{at}}{(1+d)^{t-1}}$$

$$C_{RIM} = \sum_{t=1}^{N} \frac{UIC_t + RL_t + PRC_t + INC_t}{(1+d)^{t-1}} + \sum_{t=1}^{N} \frac{RL_{at}}{(1+d)^{t-1}}$$

$$E = \sum_{t=1}^{N} \frac{E_t}{(1+d)^{t-1}}$$

Total Resource Cost Test

$$B_{TRC} = \sum_{t=1}^{N} \frac{UAC_t + TC_t}{(1+d)^{t-1}} + \sum_{t=1}^{N} \frac{UAC_{at} + PAC_{at}}{(1+d)^{t-1}}$$

$$C_{TRC} = \sum_{t=1}^{N} \frac{PRC_t + PCN_t + UIC_t}{(1+d)^{t-1}}$$

$$LCRC = \sum_{t=1}^{N} \frac{PRC_t + PCN_t + TC_t}{(1+d)^{t-1}}$$

$$IMP = \sum_{t-1}^{N} \frac{\left[\left(\sum_{i=1}^{I} \Delta EN_{it}\right), \text{ or } (\Delta DN_{it}, \text{ where } i = \text{ peak period})\right]}{(1+d)^{t-1}}$$

Program Administrator Cost Test

$$B_{pa} = \sum_{t=1}^{N} \frac{\text{UAC}_t}{(1+d)^{t-1}} + \sum_{t+1}^{N} \frac{\text{UAC}_{at}}{(1+d)^{t-1}}$$

$$C_{pa} = \sum_{t=1}^{N} \frac{\text{PRC}_t + \text{INC}_t + \text{UIC}_t}{(1+d)^{t-1}}$$

$$\text{LCPA} = \sum_{t=1}^{N} \frac{\text{PRC}_t + \text{INC}_t}{(1+d)^{t-1}}$$

Glossary of Symbols

AB_{at} Avoided bill reductions on bill from alternate fuel in year t

$\text{AC:}D_{it}$ Rate charged for demand in costing period i in year t

$\text{AC:}E_{it}$ Rate charged for energy in costing period i in year t

ARI_{RIM} Stream of cumulative annual revenue impacts of the program per unit of energy, demand, or per customer. Note that the terms in the ARI formula are not discounted, thus they are the nominal cumulative revenue impacts. Discounted cumulative revenue impacts may be calculated and submitted if they are indicated as such. Note also that the sum of the discounted stream of cumulative revenue impacts does not equal the $\text{LRI}_{\text{RIM}}\star$

BCR_p Benefit/cost ratio to participants

BCR_{RIM} Benefit/cost ratio for rate levels

BCR_{TRC} Benefit/cost ratio of total costs of the resource

BCR_{pa} Benefit/cost ratio of program administrator and utility costs

BI_t Bill increases in year t

B_j Cumulative benefits to participants in year j

B_p Benefit to participants

B_{RIM} Benefits to rate levels or customer bills

B_{TRC} Benefits of the program

B_{pa} Benefits of the program

BR_t Bill reductions in year t

C_j Cumulative costs to participants in year i

C_p Costs to participants

C_{RIM} Costs to rate levels or customer bills

C_{TRC}	Costs of the program
C_{pa}	Costs of the program
d	Discount rate
ΔDG_{it}	Reduction in gross billing demand in costing period i in year t
ΔDN_{it}	Reduction in net demand in costing period i in year t
DP_p	Discounted payback in years
E	Discounted stream of system energy sales (kilowatt hours or therms) or demand sales (kilowatt) or first-year customers
ΔEG_{it}	Reduction in gross energy use in costing period i in year t
ΔEN_{it}	Reduction in net energy use in costing period i in year t
E_t	System sales in kilowatt hours, kilowatts, or therms in year t or first-year customers
FRI_{RIM}	First-year revenue impact of the program per unit of energy, demand, or per customer.
IMP	Total discounted load impacts of the program
INC_t	Incentives paid to the participant by the sponsoring utility in year t, the first year in which cumulative benefits are *equal to or greater than* cumulative costs.
K_{it}	1 when ΔEG_{it} or ΔDG_{it} is positive (a reduction) in costing period i in year t; 0 otherwise
LCRC	Total resource costs used for levelizing
LC_{TRC}	Levelized cost per unit of the total cost of the resource
LCPA	Total program administrator costs used for levelizing
LC_{pa}	Levelized cost per unit of program administrator cost of the resource
LRI_{RIM}	Life-cycle revenue impact of the program per unit of energy (kilowatt hour or therm) or demand (kilowatt)—the one-time change in rates—or per customer—the change in customer bills over the life of the program.
$MC{:}D_{it}$	Marginal cost of demand in costing period i in year t
$MC{:}E_{it}$	Marginal cost of energy in costing period i in year t
NPV_{avp}	Net present value to the average participant
NPV_p	Net present value to all participants
NPV_{RIM}	Net present value levels
NPV_{TRC}	Net present value of total costs of the resource
NPV_{pa}	Net present value of program administrator costs
OBI_t	Other bill increases (i.e., customer charges, standby rates)
OBR_t	Other bill reductions or avoided bill payments (e.g., customer charges, standby rates)

P	Number of program participants
PAC_{at}	Participant avoided costs in year t for alternate fuel devices
PC_t	Participant costs in year t, including:

- Initial capital costs, including sales tax
- Ongoing operation and maintenance costs
- Removal costs, less salvage value
- Value of the customer's time in arranging for installation, if significant

PRC_t	Program administrator program costs in year t
PCN	Net participant costs
RG_t	Revenue gain from increased sales in year t
RL_{at}	Revenue loss from avoided bill payments for alternate fuel in year t (i.e., device not chosen in a fuel substitution program)
RL_t	Revenue loss from reduced sales in year t
TC_t	Tax credits in year t
UAC_{at}	Utility-avoided supply costs for the alternate fuel in year t
UAC_t	Utility-avoided supply costs in year t
PA_t	Program administrator costs in year t
UIC_t	Utility-increased supply costs in year t

APPENDIX c: DERIVATION OF RIM LIFE-CYCLE REVENUE IMPACT FORMULA

Most of the formulas in the chapter are either self-explanatory or are explained in the text. This appendix provides additional explanation for a specific area where the algebra was considered to be too cumbersome to include in the text.

Rate Impact Measure

The ratepayer impact measure life-cycle revenue impact test (LRI_{RIM}) is assumed to be the one-time increase or decrease in rates that will re-equate the present valued stream of revenues and stream of revenue requirements over the life of the program.

Rates are designed to equate long-term revenues with long-term costs or revenue requirements. The implementation of a demand-side program can disrupt this equality by changing one of the assumptions on which it is based: the sales forecast. Demand-side programs by definition change sales. This expected difference between the long-term revenues and revenue requirements is calculated in the NPV_{RIM}. The amount that present valued revenues are below present valued revenue requirements equals $-\text{NPV}_{\text{RIM}}$.

The LRI_{RIM} is the change in rates that creates a change in the revenue stream that, when present valued, equals the $-\text{NPV}_{\text{RIM}}$*. If the utility raises (or lowers) its rates in the base year by the amount of the LRI_{RIM}, revenues over the term of the program will again equal revenue requirements. (The other assumed changes in rates, implied in the escalation of the rate values, are considered to remain in effect.)

Therefore, the formula for the LRI_{RIM} is derived from the following equality, where the present value change in revenues due to the rate increase or decrease is set equal to the $-\text{NPV}_{\text{RIM}}$ or the revenue change caused by the program:

$$-\text{NPV}_{\text{RIM}} = \sum_{t=1}^{N} \frac{\text{LRI}_{\text{RIM}} \times E_t}{(1 + d)^{t-1}}$$

Since the LRI_{RIM} term has no time subscript, it can be removed from the summation; the formula is then:

$$-\text{NPV}_{\text{RIM}} = \text{LRI}_{\text{RIM}} \times \sum_{t=1}^{N} \frac{E_t}{(1 + d)^{t-1}}$$

Rearranging terms, we then get:

$$\text{LRI}_{\text{RIM}} = \frac{-\text{NPV}_{\text{RIM}}}{\sum_{t=1}^{N} \frac{E_t}{(1 + d)^{t-1}}}$$

Therefore,

$$E = \text{LRI}_{\text{RIM}} = \sum_{t=1}^{N} \frac{E_t}{(1 + d)^{t-1}}.$$

CHAPTER 8

Life-Cycle and Cost-Benefit Analyses of Renewable Energy
The Case of Solar Power Systems

Thomas Pastore, ASA, CFA, CMA, MBA
Maria Ignatova, CFA

Contents

8.1	Introduction	139
	Energy Challenges are Enormous	139
8.2	A Successful Energy Plan	140
8.3	Due Diligence Procedures	140
8.4	Life-cycle Analysis of a PV System from a Financial Perspective	141
	PV System Costs	141
	Available Incentives	142
8.5	Financing Structures	143
	Power Purchase Agreements	143
	Equipment Lease Agreements	144
8.6	Measuring Savings from a PV System	144
	Financial Analysis	144
	Consideration of Externalities	148
8.7	Conclusion	148

8.1 INTRODUCTION

This chapter presents financial analyses available to determining the feasibility of implementing renewable energy central plants. In particular, discussed here is the implementation of a solar photovoltaic (PV) system in southern California by a non-profit organization. However, in general these analyses can be applied to other renewable energy or hybrid systems.

Federal and state incentives along with a number of different financing structures can help make the implementation of renewable energy systems feasible.

Energy Challenges are Enormous

Multiple factors must be considered when putting together an energy consumption plan. In addition to issues regarding energy security and reliability,

Sustainable Communities Design Handbook
ISBN: 978-1-85617-804-4, DOI: 10.1016/B978-1-85617-804-4.00008-2

economic growth could affect the conditions of an energy contract. Furthermore, natural disasters could have an adverse impact on energy supply, and the environmental impact of an energy system may affect future conditions under which energy is produced. An agile energy plan incorporating renewable or green energy can address these considerations and factors.

8.2   A SUCCESSFUL ENERGY PLAN

Renewable energy sources implantation represents a paradigm shift in energy consumption planning for any organization. The following major steps need to be taken when putting together a successful energy consumption strategy:

1. Assessing the energy consumption under the existing overall infrastructure, such as building insulation, equipment age, and types of light bulbs, just to name a few. This is called a demand side audit.
2. Implementing the necessary changes, as a result of the demand side audit, to minimize energy consumption and make the overall infrastructure most efficient.
3. Installing efficient and cost-effective renewable energy central plants, including PV systems.
4. Continuously monitoring energy consumption levels and patterns.
5. Developing a curriculum program and ongoing training around the implementation and operation of a renewable energy central plant.

8.3   DUE DILIGENCE PROCEDURES

Proper due diligence procedures entail various analyses of proposed PV systems. The following analyses need to be performed to determine the feasibility of implementing a PV system:

1. Engineering analyses of design proposals, installation sites, and ongoing maintenance.
2. Financial analyses of a PV system's implementation costs, financing costs, operating costs, and maintenance costs.
3. Legal analyses of proposed contracts between a non-profit organization, the PV system installer, and the investor who becomes the owner once the PV system is energized.
4. Project management and analyses from the perspective of the non-profit organization.

8.4 LIFE-CYCLE ANALYSIS OF A PV SYSTEM FROM A FINANCIAL PERSPECTIVE

The life-cycle analysis must encompass all cash flows during the life of a PV system, from the preliminary design stage through the removal of the PV system once it ceases operations. Considerations important to this analysis include:

1. A PV system may be fully financed or upfront capital investment may be required.
2. Applications for all available incentives, both federal and state.
3. Structuring a power purchase agreement (PPA) or an equipment lease agreement with a third party that commences once the PV system is energized.
4. Maintenance of the PV system, along with production guarantees from the maintenance provider, for a negotiated time period, usually of 20 years or less.
5. Current energy costs escalated periodically to reflect expected energy costs could be used as the baseline for calculating savings during the life of the PV system.
6. Once the PV system stops operating, it has to be replaced or removed, also known as decommissioning costs.

Several different designs may be presented from the original preliminary design to the ultimate one that meets an organization's current and anticipated near future energy needs.

The period between breaking ground for construction to when the PV system is energized, may require a construction loan.

It is important to note that maintenance costs are relatively minimal since the PV panels are usually guaranteed for 20 years, and the inverters are guaranteed for 10 years. A PV system could operate for as many as 25 to 40 years.

PV System Costs

Standardized PV system designs, bulk purchases of PV panels, and uninterrupted installation schedule, could all add to cost savings.

Table 8.1 shows the costs of a hypothetical 1.0 megawatt ("MW") PV system.

As with all technology, future technological advancements are expected to make PV systems more efficient whilst costing less.

Table 8.1 PV System Costs

Type of Cost	Cost ($)	$/W	% of Total Cost
Cost of PV panels	$3,000,000	$3.00	50%
Cost of ancillary electrical equipment	$540,000	$0.54	9%
Cost of structures and installation	$1,500,000	$1.50	25%
Cost of engineering	$120,000	$0.12	2%
Cost of construction management	$240,000	$0.24	4%
Cost of general site work	$60,000	$0.06	1%
Other costs	$540,000	$0.54	9%
Total costs	$6,000,000	$6.00	100%

Available Incentives

Monetary incentives are available from both federal and state programs to assist with the cost of installing PV systems. Federal incentives are provided by the National Energy Policy Act of 2005, while state incentives are usually provided through the local utility company servicing the area and the California Public Utility Commission (CPUC).

Federal incentives include an Investment Tax Credit (ITC) or a Treasury Cash Grant (TCG) equal to 30% of eligible costs.[1] Another incentive comes from the IRS's Modified Accelerated Cost Recovery System, under which businesses can recover investments in solar, wind, and geothermal property placed in service after 1986 over a five-year schedule of depreciation deductions.[2] Since the economic life of such property is 25 to 40 years, this incentive allows for relatively rapid recovery of deductable depreciation of an investment compared to the expected economic life of the property installed.

The California Solar Initiative (CSI), which is regulated by the CPUC, offers an incentive to further reduce the cost of installing PV systems. The CSI is a performance-based incentive (PBI) that is calculated based on projected kilowatt hours produced by a PV system. Different PV system size limits exist under each utility company. In addition, the CSI is composed of a number of declining steps, where the PBI rebate rate decreases as the number of MW installed increases by certain increments.

As an example, the following summarizes incentives available through the CSI program for entities within the Southern California Edison (SCE) servicing territory (*www.GoSolarCalifornia.ca.gov*):

- For PV systems greater than or equal to 100 kilowatt (kW) in size, incentives are paid monthly based on the actual energy produced for a

[1] www.GoSolarCalifornia.ca.gov
[2] www.IRS.gov

period of five years. Systems of any size may elect to opt into the PBI program. In addition, building integrated systems, regardless of size, are required to participate in the PBI program.

- Incentives for all systems less than 100 kW initially are paid a one-time, upfront incentive based on expected system performance. Expected performance is calculated based on equipment ratings and installation factors, such as geographic location, tilt, orientation, and shading. This type of incentive is called *expected performance-based buy down*. Residential and commercial incentives are set at slightly lower rates than government and non-profit organizations incentives, which is meant to compensate their lack of access to the federal incentives.

Investing in hybrid systems that feature energy storage capacity could reduce an organization's reliance on the utility company. Without the energy storage capacity, excess power produced is either fed back into the utility company's electrical grid for a pre-negotiated credit for a set period of time, known as a feed-in-tariff, or is wasted.

Incentives may change considerably over time. It is important for the financial analyst to keep abreast of changes in incentives and formation of new incentives. Information on all federal and state incentive programs around the country is available at the Database of State Incentives for Renewables and Efficiency (www.dsireuse.org).

8.5 FINANCING STRUCTURES

Non-profit organizations are not able to benefit from any tax credit or depreciation incentives since they do not generate taxable income. For-profit third party ownership allows non-profit organizations to indirectly benefit from all available incentives that would otherwise not be available. This benefit is passed through to the non-profit organization in the form of a lower payment under the chosen financing structure, as discussed next.

Power Purchase Agreements

A PPA can be a contract between a non-profit organization and a third party, typically an investor, where the non-profit organization purchases power produced by a PV system based on a pre-determined price per unit, i.e., $/kWh produced. A PPA specifically for the purpose of providing a solar energy system is also known as a *solar service agreement*. A typical PPA term is 20 years. Such an agreement allows a non-profit organization, which cannot fully utilize all available incentives, to indirectly benefit from them through a lower PPA energy rate.

Equipment Lease Agreements

Under an equipment lease agreement, the installer sells the PV system to a third party, typically an investor, which then leases the PV system to a non-profit organization. As the PV system owner, the lessor can apply for and receive the TCG. The lease payment is a fixed amount and, unlike a PPA, does not vary with production. A typical lease term is 15 years. Tax counsel should be consulted to assure that the terms of the lease meet the criteria of an operating lease. All available incentives are reflected in the form of a lower lease payment.

8.6 MEASURING SAVINGS FROM A PV SYSTEM

Determining if a PV system is financially feasible requires a comparison of annual costs to the purchasing party, i.e., non-profit organization, over the life of the PV system to the purchasing party's offset utility costs during the life of the PV system.

The first step in calculating the utility cost that is being offset by the PV system production is establishing the appropriate utility rate per kilowatt hour, and then applying it to the PV system's kilowatt hours produced. For example, Southern California Edison utility rates include charges for energy use, by customer, and by demand. Energy use charges involve delivery service and generation charges based on time of use (TOU), customer charges and related facilities, and a power factor adjustment. Demand charges are not TOU charges. Time related demand depends on TOU during summer (12 A.M. on the first Sunday in June through 12 A.M. of the first Sunday in October) and winter (the remainder of the year). TOU rates are based on three time periods, on-peak, mid-peak, and off-peak, with maximum demand rates established for each time period based on the maximum average kilowatt input recorded during any 15-minute interval during each month. On-peak hours are noon through 6 P.M. on summer weekdays, except holidays. Mid-peak hours are 8 A.M. to noon and 6 P.M. to 11 P.M. on summer weekdays, except holidays, and off-peak hours account for all remaining hours.

Table 8.2 presents a cost comparison of a non-profit organization's annual utility costs with a hypothetical 1.0 MW PV system over an estimated 30-year economic life, financed over 20 years.

Financial Analysis

There are multiple methods available to financially analyze the feasibility of a PV system.

Table 8.2 PV System's Economic Life

Year	Solar Electricity Produced (kWh)	Utility Energy Rate ($/kWh)	PPA Energy Rate ($/kWh)	Utility Energy Cost ($)	PPA Energy Cost ($)	Net Savings (Cost) to Purchaser of Solar Energy ($)
	A	B	C	$D = A \star B$	$E = A \star C$	$F = D-E$
1	1,450,000	$0.175	$0.190	$253,750	$275,500	($21,750)
2	1,439,125	$0.184	$0.196	$264,439	$281,637	($17,198)
3	1,428,332	$0.193	$0.202	$275,579	$287,910	($12,332)
4	1,417,619	$0.203	$0.208	$287,187	$294,323	($7,136)
5	1,406,987	$0.213	$0.214	$299,285	$300,879	($1,594)
6	1,396,435	$0.223	$0.220	$311,893	$307,582	$4,311
7	1,385,961	$0.235	$0.227	$325,031	$314,433	$10,598
8	1,375,567	$0.246	$0.234	$338,723	$321,437	$17,286
9	1,365,250	$0.259	$0.241	$352,992	$328,597	$24,395
10	1,355,010	$0.271	$0.248	$367,862	$335,916	$31,945
11	1,344,848	$0.285	$0.255	$383,358	$343,399	$39,959
12	1,334,762	$0.299	$0.263	$399,507	$351,048	$48,458
13	1,324,751	$0.314	$0.271	$416,336	$358,868	$57,468
14	1,314,815	$0.330	$0.279	$433,874	$366,862	$67,012
15	1,304,954	$0.346	$0.287	$452,151	$375,033	$77,118
16	1,295,167	$0.364	$0.296	$471,198	$383,387	$87,811
17	1,285,453	$0.382	$0.305	$491,047	$391,927	$99,120
18	1,275,812	$0.401	$0.314	$511,732	$400,657	$111,075
19	1,266,244	$0.421	$0.323	$533,289	$409,582	$123,707
20	1,256,747	$0.442	$0.333	$555,754	$418,706	$137,048
21	1,247,321	$0.464	$0	$579,165	$0	$579,165
22	1,237,966	$0.488	$0	$603,562	$0	$603,562
23	1,228,682	$0.512	$0	$628,987	$0	$628,987
24	1,219,466	$0.538	$0	$655,484	$0	$655,484
25	1,210,320	$0.564	$0	$683,096	$0	$683,096
26	1,201,243	$0.593	$0	$711,871	$0	$711,871
27	1,192,234	$0.622	$0	$741,859	$0	$741,859
28	1,183,292	$0.653	$0	$773,110	$0	$773,110
29	1,174,417	$0.686	$0	$805,677	$0	$805,677
30	1,165,609	$0.720	$0	$839,616	$0	$839,616
Total	**39,084,388**	**$0.388**		**$14,747,413**	**6,847,684**	**7,899,728**

Note: For illustration purposes only.

The first is the net present value (NPV) method, which is the sum of the present values of the annual cash flows during the life of the PV system minus the present value of the investments. An appropriate discount rate accounts for the time value of money and uncertainties associated with the cash flows. This method is important, as it shows the net value of the PV system from year to year.

Another method is based on the internal rate of return (IRR), which is the discount rate that makes the project's cash flows and investments have a zero NPV. It is important to define a threshold IRR prior to evaluating the PV system. An IRR of 0% does not make a project financially feasible, as it fails to compensate an investor for the time value of money and the uncertainties associated with future cash flows.

The last method is the payback period, which is the length of time required to recover an initial investment through cash flows generated by the investment. The payback period is important when considering an organization's financial ability to implement a PV system.

To explain the strengths and weaknesses of these methods, Charles T. Horngren writes:[3]

> One big advantage of the NPV method is that it expresses computations in dollars, not in percent. Therefore, we can sum NPVs of individual projects to calculate NPV of a combination of projects. In contrast, IRRs of individual projects cannot be added or averaged to represent IRR of a combination of projects … Two weaknesses of the payback method are that (1) it fails to incorporate the time value of money and (2) it does not consider a project's cash flows after the payback period … Another problem with the payback method is that choosing too short a cutoff period for project acceptance may promote the selection of only short-lived projects. [If it uses only the payback method,] an organization will tend to reject long-run, positive-NPV projects.

The three methods just described often do not yield the same result. In Table 8.3, a financial analysis is performed based on a hypothetical 1.0 MW PV system using these methods:

At the time of the PPA's expiration, at the end of year 20 (a usual PPA term), the PV system has an NPV of $385,698, with an IRR of 25.13%. The same PV system has a payback period of just less than 10 years. The PPA terms generally assign ownership to the power purchasing entity (i.e., a non-profit organization) at expiration. If the PV system continues to operate for 30 years from today, it is projected to have an NPV of $2,395,790, with an IRR of 29.69%. The NPV dramatically increases over the last 10 years of operation because from year 20 through year 30 all power produced by the PV system is essentially free to the owner, as the contractual PPA payments cease and the non-profit organization is not incurring utility costs (see Table 8.2).

---

[3] Charles T. Horngren, Srikant M. Datar, George Foster, Cost Accounting: A Managerial Emphasis, 11th Edition, Upper Saddle River, NJ: Pearson Education, 2003, pp. 720-725.

Table 8.3 Financial Analysis of a PV System

Year	NPV	IRR	Pay Back Period
1	($20,714)	n/a	($21,750)
2	($36,313)	n/a	($38,948)
3	($46,965)	n/a	($51,279)
4	($52,836)	n/a	($58,415)
5	($54,085)	n/a	($60,009)
6	($50,868)	−55.31%	($55,698)
7	($43,336)	−27.30%	($45,100)
8	($31,636)	−11.30%	($27,814)
9	($15,911)	−0.98%	($3,419)
10	$3,700	6.06%	$28,526
11	$27,063	11.05%	$68,485
12	$54,047	14.69%	$116,943
13	$84,523	17.41%	$174,411
14	$118,369	19.47%	$241,424
15	$155,464	21.05%	$318,541
16	$195,691	22.28%	$406,352
17	$238,937	23.25%	$505,472
18	$285,091	24.02%	$616,547
19	$334,046	24.64%	$740,254
20★	$385,698	25.13%	$877,302
21	$593,585	26.52%	$1,456,467
22	$799,912	27.43%	$2,060,029
23	$1,004,693	28.07%	$2,689,017
24	$1,207,937	28.53%	$3,344,500
25	$1,409,657	28.87%	$4,027,596
26	$1,609,864	29.13%	$4,739,467
27	$1,808,570	29.33%	$5,481,326
28	$2,005,785	29.48%	$6,254,436
29	$2,201,521	29.59%	$7,060,112
30	$2,395,790	29.69%	$7,899,728

Note: For illustration purposes only.

★Expiration of PPA term.

All three methods are highly dependent on a PV system's cost (see Table 8.1), which in turn is subject to market price fluctuations of commodity type raw materials, such as PV panels and steel. If these price fluctuations cannot be controlled in the procurement process, there is the potential for a significant adverse impact on NPV, IRR, and payback period. This could make a PV system financially unfeasible.

Table 8.4 Externality Factors

Financial analysis

Net present value at year 30	$2,395,790
Add:*	
NPV of carbon credits	$500,000
NPV of renewable energy credits	$300,000
NPV of health care cost savings	$200,000
NPV of savings with externalities	$3,395,790

*The amounts shown are solely for illustrative purposes.

Consideration of Externalities

Beyond consideration of NPV, IRR and payback period analyses, both qualitative and quantitative externalities resulting from the installation of PV systems must be considered to complete the financial analyses. Additional quantitative benefits to the PV system owner include carbon credits, renewable energy credits (RECs), and possible employee health care savings as a result of a cleaner environment.

Qualitative externalities include reduction of pollution and greenhouse gas emissions, reduced dependency on utility providers, and greater control over energy price volatility. In addition, PV systems can provide power during traditional power outages, whether due to natural disasters or any other reason.

Table 8.4 presents a hypothetical analysis of externality factors to provide a quantitative concept of the effect of externalities on NPV.

Installing PV systems in parking lots and on rooftops or other existing structures provides shade while not infringing on an organization's operations and not requiring the acquisition of additional space. Finally, minimal maintenance cost is associated with PV systems, with long-term reliability of 25 to 40 years.

8.7 CONCLUSION

A traditional financial analysis is one part of evaluating the feasibility of an energy consumption plan. As outlined in this chapter, a complete financial analysis includes all factors present during the life-cycle of a PV system. These factors include, but are not limited to, the financing structure terms, investment costs, available incentives, utility energy costs, and externalities. Proper application of financial analyses to determine the financial feasibility of a PV system provides a critical portion of the overall due diligence procedures in implementing a PV system.

CHAPTER 9

Public Buildings and Institutions: Solar Power as a Solution

Legal Mechanisms for Sustainable Buildings

Douglas N. Yeoman

Contents

9.1	Alternative Energy Public Policy	149
9.2	Legal Mechanisms Facilitating Development of Alternative Energy Sources	150
	Energy Management Agreement by Community College Districts	150
	Energy Service Contract and Facility Ground Lease by Public Agencies	150
	Power Purchase Agreement by Government Agency	153
	Lease of Photovoltaic System	159
9.3	Treatment of Environmental Incentives	161

9.1 ALTERNATIVE ENERGY PUBLIC POLICY

The California Legislature in 1974 passed Public Resources Code section 25007, which established a state policy "to employ a range of measures to reduce wasteful, uneconomical, and unnecessary uses of energy, thereby reducing the rate of growth of energy consumption, prudently conserve energy resources, and assure statewide environmental, public safety, and land use goals." With continued increases in energy usage and energy costs, the legislature revised the statewide policy in 1981, adding that it was further the policy of the state to "promote all feasible means of energy and water conservation and all feasible uses of alternative energy and water supply sources [including, but not limited to, solar technologies]." (Public Resources Code section 25008.)

More recently, Governor Schwarzenegger signed the California Solar Initiative on August 21, 2006 (Senate Bill 1), which establishes a goal of the state to install solar energy systems with a generation capacity equivalent of 3000 MW, to establish a self-sufficient solar industry in which solar energy systems are a viable mainstream option for both homes and businesses in 10 years, and to place solar energy systems on 50% of new homes in 13 years.

Sustainable Communities Design Handbook
ISBN: 978-1-85617-804-4, DOI: 10.1016/B978-1-85617-804-4.00009-4

9.2 LEGAL MECHANISMS FACILITATING DEVELOPMENT OF ALTERNATIVE ENERGY SOURCES

Energy Management Agreement by Community College Districts

Education Code sections 81660 through 81662 authorize a community college district to enter into an energy management agreement for energy management systems (i.e. solar energy or solar and energy management systems) with the *lowest responsible bidder*, considering the net cost or savings to the district, less the projected energy savings to be realized from the energy management system. The maximum term of such an agreement is the estimated useful life of the energy management system or 15 years, whichever is less.

Energy Service Contract and Facility Ground Lease by Public Agencies

To implement the public policy set forth in Public Resources Code section 25008, the California Legislature in 1986 adopted and added Chapter 3.2 of the Government Code (sections 4217.10 through 4217.18), which authorizes public agencies to enter into energy service contracts for the development of energy conservation, cogeneration, and alternate energy supply sources, *without competitive bidding*. School districts, community college districts, counties, cities, districts, joint powers authorities, or other political subdivisions are included in the definition of *public agency* (Gov. Code § 4217.11, subd. (j)).

Although a direct energy service contract and related facility ground lease may be entered into without competitive bidding, a limitation to using this contracting method is the requirement that the contract involve only alternate energy equipment, including but not limited to solar, maintenance, load management techniques, or other conservation measures that result in the *reduction of energy use or makes for a more efficient use of energy* (Gov. Code, § 4217.11, subds. (a) and (c)). For example, an energy service contract may not include the installation of air conditioning where no form of air conditioning existed previously, as the addition of the air conditioning would result in an increase, not a reduction, of energy use. In this case, the air conditioning component of the project would require competitive bidding, assuming the estimated cost would exceed the bidding amount threshold of the particular public agency.

As a condition to entering into an energy service contract and any necessarily related facility ground lease, the governing board must determine that entering into such agreements are in the best interests of the public agency. Except for state agency heads, who can make the findings described later without holding a public hearing, the governing boards of all other public agencies must make the "best interests" determination at a regularly scheduled public hearing in which public notice has been given at least two weeks in advance. To support this determination, the board must find (1) that the anticipated cost to the agency for thermal or electrical energy or for the "energy conservation facility" under the contract will be less than the anticipated marginal cost to the agency of thermal, electrical, or other energy that would have been consumed by the district in the absence of those purchases; and (2) that the difference, if any, between the fair rental value for the real property subject to the facility ground lease and the agreed rent is anticipated to be offset by below-market energy purchase or other benefits provided under the energy service contract (Gov. Code, § 4217.12). The term *energy conservation facility* is defined at Government Code section 4217.11(e) to mean "alternate energy equipment, cogeneration equipment, or conservation measures located in public buildings or on land owned by public agencies."

Government Code section 4217.13 also authorizes a public agency to enter into a facility financing contract and a facility ground lease on terms determined by the board to be in the best interest of the agency if the determination is made at a regularly scheduled public hearing and if the governing body finds that funds for the repayment of the financing or the cost of design, construction, and operation of the energy conservation facility or both, are projected to be available from revenues generated from the sale of electricity or thermal energy from the facility or from funding that otherwise would have been used for purchase of electrical, thermal, or other energy required by the agency in the absence of the energy conservation facility or both. As with energy service contracts, state agency heads may make these findings without holding a public hearing.

Public agencies typically support the findings required in the preceding paragraph for entering into an energy service contract and facility financing contract upon a preliminary survey of the agency's existing energy equipment and usage conducted by the prospective contractor. Government Code section 4217.15 authorizes a public agency to base its findings on projections for electrical and thermal energy rates from the following sources: (1) the public utility that provides thermal or electrical energy to the public agency, (2) the state utilities commission, (3) the state energy resources conservation and development commission, or (4) the projections used by the department of general services for evaluating the feasibility of energy conservation facilities at state facilities located within the same public utility service area as the public agency.

Under this legislative scheme, public agencies may, "notwithstanding any other provision of law," enter into contracts for the sale of electricity, electrical generating capacity, or thermal energy produced by the energy conservation facility at rates and on such terms as may be approved by the governing board (Gov. Code, § 4217.14).

Although no competitive selection process is required, a public agency may wish to solicit proposals to ensure it is receiving the greatest available energy savings. Section 4217.16 of the Government Code provides for the option of seeking proposals in stating:

> Prior to awarding or entering into an agreement or lease, the public agency may request proposals from qualified persons. After evaluating the proposals, the public agency may award the contract on the basis of the experience of the contractor, the type of technology employed by the contractor, the cost to the local agency, and any other relevant considerations. The public agency may utilize the pool of qualified energy service companies established pursuant to Section 388 of the Public Utilities Code and the procedures contained in that section in awarding the contract.

Public Utilities Code section 388 referenced previously, which is applicable to state agencies, authorizes agencies to "enter into an energy savings contract with a qualified energy service company for the purchase or exchange of thermal or electrical energy or water, or to acquire energy efficiency and/or water conservation services, for a term not exceeding 35 years, at those rates and upon those terms that are approved by the agency."

Paragraph (b) of Public Utilities Code section 388 provides the option for state agencies and local agencies to establish a pool of qualified energy service companies that is updated at least every two years based on qualification, experience, pricing, or other pertinent factors. The paragraph further provides that energy service contracts for individual projects may be awarded through a competitive selection process to individuals or firms identified in such a pool.

Government Code section 4217.18 concludes Chapter 3.2 on energy conservation contracts by emphasizing the intended flexibility of the aforementioned sections by stating:

> The provisions of this chapter shall be construed to provide the greatest possible flexibility to public agencies in structuring agreements entered into hereunder so that economic benefits may be maximized and financing and projects may be minimized. To this end, public agencies and the entities with whom they contract under this chapter should have great latitude in characterizing

components of energy conservation facilities as personal or real property and in granting security interests in leasehold interests and components of the alternate energy facilities to project lenders.

Utilizing this statutory scheme may be advantageous when a public agency desires to implement an energy conservation project involving conservation measures where the cost of design, construction, and operation is projected to be recovered from energy savings over the life expectancy of the conservation measures. Conversely, if (1) new energy conservation measures are being considered; (2) the cost of design, construction, and operation of energy conservation measures is not projected to be recovered over the life expectancy of the energy conservation measures; or (3) the public agency either does not have the funds or does not desire to finance the new energy conservation measures, the agency may want to consider entering into a purchase power agreement as discussed nextw.

Power Purchase Agreement by Government Agency

To assist local government agencies in infrastructure financing of energy or power production projects, Assembly Bill 2660 was passed in 1996, which added Government Code sections 5956 through 5956.10 (referred to here as the *power purchase provisions*). The power purchase provisions grant the authority to a city, county, school district, community college district, public district, county board of education, joint powers authority, transportation commission or authority, or any other public or municipal corporation (collectively, *government agency*), "to utilize private investment capital to study, plan, design, construct, develop, finance, maintain, rebuild, improve, repair, or operate, or any combination thereof, fee-producing infrastructure facilities" (Gov. Code § 5956.1). The term *fee-producing infrastructure project* is defined as the "operation of the infrastructure project or facility . . . paid for by the persons or entities benefited by or utilizing the project or facility" (Gov. Code § 5956.3(c)). Any combination of private infrastructure financing, federal, or local funds may be utilized under this statutory scheme (Gov. Code § 5956.9). State agencies are specifically prohibited from utilizing the power purchase provisions (Gov. Code § 5956.10).

The power purchase provisions require that an agency solicit proposals as part of a competitive negotiation process when selecting a contractor for the studying, planning, design, developing, financing, construction, maintenance, rebuilding, improvement, repair, or operation, or any combination of these, for fee-producing infrastructure projects. *Neither competitive bidding nor compliance with any other provision of the Public Contract Code or Government Code relating to public procurements is required.* Projects may be proposed by a private entity and selected by the government agency in its discretion, subject to the following selection criteria being considered: (1) demonstrated competence and qualifications of the private entity and (2) the proposed facility must be operated at fair and reasonable prices to the user of the infrastructure facility services. The competitive negotiation process must specifically prohibit practices that may result in unlawful activity, including, but not limited to, rebates, kickbacks, or other unlawful consideration and any prohibited conflict of interest involving the employees of the government agency in violation of Government Code section 87100, which states: "No public official at any level of state or local government shall make, participate in making or in any way attempt to use his official position to influence a governmental decision in which he knows or has reason to know he has a financial interest."

The power purchase agreement to be entered into with the private entity is required by Government Code section 5956.6 to contain provisions to ensure the following:

1. Provide whether the facilities will be owned by the agency or contractor during the term of the agreement. If the facilities are leased to the contractor, the agreement must provide for a complete reversion of ownership in the facility at the expiration of the term (may not exceed 35 years), *without charge* to the government agency.

2. Compliance with the California Environmental Quality Act (CEQA) commencing at Public Resources Code section 21000 prior to the commencement of project development. Although cogeneration projects at existing facilities may be categorically exempt from CEQA if the conditions set forth at Title 14, California Code of Regulations, section 15329, are satisfied, typically, a negative declaration or mitigated negative declaration is required to demonstrate that the facility will not have a significant adverse impact on the environment.

3. Security for the construction of the facility to ensure its completion and contractual provisions that are necessary to protect the revenue streams of the project. Insurance provisions (example under item 11), hold harmless and indemnity clauses, and if appropriate, performance bonds provide such protection.

4. Adequate financial resources of the private entity to design, build, and operate the facility after the date of the agreement.

5. Authority for the government agency to impose user fees for use of the facility in an amount sufficient to protect the revenue streams necessary for projects or facilities. The user fee revenue must be dedicated exclusively to payment of the private entity's direct and indirect capital outlay costs for the project, direct and indirect costs associated with operations, direct and

indirect user fee collection costs, direct costs of administration of the facility, reimbursement for the direct and indirect costs of maintenance, and a negotiated reasonable return on investment to the private entity.

Prior to taking action to impose or increase a user fee, the government agency must conduct at least one public hearing on the proposed fee. Notice of the public hearing(s) must be given (1) by mail no less than 14 days prior to the meeting to any interested party who has requested notice of the meeting, and (2) by publication no less than 10 days prior to the meeting in a newspaper of general circulation in the jurisdiction of the government agency. All data in support of the proposed user fee must also be available for public inspection at least 10 days preceding the meeting. All costs incurred by the government agency in providing the required notice and holding the public hearing(s) may be recovered from the fees to be charged. Action to impose or increase a user fee must be taken by ordinance or resolution by the governing board of the government agency. The established fee may not exceed the estimated amount required to provide the service for which the fee is charged and a reasonable rate of return on investment.

6. Require that any revenues in excess of the actual cost and a reasonable rate of return on investment be applied by the government agency to either reduce any indebtedness incurred by the private entity with respect to the project, be paid into a reserve account to offset future operation costs, be paid into the appropriate government account, be used to reduce the user fee or service charge creating the excess, or a combination of these sources.

7. Require the private entity to maintain the facility in good operating condition at all times, including the time the facility reverts to the government agency.

8. Preparation by the private entity of an annual audited report accounting for the income received and expenses to operate the facility.

9. Provision for a buyout of the private entity by the government agency in the event of termination or default prior to the expiration of the term of the purchase power agreement.

10. Provision for appropriate indemnity promises between the government agency and private entity.

11. Provision requiring the private entity to maintain insurance with those coverages and in those amounts that the government agency deems appropriate. A sample insurance provision is as follows:

    A. **Contractor's insurance.** The Contractor shall provide and maintain insurance, at the Contractor's own cost and expense, against all claims or losses which may arise from or in connection with the performance of services by the Contractor. The obligation to maintain insurance shall not in any way affect the indemnity provided in or by Section __. District's acceptance of Contractor's insurance hereunder shall not in any way act as a limitation on the extent of Contractor's liability.

B. Coverages, subcontractor, subconsultant insurance.

(1) Contractor shall, maintain and shall require that every Subcontractor and Subconsultant, of any Tier, performing or providing any portion of the Work obtain and maintain, for the duration of its performance of the Work and for the full duration of all guarantee or warranty periods set forth in the Contract Documents (and such longer periods as required below for completed operations coverage), the insurance coverage outlined in (a) through (d) below, and all such other insurance as required by Applicable Laws; provided, however, that Subcontractors not providing professional services shall not be required to provide Professional Liability coverage and except where District has given its written approval to waive said limits for a specific Subcontractor.

(a) Commercial General Liability and Property Insurance, on an "occurrence" form covering occurrences (including, but not limited to those listed below) arising out of or related to operations, whether such operations be by the Contractor, a Subcontractor or Subconsultant, or by anyone directly or indirectly employed by any of them, or by anyone for whose acts any of them may be liable, involving damage or loss of any kind: (1) because or bodily injury, sickness or death of any person other than the Contractor's, Subcontractor's or Subconsultant's employees; (2) sustained (a) by a person as a result of an offense directly or indirectly related to employment of such person, or (b) by another person; (3) other than to the Work itself, because of injury to or destruction of tangible property including loss of use resulting therefrom; (4) because of bodily injury, death of a person or property damages arising out of ownership, maintenance or use of a motor vehicle; (5) contractual liability insurance; and (6) completed operations, with limits as follows:

$2,000,000 per occurrence for Bodily Injury and Property Damage.

$2,000,000 General Aggregate—other than Products/Completed Operations.

$1,000,000 Products/Completed Operations Aggregate for the duration of a period of not shorter than 1 year after Final Completion and Acceptance of the Project.

$1,000,000 Personal and Advertising Injury.

Full replacement value for Fire Damage.

And including, without limitation, special hazards coverage for:

$1,000,000 Material hoists

$1,000,000 Explosion, collapse and underground (XCU)

(b) Auto Liability insurance, on an occurrence form, for owned, hired and nonowned vehicles with limits of $1,000,000 per occurrence

(c) Professional Liability insurance (only to be provided by Subconsultants or Subcontractors performing professional services), written on a "claims-made" form, with limits of:
$1,000,000 per claim
$1,000,000 aggregate

(d) Excess Liability insurance, on an "occurrence" form, in excess of coverages provided for Commercial General Liability, Auto Liability, Professional Liability and Employer's Liability, with limits as follows:
$1,000,000 each occurrence (or, in the case of coverage in excess of Professional Liability, each claim).
$1,000,000 aggregate

(2) **Evidence of insurance.** Upon request of District, Contractor shall promptly deliver to District Certificates of Insurance evidencing that the Subcontractors and Subconsultants have obtained and maintained policies of insurance in conformity with the requirements of this Section ___. Failure or refusal of Contractor to do so may be deemed by District to be a material default by Contractor of the Contract.

(3) **Builder's risk "all-risk" insurance.** Builder's Risk "All Risk" Insurance will be purchased by District, which shall include primary coverage protecting the insured's interest in materials, supplies, equipment, fixtures, structures, and real property to be incorporated into and forming a part of the Project and with policy limits protecting up to the Estimated Maximum Value of the Project for any one loss or occurrence and with deductibles of between $5,000 and $25,000 per occurrence. Said Builder's Risk policy shall be endorsed to add Contractor and its Subcontractors of the first Tier and Subconsultants of the first Tier as additional named insureds, as their interests may appear, and to waive the carrier's right of recovery under subrogation against Contractor and all Subcontractors and Subconsultants whose interest are insured under such policy. If a claim results from any construction activity of Contractor or a Subcontractor of Subconsultant, then Contractor or the Subcontractor or Subconsultant having care, custody, and control of the damaged property shall pay the deductible amount. Any loss or damage covered by the Builder's Risk Policy shall be adjusted by and payable to District, or its designee, for the benefit of all Parties as their interest may appear. District shall not be responsible for loss or damage to and shall not obtain and/or maintain in force insurance on temporary structures,

construction equipment, tools or personal effects, owned, rented to, or in the care, custody and control of Contractor or any Subcontractor or Subconsultant. In the event of loss or damage caused by the acts or omissions of Contractor or its Subcontractors or Subconsultants that is not covered by the Builders Risk policy, the cost of the repair and/or replacement of such loss or damage shall be at Contractor's own expense. District, Contractor and all Subcontractors and Subconsultants each and all waive rights of subrogation against each other to the extent that said Builder's Risk policy covers property damage arising out of the perils of fire or other casualty also covered by Contractor's or a Subcontractor's or Subconsultant's insurance policy.

(4) **Policy requirements and endorsements.** Except as otherwise stated in this Paragraph 4, each policy of insurance required to be provided by Subcontractors and Subconsultants shall comply with the following:

(a) The commercial general liability insurance policy shall contain a waiver of subrogation rights against District, members of the Board of Trustees, District's Consultants, and each of their respective agents, employees, and volunteers, and the State Allocation Board

(b) The insurance policies provided for Commercial General Liability, Auto Liability, as well as any Excess Liability coverage in excess thereof shall be endorsed to include, individually and collectively, the District, members of the Board of Trustees, District's Consultants, and each of their respective agents, employees, and volunteers, and the State Allocation Board, as additional insureds.

(c) The insurance polices shall provide that the insurance is primary coverage with respect to District and all other additional insureds, shall not be considered contributory insurance with any insurance policies of the District or any other additional insureds, and all insurance coverages provided by District and any other additional insureds shall be considered excess to the coverages provided by the Subcontractor or Subconsultant.

12. In the event of a dispute, both parties shall be entitled to all available legal or equitable remedies.

13. Require that the plans and specifications for the project be constructed in compliance with all applicable governmental design standards and shall utilize private sector design and construction firms to design and construct the infrastructure facilities.

14. Comply with all applicable laws relating to public property and public works projects, including the payment of prevailing wages.

Although not required by statute, it is suggested that government agencies consider including a guarantee provision in the power purchase agreement whereby the energy provider guarantees a minimum energy output, which if not met will result in a monetary penalty on behalf of the power provider, such as requiring the energy provider to pay the difference between what the government agency is required to pay the utility company for the power shortage and what the agency would have been required to pay the energy provider had the guaranteed energy output been delivered. An example of such a provision follows:

> **Guarantee.** *Power Provider shall provide a Cumulative Output Guarantee from the Generating Facility commencing on the date of Commercial Operation and continuing until the twentieth (20th) anniversary of the Commercial Operation Date or achievement of the twentieth year cumulative output guarantee of _____kWh, whichever comes first. The guarantee is defined to be 90% of the expected annual production from the Generating Facility to be measured in kilowatt hours.*
>
> *In order to control for variations in weather, the actual output will be compared to the Cumulative Output Guarantee on a cumulative basis on the third (3rd), sixth (6th), ninth (9th), twelfth (12th), fifteenth (15th), and twentieth (20th) year during the cumulative output Guarantee Term. Actual production shall accrue to the cumulative balance each year and be compared on the anniversary dates noted above of the Commercial Operation Date to the aggregate cumulative output guarantee for the years in that measurement period as indicated in the table below [not shown]. In the event that the Guaranteed Energy Output is not achieved as described above during the term of this Agreement (the "Guaranteed Energy Output Shortage"), and Purchaser is required to purchase replacement ac kilowatt hours from Southern California Edison, then Power Provider shall refund the difference between the amount Purchaser pays Southern California Edison for the replacement power and the annual rate as specified in Exhibit __ [not shown]. The Southern California Edison replacement power price is defined as the blended average annual TOU-8 tariff for that portion of ac kilowatt hours representing the Guaranteed Energy Output Shortage. This guarantee shall immediately terminate if the Generating Facility title is transferred to Purchaser.*
>
> *Example of hypothetical shortfall payment calculation. In year 3, the governmental agency consumes 7 million kWh of electricity and pays Southern California Edison $1,050,000 for its total annual energy use under the TOU-8 rate. Therefore the blended average annual TOU-8 rate is equal to $0.15 per kilowatt hour ($1,050,000/7,000,000 kWh = $0.15/kWh). The cumulative output guarantee in year 3 is 4,193,486 kWh. The actual delivered cumulative output is 4,100,000 kWh. The shortfall is therefore 93,486 kWh. The PPA rate is $0.14333/kWh in year 3. The shortfall payment paid by Power Provider to Purchaser is $624 (93,486 kWh × [$0.15–$0.1433] = $624).*

Lease of Photovoltaic System

The Los Angeles Department of Water and Power (DWP) instituted a Solar Photovoltaic Incentive Program (the incentive program) consistent with the

California solar initiative set forth in Senate Bill 1 (SB 1, Murray), which was approved during the 2005–2006 legislative term. Public Utilities Code section 387.5(b) requires that, on or before January 1, 2008, a local publicly owned electric utility must offer monetary incentives for the installation of solar energy systems of at least $2.80 per installed watt, or for the electricity produced by the solar energy system, measured in kilowatt hours, as determined by the governing board of a local publicly owned electric utility, for photovoltaic solar energy systems. The incentive level is scheduled to decline each year thereafter at a rate of no less than an average of 7% per year.

For a local publicly owned electric utility to institute a solar energy program, Public Utilities Code section 387.5(d) requires the program to be consistent with all the following:

(1) That a solar energy system receiving monetary incentives comply with the eligibility criteria, design, installation, and electrical output standards or incentives established by the State Energy Resources Conservation and Development Commission pursuant to Section 25782 of the Public Resources Code.

(2) That solar energy systems receiving monetary incentives are intended primarily to offset part or all of the consumer's own electricity demand.

(3) That all components in the solar energy system are new and unused, and have not previously been placed in service in any other location or for any other application.

(4) That the solar energy system has a warranty of not less than 10 years to protect against defects and undue degradation of electrical generation output.

(5) That the solar energy system be located on the same premises of the end-use consumer where the consumer's own electricity demand is located.

(6) That the solar energy system be connected to the electric utility's electrical distribution system within the state.

(7) That the solar energy system has meters or other devices in place to monitor and measure the system's performance and the quantity of electricity generated by the system.

(8) That the solar energy system be installed in conformance with the manufacturer's specifications and in compliance with all applicable electrical and building code standards."

In implementing the DWP incentive program consistent with these criteria, DWP customers have been given an alternative to purchasing and owning the photovoltaic system. The customer may lease the system from a third party, provided that the following conditions are met:

1. The lease is guaranteed for at least 20 years (to cover the anticipated period of energy production on which the incentive is based).

2. The photovoltaic system is operational and operated at the expected generation capacity for a 20-year term.

3. The lease provides for customer ownership by the end of the 20-year term.
4. The lease payments may not be based on energy production from the equipment, which could be interpreted as retail sale of electricity.
5. The incentive payment is paid directly to the customer and is not assignable to a third party.

The incentive program requires that the lease agreement for the equipment be provided to DWP for review and found acceptable for the incentive payment in the sole discretion of both DWP and the Los Angeles City Attorney.

9.3 TREATMENT OF ENVIRONMENTAL INCENTIVES

When a government agency is considering use of one of the aforementioned legal mechanisms for an alternative energy program, it is important that the government agency control, if possible, as many of the environmental attributes, environmental incentives, and reporting rights as possible. As used here, the terms *environmental attributes, environmental incentives,* and *reporting rights* are defined as follows:

- *Environmental attributes* means the characteristics of electric power generation at the generating facility (the electric power generation equipment, controls, meters, etc. connected to the energy delivery point as a fixture on the site) that have intrinsic value, separate and apart from the energy output (total quantity of all actual net energy generated), arising from the perceived environmental benefits of the generating facility of the energy output, including but not limited to all environmental and other attributes that differentiate the generating facility or the energy output from energy generated by fossil-fuel-based generation units, fuels or resources, characteristics of the generating facility that may result in the avoidance of environmental impacts on air, soil, or water, such as the absence of emission of any oxides of nitrogen, sulfur, carbon, or of mercury, or other gas or chemical, soot, particulate matter, or other substances attributable to the generating facility or the compliance of the generating facility or the energy output with the law, rules, and standards of the United Nations Framework Convention on Climate Change (UNFCCC) or the Kyoto Protocol to the UNFCCC or crediting "early action" with a view thereto, or laws or regulations involving or administered by the Clean Air Markets Division of the Environmental Protection Agency or successor administrator or any state or federal entity given jurisdiction over a program involving transferability of environmental attributes and reporting rights.

- *Environmental incentives* means all rights, credits (including tax credits), rebates, benefits, reductions, offsets, and allowances and entitlements of any kind, howsoever entitled or named (including carbon credits and allowances), whether arising under federal, state, or local law; international treaty; trade association membership; or the like arising from the environmental attributes of the generating facility or the energy output or otherwise from the development or installation of the generating facility or the production, sale, purchase, consumption, or use of the energy output. Without limiting the forgoing, environmental incentives include green tags; renewable energy credits; tradable renewable certificates; portfolio energy credits; the right to apply for (and entitlement to receive) incentives under the Self-Generation Incentive Program, the Emerging Renewables Program, the California Solar Initiative, or other incentive programs offered by the State of California; and the right to claim federal income tax credits under Sections 45 and/or 48 of the Internal Revenue Code.

- *Reporting rights* means the right of the private power provider to report to any federal, state, or local agency, authority, or other party, including without limitation under Section 1605(b) of the Energy Policy Act of 1992 and provisions of the Energy Policy Act of 2005, or under any present or future domestic, international or foreign emissions trading program, that the power provider owns the environmental attributes and the environmental incentives associated with the energy output.

An example of how the environmental attributes, incentives, and reporting rights may be treated in an agreement is as follows:

(a) **Delegation of attributes to power provider.** Notwithstanding the Generating Facility's presence as a fixture on the Site, Power Provider shall own, and may assign or sell in its sole discretion, all right, title, and interest associated with or resulting from the development and installation of the Generating Facility or the production, sale, purchase or use of the Energy Output including, without limitation:

(i) All Environmental Incentives except for Solar Renewable Energy Credits associated with the Generating Facility; and

(ii) The Reporting Rights and the exclusive rights to claim that: (A) the Energy Output was generated by the Generating Facility; (B) Power Provider is responsible for the delivery of the Energy Output to the Energy Delivery Point; (C) Power Provider is responsible for the reductions in emissions of pollution and greenhouse gases resulting from the generation of the Energy Output and the delivery thereof to the Energy Delivery Point; and (D) Power Provider is entitled to all credits, certificates, registrations, etc., evidencing or representing any of the foregoing.

(b) Delegation of attributes to purchaser (government agency). Purchaser shall own, and may assign or sell in its sole discretion, all right, title, and interest associated with or resulting from the following:

(i) All Environmental Attributes and Solar Renewable Energy Credits associated with the Generating Facility; and

(ii) The Reporting Rights and the exclusive rights to claim that Purchaser is entitled to all Solar Renewable Energy Credits evidencing or representing any of the foregoing.

Based upon the public policy espoused by the California legislature over the past 30 years, a number of legal mechanisms have been authorized to provide and encourage the development and installation of alternative energy sources. With the ever-growing awareness and publicity regarding the continuing erosion of our environment, the support and advancement of solar technologies can be expected to continue for the foreseeable future.

CHAPTER 10

Seven Principles for Interconnectivity: Achieving Sustainability in Design and Construction

Christine S.E. Magar, AIA, LEED AP

Contents

10.1	Introduction	165
10.2	Some of Today's Most Influential Sustainable Design Maxims	166
10.3	A New Set of Principles	168
10.4	Principle 1. Building Independence	170
10.5	Principle 2. Building Natural Form	171
10.6	Principle 3. Building Service	172
10.7	Principle 4. Building Interconnectivity	172
10.8	Principle 5. Building Adaptability	173
10.9	Principle 6. Building Performance	174
10.10	Principle 7. Building Interdependence	175

10.1 INTRODUCTION

Envisioning a better future requires a dream, and, I believe, one that includes improving our quality of life while practicing sustainable development. Much of today's built environment is not about sustainability but about improving a bottom line or satisfying an individual's aesthetic preference. Profit and aesthetics are important, but they need not be met at the expense of human comfort, quality of life, and the health of people and the planet. The balance between an individual's preferences and a need to be good stewards of our quality of life and planet health can be achieved. Creating this balance is the responsibility of the design and engineering professional.

I offer seven principles as a path toward this balance. Best practices for future sustainable design and construction are integral to these seven principles. At the same time, they are meant to be thinking points on which to have a dialogue, debate, and finally agree on what is most appropriate in our homes, our communities, and in the practice of architecture and

Sustainable Communities Design Handbook
ISBN: 978-1-85617-804-4, DOI: 10.1016/B978-1-85617-804-4.00010-0

engineering. In reading them, you are given an opportunity to reset your priorities on development practices by putting basic shelter and comfort in partnership with nature, combined as first priority.

The principles are abstract, and so I use some tools to aid in their definitions. I include a diagram to illustrate their meaning. Various examples are offered from the small scale of a window to the large scale of a community to explain how the principle can be manifest. Finally, I also offer a means to measure each principle, as a way to create more clarity and distinction among them.

10.2 SOME OF TODAY'S MOST INFLUENTIAL SUSTAINABLE DESIGN MAXIMS

Initiated almost two decades ago, the nine Hannover principles[1] were developed for the World Expo 2000 in Hannover, Germany. These have been integral to the thinking around sustainability. They are frequently cited as starting points for activities seeking sustainability and are often the basis for sustainable design. These are an effective set of principles that address a wide spectrum of scales and environments. However, they are abstract, open to much interpretation, require much thought and work to implement, and they have no means to measure their success.

Around the same time, the U.S. Green Building Council was established and generated a set of 100 maxims organized around five categories. Named, Leadership for Energy and Environmental Design (LEED), this list evolved into a list of 69 credits—a green building rating system—and became a national (and even global) benchmarking system. Since then, LEED has become a definer of a green building, credited for major market transformations in the design, construction, and operation of buildings. It is a system that rates buildings against a consistent set of metrics, easily understood by the market. LEED was designed by a group of volunteer experts to be a market-friendly tool and has been enormously successful in shifting the paradigm in design and construction to integrate nature and technology

---

[1] Hannover principles:
1. Insist on rights of humanity and nature to coexist.
2. Recognize interdependence.
3. Respect relationships between spirit and matter.
4. Accept responsibility for the consequences of design
5. Create safe objects of long-term value.
6. Eliminate the concept of waste.
7. Rely on natural energy flows.
8. Understand the limitations of design.
9. Seek constant improvement by the sharing of knowledge.

more honestly. However, it does not easily translate to all buildings types or built environments. Neither does it reward the passive[2] building that relies on non-energy-using systems for comfort.

As a design professional seeking the most appropriate response to both our local and global concerns for the health of the earth and ourselves, I believe that an aspiration, or ideal, building and community is requisite to each region. Something that is both an abstract ideal and something that is specific and measurable. John Lyle made famous the term *regenerative design*,[3] which responds well to the definition of sustainability as defined by UN World Commission on Environment and Development "To meet the needs of the present without compromising the ability of future generations to meet their own needs." A few years ago the Living Building Challenge[4] was generated by

---

[2] Passive design: The integration in design and construction of site orientation and location, local climate conditions and changes, thermal mass, placement and design of fenestration and shading elements and solar orientation. No mechanized systems allowed.

[3] Regenerative design: Originally coined by Robert Rodale in organic farming, John Lyle expanded the term to include the provision of all necessities of daily life. "Regenerative design means replacing the present linear system of throughput flows with the cyclical flows at sources, consumption centers, and sinks." A regenerative system provides for continuous replacement, through its own functional processes, of the energy and materials used in its operation. Energy is replaced primarily by incoming solar radiation, while materials are replaced by recycling and reuse. Such a system generally has the following characteristics: operational integration with natural processes and, by extension, with social processes; minimum use of fossil fuels and humanmade chemicals except for backup applications; minimum use of nonrenewable resources except where future reuse or recycling is possible and likely; use of renewable resources within their capacities for renewal; composition and volume of wastes within the capacity of the environment to reassimilate them without damage. (John Lyle, *Regenerative Design for Sustainable Development*, New York: Wiley & Sons, p. 10.)

[4] Living building challenge:

SITES
1. Responsible site selection
2. Limits to growth
3. Habitat exchange

ENERGY
4. Net zero energy
5. Materials red list
6. Construction carbon footprint
7. Responsible industry
8. Appropriate materials/service radius
9. Leadership in construction waste

WATER
10. Net zero water
11. Sustainable water discharge

IEQ (indoor environmental quality)
12. A civilized environment
13. Healthy air: Source control
14. Healthy air: ventilation

BEAUTY AND INSPIRATION
15. Beauty and spirit
16. Inspiration and education

the Cascadia-Chapter of the USGBC, identifying 16 prerequisites that must be met to achieve the challenge. This unique challenge is defined by measurable absolutes: zero net energy and water, no toxic materials, and so forth. Although it is more difficult to achieve than LEED, it aspires to an ideal rather then settling for the minimum. From this perspective, there is no gray area, a project either achieves the challenge or it does not.

A very different kind of system came out of the New Orleans Katrina disaster, fostering an opportunity that most old communities do not have. The Make It Right[5] Program provides smart guidelines that include aggressive energy efficiency and green building goals. Most important, the design criteria require that buildings be designed to improve quality of life, support the community, and adapt to climate change.

10.3 A NEW SET OF PRINCIPLES

The following seven principles (illustrated in Figure 10.1) give focus to the important relationships between building, people, and nature. Although they are a process, they can be considered individually, keeping in mind that interconnectivity is the fulcrum on which all the other principles rely. All seven principles are met through partnership with community and the natural environment. Each is measurable, assumes that the passive-regenerative building is possible and desirable, that climate change is an opportunity to make it right and promises abundance to all when resources are shared.

1. **Architectural independence**. Buildings are independent of their systems. This is measured by occupant comfort when the utilities are turned off.
2. **Building form**. The building's form enhances and amplifies its performance. Measured by the degree of human comfort and optimal use of energy, water, and material.
3. **Building service**. The building is in service to natural systems. This is measured by the health of the natural environment.

---

[5] Make it Right (MIR9): "In December 2007, the plan to rebuild part of the flood-ruined lower Ninth Ward had the future in mind. The . . . architects . . . invited to contribute to the project . . . were asked to base their work on traditional New Orleans typologies—the 'shotgun' single family home, and the 'duplex,' a multi-family home. They were asked to make the homes green, affordable, and durable enough to weather the storms to come. They were expected to design homes that would be built using materials inspired by Cradle to Cradle™ thinking and verified to be non-toxic and reusable. And finally, they were asked to design homes that were aesthetically advanced." (www.makeitrightnola.org/index. php/building_green/architecture/.)

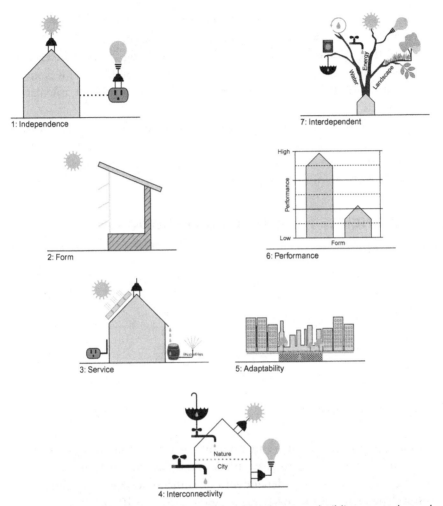

Figure 10.1 Seven principles of the relationships between buildings, people, and nature

4. **Building local interconnectivity**. A building is connected to community utilities, but in a crisis, it can disconnect and still function. Measured by building and community functionality in crisis.
5. **Building adaptability is building for the future**. The building architectural (non-energy-using) systems are adapted to the changes of climate. Climate constants are determined by geography, such as solar access, hurricanes, and earthquakes. Climate variables include solar days, rain inches, wind days, and daylight autonomy. The built environment

is designed using climate constants that can absorb the shock of its variables. This is measured by building longevity and reuse.

6. **Building performance**. Form follows building performance. The green building aspiration is a passive-regenerative building. This is measured by the degree of net-positive use of natural resources, calculated in BTUs, gallons, and tons.

7. **Natural and community interdependence**. The building architecture is inextricably integrated with its natural environment. The boundary between inside and outside is undeniably blurred and is measured by minimal use of resources.

10.4 PRINCIPLE 1. BUILDING INDEPENDENCE

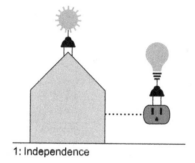

1: Independence

Figure 10.2 Building independence

Building occupants can thrive in a building because of its architecture, that is, the daylighting, the natural ventilation, the mass, orientation, and landscaping. The architect is independent of engineers so that the architecture is not dependent on the mechanized systems (Figure 10.2).

To meet the first principle, architects design buildings and communities with natural systems to meet basic comfort needs. Buildings are designed to be independent of their internal systems, such as lighting and space conditioning and water systems. It seems unimaginable today, to function in a building without its "mechanized systems." This is possible only when we design and build in partnership with nature and consider energy-using systems as supplements to natural systems. Architects can apply systems substitutions; for example, the most obvious and simplest system substitution is lighting. Space lighting is provided not only through the electrical wires but also via daylight through windows. Occupants have access to good air quality through operable windows, and space conditioning is supplemented with sweaters and tee shirts. Some substitutions to consider include all mechanized systems for natural systems, electric light for daylight, mechanical heating and cooling for passive heating and cooling (green house: orientation, mass, glass); mechanical ventilation for operable windows and the stack effect (hot air rises naturally), utility water for rainwater harvesting and gray water.

Building independence is measured by the comfort and functionality of the occupants when the utilities are turned off (See the Examples of the New Principles table at the end of the chapter for this and all illustrations of the principles described here.).

10.5 PRINCIPLE 2. BUILDING NATURAL FORM

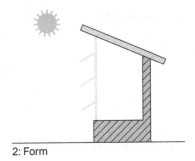

2: Form

Figure 10.3 Building natural form

The building's performance does not suffer from its form; rather its form enhances and amplifies its performance (Figure 10.3).

Its environment and its materials influence the building form; its shape can be created to amplify its performance. Frank Lloyd Wright's organic architecture defines an architecture where the building form connects directly to the natural world including its systems and materials. The building fails this principle when the form exists for itself and does not incorporate the aspects required for comfort such as daylight and ventilation. For a surprisingly successful example, Frank Gehry's Disney Hall masterfully integrates the well-disguised skylights and fenestration with the form of the building. In this case, the striking form is enhanced by daylight and saves some energy by the displaced electric lighting.

Similarly, the Disney Hall exterior can be experienced as a landscape of walls and trees, collapsing the distinction between nature and building yet highlighting the building form as an enhancer of daylight, the exterior

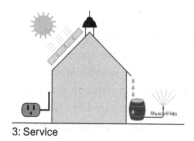

3: Service

Figure 10.4 Building service

landscape, and the view. Fenestration gives the opportunity to connect the building occupant with nature while satisfying the need for comfort. The window offers daylight, views, and natural ventilation, where all the elements can be experienced without harm.

Measured by an optimal use of energy, water, and material, this building can have any form so long as the natural systems work with the mechanical ones. A building that ignores the optimal orientation and design for daylight, space conditioning, and ventilation does not meet this criterion. See the Example table for ideas.

10.6 PRINCIPLE 3. BUILDING SERVICE

The built environment is in service to natural systems (Figure 10.4).

This principle inverts the common practice where the earth and her resources serve our needs and our buildings. From the extraction of raw materials to the increasing waste stream, we need to conceptualize the opposite pattern. We can start with a new way of thinking about our relationship to the earth, resulting in more respect for the earth's resources and therefore better resource management practices.

A building and community can be in service of their environment in many ways. If a property or community harvests and reuses rainwater for irrigation, this would serve the overall health of the regional watershed by keeping it out of the city infrastructure and in the natural aquifer.

Building service is measured by the degree of service to health of the natural environment. See the Example table for ideas.

10.7 PRINCIPLE 4. BUILDING INTERCONNECTIVITY

Building local interconnectivity acknowledges that we are inextricably linked to each other and the natural world (Figure 10.5). A building is

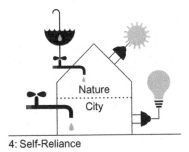

4: Self-Reliance

Figure 10.5 Building interconnectivity

connected to municipal utilities but can function independently. Yet, there is no pure independence, no self-sufficiency. A watershed demonstrates this. When there is a drought, all are affected; if the water in an aquifer is contaminated, all are affected; if the river runs low, all are affected; if my well runs dry on my property, it is likely that so has my neighbor's.

Interconnectivity also refers to the need to have some self-reliance. Although we are connected to the local grid, by generating on-site electricity, harvesting rainwater, and growing food, a building and a community can practice some autonomy. Some water must be harvested and recycled so that, in the event of an emergency, the home, building, neighborhood, or region can function on its own, meeting minimum needs.

The first step toward interconnectivity is quantifying minimum needs. Buildings serve the environment when we can measure the community (watershed, energy-shed) longevity. See the Example table for ideas.

10.8 PRINCIPLE 5. BUILDING ADAPTABILITY

Climate change is often addressed in two categories: mitigation of greenhouse gas (GHG) emissions and adaptation to the climate changes in a region. Buildings and communities can contribute to adapting to climate change (Figure 10.6). There is a lot of uncertainty in climate change. We know that the changes differ in every region and tend to be more extreme with some unpredictability. Design in a hurricane region like New Orleans may require lifting houses to avoid flood damage, design in a flood plane in Los Angeles may require storm water retention. It must be regionally based, a lifted house in Los Angeles is excessive to the thunderstorm flooding and at the same time retention basins in New Orleans does little to meet the needs for water collection during a hurricane.

5: Adaptability

Figure 10.6 Building adaptrability

Buildings and communities meet building adaptation if they survive extreme climate events. See the Example Table for ideas.

10.9 PRINCIPLE 6. BUILDING PERFORMANCE

To meet the building performance principle, design buildings and communities that give to the environment as much as they take from it; and, for this reason, establish metrics by which a net-zero building can be measured (Figure 10.7). A building incorporates natural resources in two ways: 1) the resources (materials) of which it is made and 2) the resources (fuel and water) needed to operate and maintain the building.

First, buildings are made of *materials*: concrete, wood, steel, glass and the like make up the *base* of the building; then the exterior and interior *finishes* complete the building. Base building materials are commodity-type materials usually sold by the linear feet or cubic feet. Several protocols have been developed to qualify the material's sustainability to meet the desired net-zero balance. For example, wood can be measured by its FSC[6] label to show the chain of custody. Finish materials tend to be less durable, creating a lot of waste because they have to be replaced many times during the life of the base building. One way to avoid waste is through leasing agreements, such as a carpet leasing agreement,[7] which will assure that the finish material

---

[6] Forest Stewardship Council. Established in 1993 as a response to concerns over global deforestation, FSC is widely regarded as one of the most important initiatives of the last decade to promote responsible forest management worldwide. FSC is a certification system that provides internationally recognized standard-setting, trademark assurance, and accreditation services to companies, organizations, and communities interested in responsible forestry. http://www.fsc.org/about-fsc.html

[7] Evergreen Lease™
See Evergreen Lease by Interface. Selling carpet without selling carpet—Evergreen Lease is born. Interface is one of the first companies to pioneer a product of service approach to selling carpet. One step closer to closing the loop, this program allows Interface to own the carpet, ensuring proper disposal and no carpet to landfills. Interface produces, installs, cleans, maintains, and replaces the carpet for customers. Customers lease the service of keeping a space carpeted, rather than buying carpet. They get the services of a carpet warmth, beauty, color, texture and acoustics. This is a whole new sustainable business model.

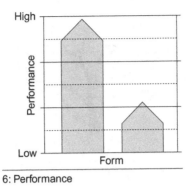

6: Performance

Figure 10.7 Building performance

(carpet) is designed to never end up in the dump. The leasing agreement requires a community of carpet users, so that the carpet material can be recycled and reused, and then be calculated at a net zero-balance.

Second, building operation requires water and various sources of energy: electricity, gas, propane, and others. A high-performing building uses a minimum of water and energy and ideally generates the energy used on site and recycles all the water. To achieve a net-positive balance in water and energy, we need to follow two steps. First, we minimize the need for water and energy. Once we reduce the need for water, we then recycle by irrigating with gray water and harvest rainwater. Similarly, in energy, we reduce the load and generate energy on-site to achieve net-zero energy and carbon neutral building.

A high-performing building is achieved when the net material use, net carbon, net energy use, and net water use are at least net zero[8] with an aspiration to net positive. See the Example table for ideas.

10.10 PRINCIPLE 7. BUILDING INTERDEPENDENCE

The building architecture is integrated with its natural environment (Figure 10.8). The boundary between inside and outside is undeniably blurred. This is true at every scale, from each building component and system to buildings and neighborhoods and regions. Independence appears in many forms, such as double or triple programming something like a rain planter

---

[8] Net Zero Energy Building (ZEB) is defined as a building that produces at least as much energy on its site as it uses.

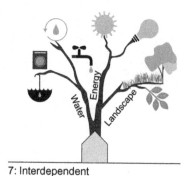

7: Interdependent

Figure 10.8 Building independence

that provides a place for rainwater to drain, for street beautification, and irrigation storage. Most important, this requires that we let nature do most of the work and give her what she needs. For example, instead of letting the rainwater collect through a concrete and mechanized storm water infrastructure, each parking lot, each garden and park, each sidewalk can play a role in retaining the rainwater so it can be absorbed into the earth or stored for future irrigation. Independence is measured by minimal use of resources. See the Example table for ideas.

Examples of the New Principles

	Small Scale	Medium Scale	Large Scale
#1 Dependence	All Scales: GEOs Net-Zero Energy Mixed-Use Neighborhood in Arvada, Colorado. Orient and design buildings to optimize building comfort through the architecture, where, for example autonomous daylight can be achieved (100% daylight when the sun is shining). A good example of this is the Colorado project called GEO[1], considered to be the largest "net-zero energy, urban mixed-use neighborhood in the United States." It achieves this metric by establishing a "symbiotic [relationship] between urban design, landscape architecture, and architecture at all scales."		
#2 Form	Any type of shading device, such as a deep overhang, or a light reflector, such as a light shelf, meets this principle.	The Disney Hall form disguises the saturation of daylight and the daylight enhances the form. In this way the form does not rely only on the daylight performance but is also sculptural.	A Master plan can incorporate building orientation for the best sun, wind, and view access. Often preferred view orientations work against this. Design ingenuity can always optimize all three. GEO 17 is a good example of this.
#3 Service	A building or community in service of nature is as much a state of mind as it is physically manifest.		
#4 Interconnectivity	With 12 inches of rain on average a year, a Los Angeles 5000ft2 property produces about 33,000 gallons of rain per year. However, a year's worth of utility bills may show a use of over 100,000 gallons per year. Self-reliance would require a new calculation so that the minimum need can be met if the water utility were to be turned off.		District energy, water, and waste provisions are at a local level and thus dependence on large centralized infrastructure can be deemphasized. Dockside Green is a good example of this where they used on-site generation from solar, wind, and geothermal as well as burning waste from local industry. See discussion in example for #6.

(Continued)

Examples of the New Principles

	Small Scale	Medium Scale	Large Scale
#5 Adaptability	Fire is a grave concern for climate adaptation in the southwest. A smart solution to fire protection for a single family dwelling is a border of succulent plants that saved a house in a Santa Barbara neighborhood that was almost entirely in ashes after the 2008 fire[2].	Make it Right[3], New Orleans. This non-profit foundation was established and hired an architect to create some basic principles of residential design to be met for climate adaptation, human comfort, and community. At least 150 homes will be built based on these principles. These homes follow principles that I believe are appropriate for many communities in the region.	Building adaptability in existing urban fabric is not easy and can be addressed from various scales. In Utah[4], Sandy City built High Point Park, which is a large storm water retention basin as a way to deal with flooding during thunderstorms. Los Angeles has recently changed its ordinances to allow for stormwater retention on single-family home sites. The combination of addressing stormwater on a site, neighborhood, and municipal level makes sense.
#6 Performance	A simple log of utility bills to compare seasonal and yearly uses of water and electricity can be helpful in resource reduction goals.	"The general layout BedZED[5], Beddington, England takes advantage of solar gain and light by facing all of the homes to the south with terraces and large glass openings to maximize solar gain. The housing units are, in many cases, double height loft spaces to bring light deep into the space and make use of heat from solar radiation. The buildings make extensive use of their roofs and terraces to provide space for green/planted roofscapes, solar panels, garden terraces, and bridges. On the north side of every housing block is an office space. The office space is placed in the shadow created by the south facing housing unit and receives mostly indirect diffused light that is better	On-site energy generation or water harvesting relies heavily on region and climate. To bring more consistency, district heating and cooling can be established in a neighborhood. In this way, a shared electric grid can optimize roof-solar-generated electricity from unoccupied residences for office air conditioners. Similarly, water used from residences can be recycled and contribute to water needs in workplaces. Dockside Green[7] in Victoria, British Columbia, Canada is a good example of this.

#7 Independence	Portland, Oregon rain-planters that are installed in street right-of-ways have at-least three purposes. They serve as rainwater catchments, irrigation storage, and street beautification.	A multifamily fifteen-story building and its adjoining park are symbiotically related. The park serves as open space for the building's residents and the building provides greywater and rainwater runoff for irrigation to the park.	for working. The illustration for this principle shows the massing of the project and its relationship to the project's solar strategy. Dark masses are housing. Lighter masses are office space tucked behind the housing."[6]	A watershed is an eco-system that includes the built environment. A healthy watershed relies on smart urban rainwater management. The Sun Valley Park Multiuse Project was completed in 2006 as part of the Sun Valley Watershed Management Plan. TreePeople is working in partnership with the Los Angeles County Department of Public Works, the City of Los Angeles, and other local stakeholders to create a large-scale sustainable watershed management demonstration project in a 2,700-acre San Fernando Valley watershed. http://www.treepeople.org/sun-valley-watershed

[1] GEOs Net-Zero Energy Mixed-Use Neighborhood Arvada, Colorado. http://www.michaeltavelarchitects.com/GEOS%20Presentation.pdf

[2] Fred Ashtiani, Tea Fire that burned 33 houses in Montecito County. http://freshdirt.sunset.com/2008/11/california-wild.html

[3] Make it Right. http://www.makeitrightnola.org/

[4] High Point Park, Sandy City, Utah. This dual-purpose park has transformed the storm water retention basin into a beautiful park feature for local residents. Containing 1 softball field and utility play field, 2 tennis courts, a playground and basketball half court, the park is used frequently. The park has an outdoor-lighted pavilion, containing twelve picnic tables.

[5] http://www.zedfactory.com/projects.html

[6] BedZED. http://greenlineblog.com/2007/11/bedzed-beddington-zero-energy-development/

[7] Dockside Green. http://docksidegreen.com/index.php?option = com_frontpage&Itemid = 1

CHAPTER 11

The Los Angeles Community College District

Establishing a net-zero energy campus

Calvin Lee Kwan[a] and Andrew Hoffmann[b]

[a]Environmental Science and Engineering Program, School of Public Health, 650 Charles E. Young Drive South University of California, Los Angeles, CA 90095, USA. Email: clkwan@ucla.edu
[b]BuildLA, 915 Wilshire Blvd., Suite 810, Los Angeles, CA 90017, USA. Email: Andrew.Hoffmann@ build-laccd.org

Contents

11.1	Introduction	181
11.2	Background	182
11.3	Goal and Objectives	184
11.4	Importance of this Study	184
11.5	Renewable Energy Options	185
	Wind	186
11.6	LACC Current Situation	188
	Energy Demand vs. Energy Consumption	190
	Electric Utility Rates	190
	City College Campus Energy Consumption and Demand	191
	City College Campus Growth and Demand Side Management	194
11.7	Projected 2015 Campus Energy Demand and Consumption	201
11.8	LACC Solar Insolation	202
11.9	Solar PV Array and Setup	204
	PV Array Size	209
11.10	Discussion	209
11.11	Conclusion	210
Appendix a	Map of LA City College Campus Indicating Previous, Current and Planned Renovations/construction	212
Appendix b	LADWP Energy Rates as of October 1, 2009 – Specific for LACC Operations	213
References		214

11.1 INTRODUCTION

With the distinct advantage of having "unique academic freedom, critical mass and a diversity of skills to develop new ideas" (Calhoun and Cortese 2005), tertiary institutions are ideal locations to promote society changing technology, concepts and practices. Industry, government and private citizens have often contributed endowments, grants and monetary gifts to support research and various functions of tertiary institutions. Universities and colleges are essentially microcosms of society – miniature communities complete with power generating facilities, transportation operations, residential and business functions as well as food and waste services. Thus, many of the challenges that the greater society faces can often be observed or mimicked in the university setting albeit in a more controlled environment. This provides an excellent opportunity and platform to develop, test and refine solutions aimed at addressing societal challenges – including those associated with climate change.

11.2 BACKGROUND

Very few institutions in the United States – tertiary, secondary or primary – have met more than 30% of its energy demand through renewable technologies. The best example to date is a small liberal arts campus located in Morris, Minnesota, part of the University of Minnesota system. The campus is able to meet 60% of its energy demand through its wind turbine farm[1]. The school is able to do this because of several factors including the non-energy intensive nature of the programs offered, the small total campus population and abundance of open land available for installing renewable energy resources. This combination of factors is not typical of a tertiary institution, let alone a city. Regardless, this is still a feat. However, the Los Angeles Community College District (LACCD) aims to set the standard even higher by making all nine of its campuses operate at net-zero energy.

With over 180,000 students enrolled across nine colleges, the Los Angeles Community College District is the largest community college system in the United States. The LACCD has taken a major step along with the American Association of Sustainable Higher Education (AASHE) with its goal to advance the utilization of advanced energy technologies to make all college campuses operate at "net-zero energy consumption" and "climate neutral". This ambitious climate change policy was made feasible with the passing of Measure J in the November 2008 California State elections. This measure allocated $3.5 billion dollars to LACCD projects including "modernization, renovation, improvement and new construction projects, constructing energy infrastructure improvements and upgrading technology systems[2]." Of this funding, just over $200 million has been earmarked specifically towards making the District's goal of being energy independent a reality. What makes

[1] University of Minnesota, Morris Campus. http://www.morris.umn.edu/about/
[2] **Measure J**: Community College Classroom Repair, Public Safety, Nursing & Job Training." http://www.smartvoter.org/2008/11/04/ca/la/meas/J/

this goal especially unique is that the District has decided to actually produce the electricity by installing renewable energy technology on its campuses, as opposed to purchasing renewable energy certificates which many other tertiary institutions currently do (AASHE).

A net-zero energy campus is defined as a one where, over the course of a specified time, the amount of energy provided by on-site renewable energy resources is equal to the amount of energy used. This is inherently different from an "off-the-grid" concept where at *any* point in time, *all* campus energy demand is met through renewable resources. Despite having strong financial support, whether the LACCD can actually realize its goal of developing a net-zero energy school system has yet to be determined.

Using renewable energy resources to generate a portion of a campus energy demand is already difficult, let alone trying to establish a net-zero energy operation. Currently, a number of viable renewable energy technologies exist, including solar PV, wind, hydroelectric and to a certain extent microturbines. However, one of the biggest concerns when considering installing renewable energy systems (RES) – and often times the deciding criteria of such projects – is the duration of time required before a return on investment (ROI) is observed. The limited number of case studies of tertiary institutions with RES installed and the relatively recent trend of installing such systems makes it difficult to accurately determine factors such asproject feasibility, payback period and even product performance. Further complicating these calculations are government incentives, rebate programs and initiatives that vary state to state. As shown in Table 11.1, studies investigating small-scale (>1 MW) RES estimate ROI times anywhere between 4 – 30 years,

Table 11.1 Summary of estimated ROI times for various RES projects

Study	Type of RES	Size of System*	Estimated ROI
(Dalton, Lockington et al. 2008)	Wind/Diesel Hybrid	1.8 MW	4.3 years
(Rhoads-Weaver and Grove 2004)	Wind	100 kW	7 years
(Yang 2004)	Solar PV	—	11.2 years
(Smestad 2008)	Solar PV	1 GW	15 years
(Edwards, Wiser et al. 2004)	Wind	3.6 GW	20 years
	Solar PV	3 MW	30 years
	Biomass	1000 L of Ethanol	11 years
(Yue and Yang 2007)	Wind	10 kW	12 years
(Forsyth, Tu et al. 2002)	Wind/Diesel Hybrid	1.8 MW	4.3 years

*MW = megawatt = 1,000,000 W
GW = gigawatt = 1,000,000,000 W

though these are largely theoretical calculations. There are cases of deploying renewable energy technologies, but the approaches have largely been used to demonstrate technological capability rather than providing economic analyses.

As organizations, businesses and nations begin focusing more attention on the role of RES in their energy portfolios, a greater understanding of the decision-making process and economics associated with RES projects is needed. This chapter aims to do just that by using the Los Angeles Community College District's net-zero energy initiative as a case study.

For this study, the City College (LACC) campus of the LACCD was selected to determine whether a net-zero energy campus can be achieved through a combination of renewable energy technologies and demand side management. City College was chosen specifically because it is situated in a dense urban area in the heart of Los Angeles. The campus itself mimics many cities – clusters of buildings up to 10 stories high situated in very developed areas with little open space available. This puts a unique challenge in identifying suitable locations for installation of various renewable energy technologies as well as identifying new types of technologies to implement in the project.

11.3 GOAL AND OBJECTIVES

The goal of this study was to evaluate whether the LACCD net-zero energy campus concept is currently feasible both technologically and economically. This was accomplished by meeting the following objectives:

1) Develop an understanding of the current energy demand of LACC including daily and annual fluctuations.
2) Complete a comprehensive energy audit and identify areas where energy demand and consumption can be reduced.
3) Project the future energy demands of LACC.
4) Identify the minimum PV array grid size that will satisfy current and year 2020 LACC energy demand.

11.4 IMPORTANCE OF THIS STUDY

Although installing renewable energy technologies is not a new concept, this study certainly adds new information to a rapidly growing field. These include:

1) *Establishing a clear guideline for organizations such as academia, towns and cities to follow with the aim of becoming net-zero energy consuming.* Currently, many other universities, cities and towns have made active commitments

towards increasing their renewable energy portfolio. However, to this date, these efforts have largely involved purchasing renewable energy credits, as opposed to actually using energy generated from renewable energy technologies. Schools and towns that have implemented renewable energy technologies have largely done so on a haphazard basis, installing randomly sized systems without the intention of fulfilling a particular percentage of their energy consumption or demand. Although the installed systems have yet to generate enough energy to completely fulfill an organizations total energy demand, this disorganized approach is inappropriate. By providing information on demand side management and methods to calculate the size of renewable energy technology arrays, organizations will be able to take a more quantified approach to establishing a renewable energy infrastructure suitable for their operations. This work will also be valuable for large organizations and businesses such as convention centers, shopping centers and grocery stores, particularly information on demand side management used to minimize energy consumption.

2) *Providing a model for cities to follow.* The deliberate choice of City College for this study is significant. Instead of choosing other LACCD campuses that have more available open land, consequently making it easier to install enough solar PV to meet energy demands, City College's limited area requires innovative thinking to identify and install renewable energy technologies. Rooftops, walkways and building sides will need to be utilized to produce enough energy, all without affecting the aesthetics of the campus. This is similar to the challenges that many towns and even other academic institutions face – a strong desire to increase renewable energy, but with limited space available. This study seeks to provide options for organizations to consider when establishing renewable energy programs.

11.5 RENEWABLE ENERGY OPTIONS

Renewable energy resources such as geothermal, biomass, hydropower and ocean energy are unavailable to LACC due to a combination of geographic restrictions, city ordinance codes and building restrictions. One possibility is utilizing hydrogen, whichhas tremendous potential as a fuel and energy resource. However, the technology needed to realize that potential is still in early developmental stages (Dillon, Nelson et al. 2006; Edwards, Kuznetsov et al. 2008; Page and Krumdieck 2009). This leaves wind and solar as the most feasible and likely renewable energy resources for LACC to utilize.

Wind

Wind energy is typically captured by wind turbines, which are then used to generate electricity. The amount of electricity generated from wind is proportional to the cube of its velocity as seen in equation 1. This means that doubling the wind speed increases the available power for capturing by a factor of 8.

Equation 1: Wind power experienced across the area swept by a wind turbine rotor

$$P = 0.5 \times \rho \times A \times V^3 \qquad (11.1)$$

Where:

P = power in watts (746 watts = 1 hp) (1,000 watts = 1 kilowatt)

ρ = air density (~1.225 kg/m^3 at sea level)

A = rotor swept area, exposed to the wind (m^2)

V = wind velocity in meters/sec

Thus, minor fluctuations in wind speed can result in a large difference in available energy and in electricity produced, and consequently, a large difference in the cost of the electricity generated. A wind turbine operating in 5 mph wind can generate over 30% more electricity than one in 4 mph wind because the cube of 5 (125) is 31% larger than the cube of 4 (64). However, this also suggests that there is very little energy that can be captured in environments with low wind speeds. It is also important to keep in mind that actual electricity generation will also depend on the efficiency of the turbine that is installed.

A wind profile for LACC was constructed to evaluate the feasibility of utilizing wind as a consistent renewable energy resource. Daily average wind speed data recorded over a 10-year period were collected from the National Oceanic and Atmospheric Administration's (NOAA[3]) USC Downtown LA weather station (WBAN ID: 93134). The USC Downtown LA weather station is located 5 miles away from the LACC campus and is the closest available weather station. Although microclimate differences between the two sites are expected, the close proximity and geographic similarity between the USC and LACC campuses support using data from the USC weather station.

Using historic data, annual average daily wind speeds since 2000 were graphed and are shown in Figure 11.1. From 2000 through 2005, the annual average daily wind speed remained constant at around 2.5 mph. However, a 30% drop in average daily wind speed to 1.72 mph was observed in 2006,

[3]NOAA Climate Data Online. http://www7.ncdc.noaa.gov/CDO/cdo

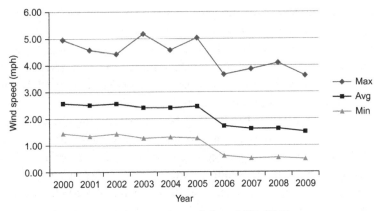

Figure 11.1 Annual average daily wind speeds from 2000–2009.

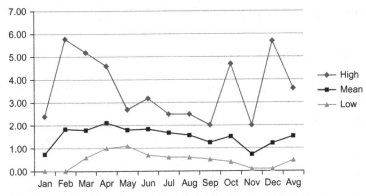

Figure 11.2 2009 Monthly average daily wind speeds.

and has been sustained at approximately that level since. Although climate change has been linked to decreased wind speeds (Pryor and Barthelmie; Segal, Pan et al. 2001; Pryor, Barthelmie et al. 2009), the sudden drop is more likely attributed to a change in the weather station's surroundings such as construction of new buildings.

In 2009, the monthly average daily wind speed recorded in Los Angeles was 1.51 mph, while the highest and lowest average daily wind speeds were 3.61 and 0.57 mph respectively. More detailed analysis reveals that the monthly average daily wind speed remained between approximately 1 and 2 mph. There was greater variation in the monthly highest average daily wind speed as seen in Figure 11.2. This was likely due to seasonal variations in storms, Santa Ana winds and inclement weather conditions.

Although there is clearly a constant flow of wind throughout the year, due to itsconsistentlow velocity, it is unlikely that wind energy will be a major contributor to the LACCD's net-zero energy goal. Wind speeds of less than 9.8 mph are classified as wind power class 1. Class 1 winds are 'generally not suitable [for wind turbine applications], although a few locations with adequate wind resource for wind turbine applications may exist' (NREL 1986). Typically a minimum of wind power class 3 (having velocity of 12.5 mph or greater) is required for large-scale wind turbine applications.

The wind resources at LACC may be appropriate for micro-scale, less energy intensive operations such as providing energy for aesthetic or accent lighting. However, for more energy intensive processes, a more reliable and consistent renewable energy resource with greater power generating capability is needed. This leaves solar energy. With an average of 3,200 hours of sunshine each year, solar radiation will be the most likely source of renewable energy available to LACC. The rest of this chapter will focus on utilizing solar RES systems to meet the LACCD net-zero energy campus concept.

11.6 LACC CURRENT SITUATION

The City College campus is built on 49 acres of land. As of October 1, 2009, there were 21 existing buildings, 7 planned new building constructions and 2 large parking lots. According to the LACC long-range developmentplan, 8 buildings on campus are scheduled to undergo complete renovation over the next five years. There are currently 10 types of buildings on the LACC campus and these are listed in Table 11.2. A map of the campus can be found in Appendix a.

Like many towns and cities, open, undeveloped land on the LACC campus is a premium commodity. Given the lack of open space and the presence

Table 11.2 List of building types found on LACC campus

Building Type
Cafeteria/Food services
Child Development Center
Classrooms
Classrooms w/ Lab
Gymnasium
Health Center
Industrial Trade
Library
Office/Administrative
Theater

of tall surrounding buildings that may impede wind exposure and solar radiation at ground level, it is difficult to install large renewable energy systems on the ground. Therefore any RES installation will likely utilize existing rooftops and carports. Even so, not all rooftop areas are suitable for RES installation, as existing heavy equipment, building design and orientation, and rooftop structure will affect the installation, efficiency and effectiveness of a RES. Aerial maps of LACC and its local vicinity coupled with an analysis of existing building rooftops and carports were used to determine the total suitable area for installing a RES. In order for a rooftop or open carport to be suitable for installing RES, the following preliminary criteria had to be met:

Suitable rooftops must:
- have a clear view of southwestern sky;
- be flat or slanted at no greater than 30° tilt;
- not have trees higher than ¾ the height of the building growing within 5 meters of the building;
- have open access to a water supply within 5 meters of the building.

Figure 11.3 illustrates how suitable rooftop areas for solar RES installations were identified.

Suitable carport areas must:
- have an unobstructed view of southwestern sky;
- not have trees growing within 5 meters;
- have open access to a water supply.

Table 11.3 summarizes the findings of this analysis.

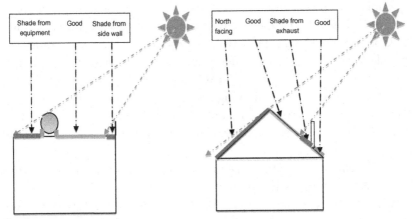

Figure 11.3 Illustration of available rooftop space for solar RES installation on flat and slope topped buildings.

Table 11.3 Summary of total and available area for installation of RES

Location	Total Space (m²)	Available space for solar (m²)
Rooftop	61,910	18411
Carports	8,163	6,560
	TOTAL	24,971

Energy Demand vs. Energy Consumption

A key component of this project was analyzing and understanding LACC's electrical energy demand and electrical energy consumption. Although the two parameters are related, they are not the same. Most consumers will be familiar with electrical energy *consumption*, which is a measurement of total electrical energy that is used. The unit of this is kilowatt-hours (kWh) and is typically reported on electrical utility bills. Electrical energy *demand* is a measurement of the rate of consumption of total electrical energy and is measured in kilowatt-hours per hour, or simply, kilowatts (kW). For example, if a building consumed 24 kWh of energy over a 24-hour period, its demand would be 1 kW. However, if the consumption occurred in 12 hours, demand would be 2 kW, likewise 3 kW if consumed over 8 hours. Peak demand refers to the maximum electrical energy demand (W) experienced during any one point of the day.

Electric Utility Rates

The LACC campus is located within the service area of the Los Angeles Department of Water and Power (LADWP), the largest municipal utility in the United States. Under the LADWP time-of-use electrical tariff scheme, LACC is considered a Large General Service and is charged according to Schedule A-3A. Electrical rates differ depending on the time of year and time of day, varying between $0.02197/kWh up to as high as $0.04390/kWh. The calendar year is divided into high season starting from June 1st through September 30th and low season, from October 1st to May 31st. The daily rate period breakdown is summarized in Table 11.4. Full details of the tariff can be seen in Appendix a.

By implementing a differential rate scheme there is financial incentive for LADWP consumers to minimize electrical energy consumption particularly during peak periods. This poses a challenge for LACC because the core of its operations take place during peak periods and are difficult to reschedule. Installing a RES can be used as a method to reduce energy

Table 11.4 Summary of daily electric rate time schedule

Charge	Time
TOU 1: High Rate	Monday–Friday, 13:00–16:59
TOU 2: Low Rate	Monday–Friday 10:00–12:59, 17:00–19:59
TOU 3: Base Rate	Monday–Friday 8:00–9:59 Saturday and Sunday All day

consumption and demand during peak usage periods in a process known as peak shaving. In this process, electricity generated from RES is used to meet a portion of electricity needs thereby reducing the energy demand drawn from the utility provider. This is particularly useful if the RES is large enough to reduce maximum electrical energy demand as opposed to simply consumption. However, the ability of a RES to maximize peak shaving depends heavily on its design.

City College Campus Energy Consumption and Demand

Renewable energy systems are not cheap and designing the system can be complicated – if it is too small, future growth may be compromised. Constructing larger systems that initially generate excess energy in anticipation of future growth may be financially unappealing and is even discouraged. Currently in California there are two options for excess electrical energy; collect and store the electrical energy using batteries and energy storage systems, or feed the energy back into the grid. In the first option, batteries and storage devices are bulky and costly - finding additional land to build energy storage centers is also a whole other consideration. In the second scenario, while many electric utility companies welcome feeding excess electrical energy back into the grid, they do not compensate the producers for the surplus electricity they send back into the grid. This discourages consumers from fully utilizing all spaces to install RES, and can even encourage consumers to waste energy in order to avoid giving it free to utility companies. Therefore, it is critical that a detailed understanding of the campus' energy demand is developed. By doing so the correct sized RES can be identified and installed.

To understand the current energy demand and consumption of LACC, energy bills between August 1st, 2007 and July 31st 2008 were collected and summarized and shown in Table 11.5. In 2008, LACC spent just over $800,000 purchasing over 8,594,000 kWh of electricity from LADWP. A breakdown of energy consumption by specific areas and buildings on campus were unavailable due to LACC having only one electric meter.

Table 11.5 LACC 2008 Monthly Energy Demand

Month	Peak Demand	TOU1 (kW)	TOU2 (kW)	TOU3 (kW)	Energy Cost	Facility Cost	Demand Charge	Utility Bill
1	1342.0	1342.0	1224.0	1230.0	$39,861.74	$5,368.00	$5,293.00	$50,672.74
2	1421.0	1421.0	1418.0	1239.0	$40,368.10	$5,684.00	$5,609.00	$51,811.10
3	1545.0	1532.0	1545.0	1391.0	$47,287.64	$6,180.00	$6,053.00	$59,670.64
4	1439.0	1439.0	1429.0	1220.0	$43,198.64	$5,756.00	$5,681.00	$54,785.64
5	1872.0	1872.0	1812.0	1490.0	$51,615.08	$7,488.00	$7,413.00	$66,666.08
6	1709.0	1709.0	1686.0	1569.0	$48,139.72	$6,836.00	$20,364.00	$75,489.72
7	1805.0	1790.0	1805.0	1572.0	$55,654.10	$7,220.00	$21,450.00	$84,474.10
8	1857.0	1836.0	1857.0	1648.0	$56,341.74	$7,428.00	$22,020.00	$85,939.74
9	2142.0	2142.0	2100.0	1856.0	$55,287.33	$8,568.00	$25,503.00	$89,508.33
10	1782.0	1782.0	1744.0	1436.0	$56,698.48	$7,128.00	$7,053.00	$71,029.48
11	1637.0	1637.0	1600.0	1325.0	$46,748.25	$6,548.00	$6,473.00	$59,919.25
12	1465.0	1465.0	1448.0	1276.0	$40,909.59	$5,860.00	$5,785.00	$52,704.59
Total					$582,110.41			$802,671.41

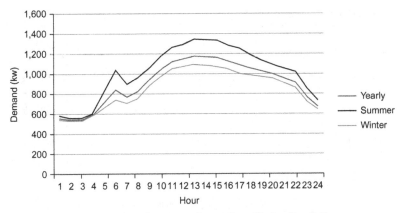

Figure 11.4 2008 Average hourly energy demand profile for City College.

Electrical energy consumption and demand was analyzed by collecting fifteen-minute interval utility energy consumption data. Data were categorized into winter (November – February) and summer (June – September)months as we expected the greatest energy demand variance between these two groups. Average hourly energy demand profiles for 2008 were then produced and are shown in Figure 11.4.

As expected, summer months have higher energy demands than winter months, using on average 20% more energy throughout the day. This can likely be attributed to the increased need for heating, ventilation and cooling in the warmer months.

All three profiles exhibited the same characteristics. Average hourly energy demand reached a peak at noon each day while average daily minimum energy demand for City College was approximately 540 kW regardless of season. This represents the baseline energy demand of City College, meaning that at any point in time, the lowest campus energy demand is 540 kW. A sudden spike in energy demand is observed at approximately 6 AM each day, and can be attributed to the coordinated startup of campus equipment such as elevators, HVAC systems, lighting and computers in preparation for daily operation.

A second graph was developed to identify how LACC's energy demand and consumption patterns fluctuated during the course of a year. These are summarized in Figure 11.5. As expected, electrical energy consumption reached a peak in July with over 730,000 kWh used that month. This is likely due to the increased electricity needed to operate air conditioning during the summer months. LACC electrical energy demand

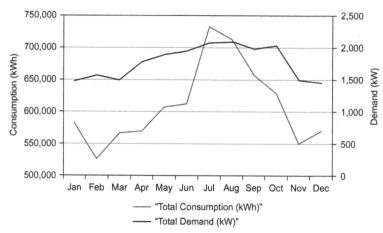

Figure 11.5 2008 Average monthly energy consumption and demand.

reached a maximum in August, drawing 2,084 kW, though demand in July was almost the same, at 2,071 kW. The 2008 electrical energy demand of LACC ranged between 1,447 kW and 2,081 kW.

City College Campus Growth and Demand Side Management
Campus Growth
The long range development plan for LACC projects that the campus will continue to grow and expand into 2020. This includes hiring more faculty and staff, enrolling more students and retrofitting and constructing buildings. In support of this, since 2007, the LACCD has implemented two major programs that will significantly alter each campus' current hourly energy demandand consumption profiles. The first program consists of construction bond measures A/AA/J that are responsible for both campus retrofits as well as significant build out.

From these measures, LACC is expected to construct seven new buildings and retrofit another eight. These projects will increase the total available floor area of LACC from its current total of 67,040 m² to 96 849 m² – an increase of over 44%. Once these building projects are complete, the electrical energy demand and total consumption of the campus will also increase. However, it is uncertain exactly how much the energy consumption and demand will increase by, making it difficult to account for when designing the size of the RES. We can only estimate the future energy consumption and demand.

The planned new buildings are similar to the existing campus buildings in that they are either administrative, classroom or computer laboratory rooms. It is common at research-oriented institutions that laboratory buildings occupy less than 25% of a campus' gross area, but account for over 50% of the total energy consumption. There are no plans for constructing energy intensive buildings such as technical or research laboratories on the LACC campus. Given this, it is reasonable to assume that energy consumption associated with the new buildings would result in a linear increase to LACC's energy demand that is proportional to the building type and square meter increase planned by the end of the bond programs.

Without individual metering, it was impossible to determine the cooling requirements and average energy consumption and demand of various building types on the LACC campus. Therefore, in order to estimate future energy demand, data on existing and future LACC buildings were collected. This included the building type and its total usable area. These are shown in Table 11.6.

Without LACC-specific electrical end use data, we referred to the 2006 California Energy Commission's (CEC) 'California commercial end-use

Table 11.6 Existing and planned buildings on LACC campus

Project ID	Building	Type	Area (m²)
ID	**Existing Building**		
01C-109	FRANKLIN HALL	Classroom	9,555
01C-106	ADMINISTRATION	Office/ Administrative	7,947
01C-134	MLK LIBRARY - To become SS Building	Library	6,537
01C-136	COMMUNICATIONS	Classroom	6,119
01C-108	DA VINCI HALL	Classroom	5,875
01C-107	CLAUSEN HALL	Classroom	5,634
01C-111	JEFFERSON HALL	Classroom	4,675
01C-133	THEATER ARTS	Theater	4,634
01C-150	CHEMISTRY	Classroom w/Labs	3,450
01C-148	GYM-WOMEN	Gymnasium	3,065
01C-110	HOLMES HALL	Classroom	2,848
01C-151	LIFE SCIENCES BLDG	Classroom	2,094
01C-132	CAFETERIA	Cafeteria\Food services	1,758
	HEALTH, FITNESS, PE BLDG	Health Center	1,152
	RADIOLOGIC TECH	Classroom w/Labs	446
	UTILITY BLDG	Other	358
	CARPENTER SHOP	Other	279
	WOMENS-DRESSING-ROOM	Gymnasium	242

(Continued)

Table 11.6 Continued

Project ID	Building	Type	Area (m²)
	BUNGALOWS X & Y	Other	214
	PLUMBER SHOP	Other	80
	BUNGALOW Z-1	Other	80
	TOTAL		**64,431**
	NEW BUILDINGS PROP A/AA		
01C-101	Science and Technology Building	Classroom w/Labs	7,915
01C-115	Health, Fitness, PE Building	Health Center	3,689
01C-122	Child Development Center	Child Development Center	2,412
01C-131	NEW MLK LIBRARY – LRC	Library	5,806
	TOTAL		**19,822**
	NEW BUILDINGS MEASURE J		
01C-146	Physical Plant (M&O Building)	Other	1,579
01C-147	Learning Support Center	Classroom	2,648
01C-145	Green Technology Student Union Building	Classroom w/Labs	5,760
	TOTAL		**9,987**
	EXTERIOR LOTS & ACTIVITIES		
01C-116	Parking Structure	Other	2,430
	TOTAL		**2,430**

survey'[4] to calculate projected energy consumption and demand. The end use survey contained data on average energy consumption by different building types. It estimates that for college buildings in Southern California, the average energy consumption per square meter is 146.6 kWh/year. Over 70% of electricity consumption in a college building is dedicated to heating, ventilation and cooling (HVAC) and indoor lighting. A complete breakdown of the California End Use Survey for college buildings can be found in Table 11.7. The end use survey also estimated that parking lots and structures consume 32.29 kWh/m²/year.

To assess the validity of the CEC's end use survey figure, we estimated the LACC energy consumption per square meter. Using 2008 data, total campus electrical consumption (8,594,400 kWh) was divided by the campus'

[4]California Energy Commission. "California Commercial End-Use Survey." http://www.energy.ca.gov/ceus/

Table 11.7 College and University Electric End Use (Itron 2006)

Component	Electrical EI (kWh/m²/yr)	Electrical EI (%)
Heating	12.70	8.66%
Cooling	22.93	15.63%
Ventilation	21.42	14.60%
Water Heating	1.51	1.03%
Cooking	1.72	1.17%
Refrigeration	3.34	2.27%
Interior Lighting	48.55	33.09%
Office Equipment	12.16	8.29%
Exterior Lighting	7.32	4.99%
Miscellaneous	5.60	3.82%
Process	0.32	0.22%
Motors	6.67	4.55%
Air Compressors	2.48	1.69%
TOTAL	**146.60**	

gross area excluding parking lots ($64{,}431\,m^2$). We found that in 2008, LACC's electric end use rate was $133.36\,kWh/m^2$/year, approximately 10% less than the CEC's estimated value. The lower LACC value is expected given the institutions' primary focus being on student education as opposed to research. This means there are fewer energy intensive operations compared to at research universities, technical colleges and professional schools – all of which are also included in the CEC's college end use estimate.

Using both the current LACC and CEC electrical end usefigures, we estimate that after completion of all building retrofits and new construction projects, the total campus energy consumption based on current patterns will be between 12,920,000 kWh and 14,200,000 kWh. With all new campus buildings and retrofits targeted to be LEED Gold or LEED-EB Gold certified, coupled with LACC's continued focus on education instead of research, it is likely that future campus energy consumption will be closer to the lower estimate.

Demand Side Management

The second program implemented at LACC was a demand side management initiative aimed at reducing the current load burden. Cost savings achieved from the program were used to pay for capital investment of the energy saving technologies. This involved performing a comprehensive energy audit for LACC and making recommendations on how to reduce the energy demand through infrastructure upgrades. The audit focused on developing the following items:

- Energy conservation measures for LACC
- Utility analysis demonstrating effectiveness of conservation measures
- Financial analysis of conservation measures

All buildings on the LACC campus were audited. The following is a list of minimum investigation areas performed for each building:

- Age
- Condition
- Performance
  - Lighting system – including controls and fixture retrofits
  - Building automation
  - HVAC systems
  - Equipment upgrades
  - Ability to connect to a central plant
  - Structural upgrades

The results of the audit recommended a list of energy conservation measures with the cost, savings and ROI listed. These are summarized in Table 11.8 and briefly described below. By implementing these measures, it

Table 11.8 Summary of proposed electrical energy control measures for LACC

Control Measure	Electricity Savings (kWh)	Utility Savings	Rebates	Cost	ROI
ECM 1: Interior lighting upgrades	519,259	$72,896	$67,590	$1,222,722	15.8 y
ECM 2: Installation of lighting controls	116,226	$14,347	$7,340	$167,028	11.1 y
ECM 3: Window glazing	61,468	$10,535	$6,723	$156,482	14.2 y
ECM 4: HVAC fan and motor upgrades	19,843	$2,275	$1,759	$35,308	14.7 y
ECM 5: Upgrade and integrate with EMS	845,481	$250,551	$27,141	$1,996,901	7.9 y
ECM 6: Installation of utility sub-meters	0	$0	$0	$498,416	n/a
TOTALS	1,562,277	$350,604	$110,553	$4,076,857	12.7y
Savings from 2008	18.2%				

is estimated that LACC can reduce its current energy consumption by up to 18.2%, resulting in over $350,000 of savings each year. Simple payback periods were calculated using estimated electrical savings, estimated rebates and incentives and current LADWP electrical rates. All energy conservation measures (ECM) except ECM 5 had ROIs of greater than 11 years. However, it is likely the payback period is even lower once additional factors such as annual electrical rate increases (estimated to be approximately 3% each year), inflation, compounded savings, replacement costs and operations and maintenance costs (or lack thereof) are taken into account.

ECM 1: Interior Lighting
The survey identified 12 locations on the LACC campus for lighting retrofits and upgrades. A total of 6,252 light fixtures were identified for replacement or retrofitting. The recommended retrofits are:

- *Replace 32 W T-8 Fluorescent Lamp Fixtures* with high color rendering 28W T-8 lamps and electronic low power ballasts.
- *Replace T-12 Fluorescent Lamp Fixtures* with 28W T-8 lamps and electronic low power ballasts.
- *Replace Metal Halide Fixtures* with high bay, T-8 4 lamp fixtures with full power ballast.
- *Replace High Pressure Sodium Fixtures* with t-8 lamp fixtures with step-dimming ballast and wire cages. Provide occupancy and daylighting sensors as part of individual fixtures.
- *Replace Incandescent Fixtures* with compact fluorescent lamps.
- *Replace Incandescent Exit Signs* with LED exit signs.

Total cost of this ECM is $1,222,722, with a projected annual electrical savings of $72,896 and 519,259 kWh.

ECM 2: Installation of Lighting Controls
Seven locations on the LACC campus were identified for lighting control upgrades. Buildings were analyzed by space types with the following recommended ECM installation solutions:

- *Private Offices:* Lighting control by wall switch with occupancy sensors.
- *Meeting Rooms:* Lighting control by wall or ceiling mounted infrared or dual technology occupancy sensors.
- *Open Spaces:* Lighting control by ceiling mounted infrared or dual technology occupancy sensors.
- *Restrooms:* Lighting control by wall switch or ceiling mount with occupancy sensors.

- *Storage areas:* Lighting control by wall switch with occupancy sensors.
- *Hallways:* Lighting control by ceiling mounted infrared occupancy sensors.
- *Vending Machines:* Occupancy sensors to control vending machine lights and refrigeration.

367 areas were identified for lighting control upgrades. The total cost of this ECM is $167,028, with a projected annual energy savings of $14,347 and 116,226 kWh.

ECM 3: Solar Film Application

Applying solar film is an effective method to reduce solar heat gain in the summer and heat loss in the winter thereby reducing electricity consumption associated with HVAC systems. Solar film is most effective when installed on east or west facing windows that receive direct sunlight. The energy audit identified 1,229 m² of window area for solar film application.

The total cost of this ECM is $156,482, with a projected annual energy savings of $10,535 and 61,468 kWh.

ECM 4: HVAC Fan and Motor Upgrades

Four locations on the LACC campus were identified as having HVAC fans or motors that were inefficient or the incorrect size for its current application. Replacement with more efficient and better performing motors was recommended.

The total cost of this ECM is $35,308, with a projected annual energy savings of $2,275 and 19,843 kWh.

ECM 5: Upgrade and Integrate Energy Management Services

Although LACC has a centralized energy management system some buildings are not integrated into this system, and thus cannot be remotely controlled. This makes it difficult to monitor and control building systems and energy use. Energy management systems are used to provide centralized oversight and remote control of HVAC systems, lighting and other campus building systems. This can help monitor and integrate functions of multiple buildings from a central location, allowing better control and analysis of energy use and costs.

The audit identified ten buildings on the LACC campus to be integrated into the energy management system. This ECM will cost $1,996,901 with a projected annual energy savings of $250,551 and 845,481 kWh.

ECM 6: Installation of Sub-metering

The LACC campus has only 1 electric meter. This makes it difficult to construct energy profiles of different operations on campus and identify areas for energy management improvement. Installation of submeters will allow for a better and more detailed understanding of LACC's energy consumption in order to establish an effective energy management policy.

The cost of this ECM is $498,416. There are no direct energy savings associated with this ECM.

11.7 PROJECTED 2015 CAMPUS ENERGY DEMAND AND CONSUMPTION

If all ECM recommendations are adopted, we project that the LACC annual energy consumption, before completion of building retrofit and new construction projects, will be reduced to just over 7,000,000 kWh, resulting in over $350,000 of savings. Upon completion of all campus construction and building retrofit projects, we estimate that total annual LACC campus energy consumption will increase to between 11,300,000 and 12,600,000 kWh per year by 2015. Figure 11.6 shows a comparison of 2008 energy demand vs. the projected 2015 energy demand at the conclusion of all building retrofit and construction projects. We project that peak energy demand at noon each day will increase from 1177 kW in 2008 to 1950 kW by 2015, an increase of 66%. Similarly, the baseline energy demand will increase from 540 kW in 2008 to 895 kW by 2015.

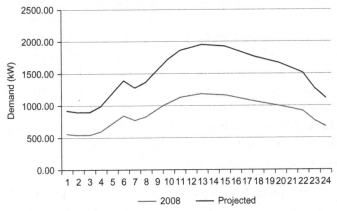

Figure 11.6 Comparison of 2008 and projected 2015 hourly energy demand.

This projection assumes that all building retrofits and new construction projects will, at a minimum, utilize the same technology currently used in existing buildings and that the current campus energy consumption pattern will not change significantly in the future. It is possible that actual future energy consumption will be lower as improved or newer energy savings technologies develop and are incorporated into LACC operations. This includes the development of more efficient chillers for ventilation and cooling services, improved performance of LEDs for lighting purposes and better control of HVAC systems.

11.8 LACC SOLAR INSOLATION

Maximum daily solar insolation absorbed by an installed solar PV array will vary throughout the year. In Los Angeles, maximum solar insolation occurs in the summer months between June and July. In the winter months, the angle of the sun is lower, as well as the number of hours of sunlight each day resulting in less daily solar insolation.

Hourly solar insolation data for the year 2005 were obtained from the National Renewable Energy Laboratory's solar radiation database[5]. Data beyond 2005 were unavailable. The data were collated and analyzed using NREL's PV Watts v2 program. Using the data obtained from NREL, the program determines the solar radiation incident of the PV array at the given array and azimuth tilt angles. To generate solar radiation data of hourly resolution, another NREL program HOMER v2.67 Beta[6] was employed to synthesize the desired profiles. HOMER uses an algorithm based on the work of (Graham and Hollands 1990). The algorithm produces synthetic solar data with specific statistical properties that results in a data sequence that has realistic day-to-day and hour-to-hour variability and autocorrelation.

Average hourly solar insolation was calculated for the entire year, winter months (November–February), and summer months (May – August). From Figure 11.7 we see that in all three cases hourly average solar insolation is greatest at approximately 12:00 PM each day. This is expected as the sun is in its highest position at noon each day, providing the most direct solar insolation. The figure also shows that at midday solar PV panels installed at LACC will generate on average $0.30 \, kW/m^2$ more energy in the summer months than in winter months.

[5]National Renewable Energy Laboratory: Solar Radiation Database. http://rredc.nrel.gov/solar/old_data/nsrdb/
[6]National Renewable Energy Laboratory: HOMER https://analysis.nrel.gov/homer/

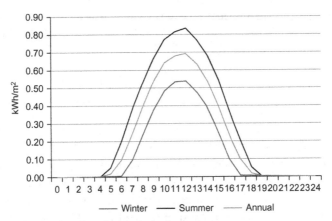

Figure 11.7 Average hourly solar insolation.

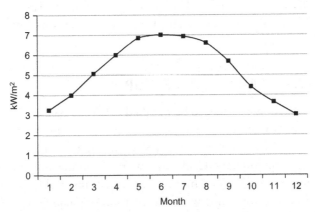

Figure 11.8 Average daily solar insolation recieved by one m² solar PV panel.

As shown in Figure 11.8 and summarized in Table 11.9, average daily solar insolation per m² at LACC is almost twice as much in the summer months (6.85 kW/m²) as compared to winter months (3.48 kW/m²). This indicates that an installed solar PV array at LACC could produce almost twice as much energy in the summer as it would in the winter.

The difference in received solar insolation between the summer and winter months presents a challenge in determining an appropriate solar PV array size. An array that entirely meets summer energy consumption will likely not be able to generate enough energy to meet consumption in the winter months. Building a PV array that entirely meets winter energy consumption may generate excess energy in the summer and depending

Table 11.9 Average daily insolation received by 1 m² of solar PV

Time Period	Average Daily Solar insolation (kW/m²)
Annual	5.21
Summer	6.85
Winter	3.48

on whether the excess energy can be sold back to energy service providers, this may not be a fiscally responsible option. However, this is an issue that is applicable only to off-grid or energy independent scenarios. Given that the goal of this project is to examine whether City College can establish a net-zero energy campus we must simply identify a PV array size that, over the course of a year, will generate a quantity of electricity equal to total annual electrical consumption. Therefore we willuse the value for annual average daily solar insolationreceived by an area of one m² (kW/m²) in all further calculations.

11.9 SOLAR PV ARRAY AND SETUP

The renewable energy system to be installed at LACC consists of PV panels, an inverter, a controller, and other essential cables and components. The performance of solar PV panels vary from 50W/m² up to as high as 300W/m² (CEC 2009). As of 2009, the California Energy Commission approved over 1,800 different solar PV panel modules for use in renewable energy installations. For our calculations, we selected monocrystallineSOLON Black 280/09/01 300W panels manufactured by Solon FuerSolartechnik. The modules have a STC rated power of 300W at a cell temperature of 25 °C and solar insolation of 1 kW/m2 and a PTC rating of 250W. The panels have a dimension of 1.90 m in length and 0.99 m in width, with a quoted cell efficiency of approximately 18%.

Solar inverters are a critical component to any solar power system. Inverters change DC (direct current) from the solar photovoltaic array, into AC (alternating current) for use on the campus. Two key categories of solar inverters are used in solar energy systems. In systems that are not connected to the utility grid the inverter takes DC current from the PV array and converts it to AC current to power a batteries or other energy storage systems from which campus power demand is drawn. In the case of this "off-grid inverter" the PV array and inverter are essentially charging batteries to keep power supplied to the building.

Table 11.10 Summary of derate vales for solar PV system (NREL 2009)

Component	Range of Acceptable Values	Typical Value
PV module efficiency	0.80–1.05	0.95
Inverter & transformer	0.88–0.98	0.92
Mismatch	0.97–0.995	0.98
Diodes and connections	0.99–0.997	0.995
DC wiring	0.97–0.99	0.98
AC wiring	0.98–0.993	0.99
Soiling of PV panels	0.30–0.995	0.95
System availability	0.00–0.995	0.98
Shading	0.00–1.00	0.98
Sun-tracking	0.95–1.00	1.00
Age	0.70–1.00	1.00

The second category ties into the utility grid and again converts DC to AC. However, with a "grid-tie inverter" the power is first supplied to the building and any remaining energy is sent back into the LADWP power grid. As of November 28th, 2009, any electricity sent back into the power grid is done so with no compensation from LADWP or any other energy service provider. LACC will essentially be providing free clean electricity to the energy service provider. The California legislature is currently considering passing two bills, SB 32 and AB 920, that would require energy service providers to reimburse customers for any excess electrical energy that is sent into the grid from renewable energy installations.

As with all electrical systems, efficiency of the solar PV array is affected by temperature, electrical resistance and losses due to connections and wiring. Therefore when determining the size of a PV array, a derate factor that takes into account these inefficiencies must be used. Table 11.10 summarizes the typical derate factor and its acceptable range for various components of a solar PV array according to NREL.

The overall derate factor is calculated by multiplying the various component derate factors. The standard NRELderate factor is 0.77, with the best possible efficiency being 0.94. Thus a 1 kW solar PV array will have an effective power of 770W, and as high as 940W.

The long term efficiency of the solar PV array will depend on the frequency of routine maintenance on and around the system as well as its age. Soiling of the solar PV panels due to accumulation of dirt or other foreign matter on the surface of the PV module will reduce the efficiency of the

Table 11.11 Summary of minimum PV array size required to generate various energy targets

Time	2010 7,000,000 kWh (kW)	2010 8,500,000 kWh (kW)	2015 11,300,000 kWh (kW)	2015 12,600,000 kWh (kW)
Annual	4601	5,587	7,427	8,282
Summer	3,499	4,249	5,649	6,299
Winter	6,888	8,364	11,120	12,399

panel, and can account for up to 25% loss in power generated. Shading of the solar PV array will also reduce the efficiency. As the solar PV array ages, it is expected to lose approximately 1% efficiency each year.

Using this information, we determined the minimum size of the solar PV array required to meet LACC's annual energy demand. There will be two main configurations for PV arrays at City College: roof mounted structures and carport structures. The current standing design has a panel slope of 5° and an azimuth of 180°. For this study it was assumed roof top units will be designed with a 20° and 180° panel slope and azimuth respectively, and the ground or carport mounted units will have a slope equal to the latitude, approximately 34° with an 180° azimuth. The array tilt can be increased up to 15° past the location's latitude in order to maximize winter production from the ground mounted system. The minimum solar PV array size for various annual energy consumption scenarios was calculated using equation 2, the data in Table 9 and the solar PV specifications outlined above. The results are summarized in Table 11.11. PV array sizes for meeting only summer (June – September) or winter (November – February) electrical consumption patterns were also calculated for reference.

$$Y_{PV} = \frac{Y_{max}}{e_{PV} G_{TD} 365} \qquad (11.2)$$

Where
Y_{max} is the desired maximum total AC output of the array (kWh)
e_{pv} is actual efficiency of the solar panel (%)
G_{TD} is the total daily solar insolation (kWh/m^2)
Y_{PV} is the rated capacity of the PV array (kW)

The results indicate a solar PV array with a rated capacity of 4,601 kW is needed to generate the required 7,000,000 kWh of annual electricity

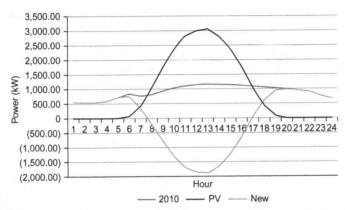

Figure 11.9 Comparison of 2010 campus energy demand before and after 4,601 kW PV array installation.

consumption after taking into account all suggested ECM measures. This increases to 5,587 kW if no ECM measures are implemented. The array size increases to at least 7,427 kW if the campus is to maintain its net-zero energy concept in 2015 after all building retrofits and construction projects are complete.

The following equation was used to calculate the output of the PV array [kW] at any given time step:

$$P_{\mathrm{PV}} = Y_{\mathrm{PV}} f_{\mathrm{PV}} \left(\frac{G_T}{G_{T,\mathrm{STC}}} \right) [1 + \alpha_P (T_c - T_{c,\mathrm{STC}})] \qquad (11.3)$$

Where:

Y_{PV} is the rated capacity of the PV array (kW)

f_{PV} is the derate factor (%)

G_T is the solar insolation incident on the PV array in the current time step (kW/m²)

$G_{T,\mathrm{STC}}$ is the incident insolation at standard test conditions (1 kW/m²)

α_P is the temperature coefficient of power (%/°C)

T_c is the PV cell temperature in the current time step (°C)

$T_{c,\mathrm{STC}}$ is the PV cell temperature under the standard test conditions [25°C]

A derate factor of 0.77 was assumed for the system. Figure 11.9 illustrates how the 2010 LA City College energy demand will change over the course of a day as a result of energy produced by the 4,601 kW PV array. We can see from the graph that the net campus energy demand reaches 0 at approximately 8 AM, and remains negative – that is the campus does

Table 11.12 Estimated monthly electrical production (kWh) by various sized PV arrays

PV Array Size (kWh)

Month	4601	5,587	7,427	8,282
January	483,528	587,149	780,518	870,372
February	493,860	599,695	797,196	888,969
March	600,418	729,089	969,204	1,080,779
April	612,800	744,125	989,191	1,103,068
May	666,512	809,346	1,075,893	1,199,750
June	666,958	809,888	1,076,614	1,200,553
July	675,372	820,105	1,090,196	1,215,700
August	685,397	832,278	1,106,377	1,233,743
September	616,646	748,795	995,400	1,109,990
October	585,676	711,187	945,406	1,054,242
November	522,977	635,076	844,230	941,418
December	462,292	561,362	746,238	832,146
Annual Total	7,072,457	8,588,094	11,416,463	12,730,730
Required	7,000,000	8,500,000	11,300,000	12,600,000
Difference	+72,457	+88,094	+116.463	+130,730

not require any electricity from the grid – until 17:15 PM. At noon, when the sun is highest during the day and solar energy production is at its peak, the solar PV array is producing just over 3,000 kW of energy, almost twice what the actual peak demand is.

The energy generated by the PV arrays ($E_{G(t)}$) over a specified period of time (t) is expressed as follows:

$$E_{G(t)} = \sum_{i=1}^{3} \left(N_{PV} E_{PV(t)} \right)_i \qquad (11.4)$$

Where:

$E_{PV(t)}$ is the energy by PV array i during time step t [kWh]
N_{PV} is the number of PV modules in PV array i

Using equation 4 we estimated the monthly energy produced by the various PV array sizes. Table 11.12 summarizes these results. In each case we found that the proposed PV array size would produce enough energy to meet the required annual electrical consumption and to establish a net-zero energy operation.

Table 11.13 Minimum PV array area for various PV array sizes

	PV Array Sizes			
	4,601 (kW)	5,587 (kW)	7,427 (kW)	8,282 (kW)
Minimum area (m²)	36,808	44,696	59,416	62,313

PV Array Size

The actual area the solar PV array will occupy was estimated using the following formula:

$$A = \frac{Y_{PV}}{P_{PTC} \, P_A} \tag{11.5}$$

Where:
A = area of entire PV array (m²)
Y_{PV} = rated capacity of the PV array (kW)
P_{PTC} = power rating under PTC
P_A = area of one panel (m²)

The summary of results is shown in Table 11.13.

In each PV array scenario, the minimum required area to establish a net-zero operation at LA City College using only solar PV is currently greater than the available area for solar panel installation.

11.10 DISCUSSION

Despite being able to reduce current energy demand by over 18% through a series of demand side management measures, the LA City College campus still requires at least 7,000,000 kWh of energy each year. The current campus configuration cannot accommodate such a large PV array.

The greatest limitation to using solar PV to provide power is that the system can only generate electricity in sunlight, which in Los Angeles on average is about 6 hours of usable sunlight a day. This limitation is exacerbated in net-zero or off the grid applications, which require that larger PV arrays be built in order to maximize capture of sunlight and generation of electricity. This may require another source of renewable energy. It is possible in the future that fuel cells may be able to generate the quantities of energy needed for LA City College to operate at net-zero energy. Fuel cells occupy less space, generate more energy, and can produce electricity on a 24/7 basis.

However, fuel cell technology is still relatively undeveloped and requires further testing. Thus solar PV, at least in the case of LA City College, is the only feasible and reliable renewable energy option available in the near future.

A concern of installing such large PV arrays is the costs associated with the system. Although the price per installed watt of solar PV has dropped over the last ten years, at $4.23/watt it still remains relatively expensive, particularly if self-financed. Since City College cannot accommodate the necessary PV to achieve net-zero operations, a financial analysis of the system was not performed. However, this should be a critical component in all other feasible projects. While City College has allocated $20 million to the renewable energy project, spending it entirely on installing solar PV may not be the optimal design. In fact financial analyses such as return on investment (ROI), internal rate of return (IRR), simple payback period, years to positive cash flow, net present value and cost of energy per kWh produced should be performed on a variety of scenarios in order to identify the optimal solar PV array size.

It is also important to understand the influence of government polices and regulations on the proliferation of renewable energy. GHG emission targets have been established by California AB32 and SB 107 mandates that all municipal utility providers have at least 20% of their electricity from renewable energy. The aim of both of these regulations is to increase the development of renewable energy in California. Government incentives and initiatives – both on a state and federal level - have played a significantrole in propagating the LACCD vision. The question is how significant have government efforts been in promoting installation of RES among organizations that may not have such forward thinking? Clean and renewable energy must be a major component of future energy production if we are to reduce GHG emissions. This would require governments to develop renewable energy friendly policies, incentives and initiatives. An evaluation of the impact of California's renewable energy initiatives[7] on the LACCD net-zero energy project would provide governments around the world with better information on how to develop similar effective programs.

11.11 CONCLUSION

Although this project evaluated and concluded that the net-zero energy concept will not work on the Los Angeles City College Campus, it has provided valuable information including:

[7]See California Executive Order S-14-08. http://gov.ca.gov/executive-order/11072

- Significant energy consumption reduction can be achieved by performing a comprehensive audit to identify areas for improvement. By doing so we were able to reduce the current LA City College energy consumption by 18%. This is a useful process for any operation in order to minimize electrical consumption and therefore costs associated with purchasing energy.
- When considering installing solar PV arrays, it is extremely important that robust and comprehensive data are available. This will allow users to produce detailed solar, energy demand and consumption profiles to design an optimal renewable energy system.
- Net-zero or off-the-grid ambitions are very difficult to achieve particularly in developed neighborhoods. Such goals may require the installation of other renewable energy technologies.
- We still face many challenges before we can fully utilize solar PV to supply large quantities of clean energy. The current design of solar PV still requires significant amounts of open space with unobstructed views of the sky – a combination that is not readily available in large metropolitan cities.

The LACCD is truly a visionary in establishing a net-zero energy goal for its nine campuses. Although the current design of LA City College will not accommodate a net-zero operation, it does not mean the District should abandon the project nor should it discourage other organizations from engaging in similar aspirations. In fact, there are still several areas that can be utilized for solar PV installation on the LA City College campus such as walkways and to some degree, south facing vertical walls. Furthermore the advancement of solar PVs in the future will undoubtedly result in panels with higher power rating, greater efficiency and better performance.

It is also important to realize that LA City College is one of the smaller LACCD campuses in terms of land available. Unlike the other LACCD campuses, LA City College cannot expand beyond its current area because of it being located in the middle of a densely populated and built area. This will make it very unlikely that City College will ever be able to install enough solar PV on its campus to operate at net-zero energy. However, the District can still achieve its overall net-zero energy goal by installing larger PV arrays on other campuses with less land restriction in order to offset the shortages at LA City College.

APPENDIX a: MAP OF LA CITY COLLEGE CAMPUS INDICATING PREVIOUS, CURRENT AND PLANNED RENOVATIONS/CONSTRUCTION

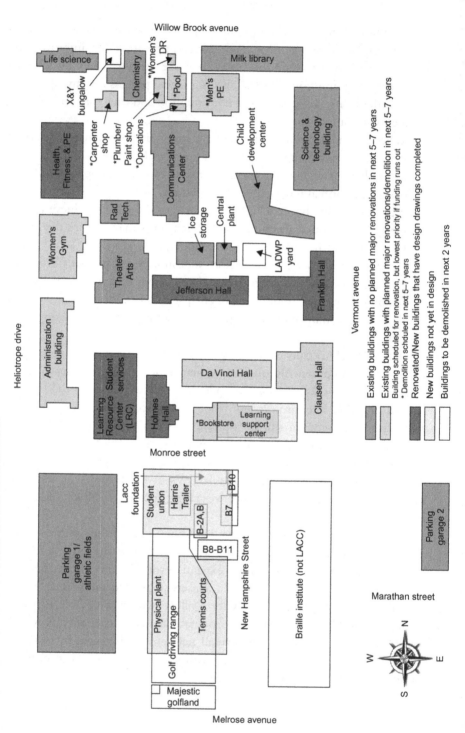

Willow Brook avenue

Life science

X&Y bungalow

Chemistry

*Women's DR

Milk library

*Pool

*Men's PE

*Carpenter shop

*Plumber/ Paint shop

*Operations

Child development center

Science & technology building

Health, Fitness, & PE

Communications Center

Women's Gym

Rad Tech

Ice storage

Central plant

LADWP yard

Theater Arts

Jefferson Hall

Franklin Hall

Heliotrope drive

Administration building

Da Vinci Hall

Clausen Hall

Learning Resource Center (LRC)

Student services

Holmes Hall

*Bookstore

Learning support center

Vermont avenue

Existing buildings with no planned major renovations in next 5–7 years

Existing buildings with planned major renovations/demolition in next 5–7 years

Building scheduled for renovation, but lowest priority if funding runs out

*Demolition scheduled in next 5–7 years

Renovated/New buildings that have design drawings completed

New buildings not yet in design

Buildings to be demolished in next 2 years

Monroe street

Lacc foundation

Student union

Harris Trailer

B-10

B7

B-2A,B

B8-B11

New Hampshire Street

Parking garage 1/ athletic fields

Physical plant

Golf driving range

Tennis courts

Braille institute (not LACC)

Parking garage 2

Marathan street

N
W   E
S

Majestic golfland

Melrose avenue

APPENDIX b: LADWP ENERGY RATES AS OF OCTOBER 1, 2009 - SPECIFIC FOR LACC OPERATIONS

ELECTRIC TARIFF STRUCTURES			
Name of Utility:	LADWP		
Rate Schedule:	A-3A – Large General Service		
Component	Charge	Unit	Description
Customer Charge:	$75	Per Month	
Demand Charge:	$4.00	Per Monthly kW	Low Season, High Peak
	$3.00	Per Monthly kW	High Season, Low Peak
	$9.00	Per Monthly kW	High Season, High Peak
	$4.00	Per Billed kW	Facilities Charge
	$0.46	Per Billed kW	Electric Subsidy Adjustment (ESA)
	$0.96	Per Billed kW	Reliability Cost Adjustment (RCA)
Consumption Charge:	$0.03863	Per kWh	Low Season, Low Peak
	$0.03863	Per kWh	Low Season, High Peak
	$0.02197	Per kWh	Low Season, Base Peak
	$0.03764	Per kWh	High Season, Low Peak
	$0.04390	Per kWh	High Season, High Peak
	$0.01755	Per kWh	High Season, Base Peak
	$0.0499	Per Total kWh	Energy Cost Adjustment (ECA)
Power Factor Correction: 0.9<pf<0.949	$0.00059	Per kVARH	High Season, Base Peak
	$0.00113	Per kVARH	High Season, Low Peak
	$0.00164	Per kVARH	High Season, High Peak
	$0.00073	Per kVARH	Low Season, Base Peak
	$0.00145	Per kVARH	Low Season, Low Peak
	$0.00145	Per kVARH	Low Season, High Peak
Other Charges:	$0.00022	Per Total kWh	State Energy Surcharge (SES)
Determination of Billed Demand:		kW	Maximum of greatest demand from last 12 months or 500 kW
Other Rate Details:	Current RCA charge is $0.51/kW; rate will change to $0.96/kW in July 2009		
	Current ECA charge is $0.0444/kWh; rate will change to $0.0454/kWh in July 2009; rate will change to $0.0499/kWh in October 2009		
	*High Season is from June 1ˢᵗ to September 30ᵗʰ.		
	*Low Season is from October 1ˢᵗ to May 31ˢᵗ.		
	*High Peak occurs Monday through Friday, 1:00 pm – 4:59 pm.		
	*Low Peak occurs Monday through Friday, 10:00 am – 12:59 pm, 5:00 pm to 7:59 pm.		
	*Base Period occurs Monday through Friday, 8:00 pm – 9:59 am and Saturday through Sunday, all day.		

ELECTRIC TARIFF STRUCTURES (continued)			
Name of Utility:	LADWP		
Rate Schedule:	A-2B –General Service		
Component	Charge	Unit	Description
Customer Charge:	$28	Per Month	
Demand Charge:	$4.25	Per Monthly kW	Low Season, High Peak
	$3.25	Per Monthly kW	High Season, Low Peak
	$9.00	Per Monthly kW	High Season, High Peak
	$5.00	Per Billed kW	Facilities Charge
	$0.46	Per Billed kW	Electric Subsidy Adjustment (ESA)
	$0.96	Per Billed kW	Reliability Cost Adjustment (RCA)
Consumption Charge:	$0.02252	Per kWh	Low Season, Base Peak
	$0.04045	Per kWh	Low Season, Low Peak
	$0.04045	Per kWh	Low Season, High Peak
	$0.01879	Per kWh	High Season, Base Peak
	$0.03952	Per kWh	High Season, Low Peak
	$0.04679	Per kWh	High Season, High Peak
	$0.0499	Per Total kWh	Energy Cost Adjustment (ECA)
Other Charges:	$0.00022	Per Total kWh	State Energy Surcharge (SES)
Determination of Billed Demand:		kW	Greatest demand from last 12 months
Other Rate Details:	Current RCA charge is $0.51/kW; rate will change to $0.96/kW in July 2009		
	Current ECA charge is $0.0444/kWh; rate will change to $0.0454/kWh in July 2009; rate will change to $0.0499/kWh in October 2009		
	*High Season is from June 1st to September 30th.		
	*Low Season is from October 1st to May 31st.		
	*High Peak occurs Monday through Friday, 1:00 pm – 4:59 pm.		
	*Low Peak occurs Monday through Friday, 10:00 am – 12:59 pm, 5:00 pm to 7:59 pm.		

REFERENCES

AASHE. Mandatory Student Fees for Renewable Eneergy and Energy Efficiency. from: <http://www.aashe.org/resources/mandatory_energy_fees.php>.
Calhoun, T., Cortese, A.D., 2005. We Rise to Play a A Greater Part: Students, Faculty, Staff and Community Converge in Search of Leadership from the Top. S. A. a. R. Panel. Society for College and University Planning, Ann Arbor, Michigan.
CEC. (2009). List of Eligible SB1 Guidelines Compliant Photovoltaic Modules. Retrieved December 24, 2009, from: <http://www.gosolarcalifornia.org/equipment/pvmodule.html>.
Dalton, G.J., Lockington, D.A., et al., 2008. Feasibility analysis of stand-alone renewable energy supply options for a large hotel. Renewable Energy 33 (7), 1475–1490.
Dillon, A., Nelson, B., et al., 2006. Importance of Turning to Renewable Energy Resources with Hydrogen as a Promising Candidate and on-board Storage a Critical Barrier. T. M. R. Society.
Edwards, J., Wiser, R., et al., 2004. Building a market for small wind: The break-even turnkey cost of residential wind systems in the United States. Lawrence Berkeley National Laboratory: Lawrence Berkeley National Laboratory.

Edwards, P.P., Kuznetsov, V.L., et al., 2008. Hydrogen and fuel cells: Towards a sustainable energy future. Energy Policy 36 (12), 4356–4362.

Forsyth, T., Tu, P., et al., 2002. Economics of grid-connected small wind turbines in the domestic market, NREL.

Graham, V.A., Hollands, K.G.T., 1990. A method to generate synthetic hourly solar radiation globally. Journal Name: Solar Energy (Journal of Solar Energy Science and Engineering); (USA); Journal vol. 44:6: Medium: X; Size: pp: 333–341.

Itron., 2006. California Commercial End-Use Survey. California Energy Commission.

NREL., 1986. Wind Energy Resource Atlas of the United States. N. R. E. Laboratory, Richland, WA. U.S. Department of Energy.

NREL., 2009. PVWatts v2.0. from, <http://www.nrel.gov/rredc/pvwatts>.

Page, S., Krumdieck, S., 2009. System-level energy efficiency is the greatest barrier to development of the hydrogen economy. Energy Policy 37 (9), 3325–3335.

Pryor, S.C., Barthelmie, R.J., Climate change impacts on wind energy: A review. Renewable and Sustainable Energy Reviews, 14 (1), 430-437.

Pryor, S.C., Barthelmie, R.J., et al., 2009. Wind speed trends over the contiguous United States. J. Geophys. Res. 114 (D14), D14105.

Rhoads-Weaver, H., Grove, J., 2004. Exploring joint green tag financing and marketing models for energy independence. Global Windpower.

Segal, M., Pan, Z., et al., 2001. On the potential change in wind power over the US due to increases of atmospheric greenhouse gases. Renewable Energy 24 (2), 235–243.

Smestad, G., 2008. The Basic Economics of Photovoltaics. Optical Society of America, San Jose.

Yang, D., 2004. Local photovoltaic (PV) wind hybrid system with battery storage or grid connection. NREL.

Yue, C.D., Yang, G.G.L., 2007. Decision support system for exploiting local renewable energy sources: A case study of the Chigu area of southwestern Taiwan. Energy Policy 35 (1), 383–394.

CHAPTER 12

Transformational Relationship of Renewable Energies and the Smart Grid

A.J. Jin, Ph.D

Contents

12.1	Introduction	217
12.2	Solar Electricity Systems and their Relationship with the Grid	219
12.3	Wind Power	223
12.4	Data Response and Power Transmission Lines	226
12.5	The Smart Grid and a Market Solution	228
Acknowledgments		231
References		231

12.1 INTRODUCTION

The increase of greenhouse gas emission is creating numerous problems for both human health and a stable global climate. Growing energy consumption and fluctuating oil prices also cause increased national security concerns. The scope of the challenges in both the energy and climate sectors is far-reaching and directly relates to our dependence on traditional carbon-based fossil fuel. This situation is currently thought of as at the heart of our energy crisis.

There is no shortage of energy flowing to the earth, since the sun radiates an enormous amount of power (170,000 TW) onto the earth's surface. Although most of the solar renewable energy is not available to us, acquiring only about 0.01% solar energy is sufficient to meet the world's needs today. As we discuss soon, the target of using a portfolio of renewable energies is gaining important government support and attracting significant private investment. Renewable energy technology is currently developing fast and is projected to soon become economically competitive.

President Barack Obama championed renewable energies as well as the smart grid, and he announced several billion dollars in U.S. government support along with private investment toward that end (Associated Press

Sustainable Communities Design Handbook
ISBN: 978-1-85617-804-4, DOI: 10.1016/B978-1-85617-804-4.00012-4

2009). The president encouraged the new grid system to be smarter, stronger, and more secure in future. The following sections discuss how these targets can be achieved through a transformational relationship among the advanced energy technologies, including both renewable energies and the smart grid.

To date, the advanced energy technologies have shown us that the development of these technologies could become the linchpin of a new modern energy infrastructure. Sections 12.4 and 12.5 show that the new system is smarter due to its ability to monitor and control energy consumption comprehensively in real time. The new system is beneficial in terms of demand response, energy efficiency, and compatibility with the large supply of renewable energy sources. The aforementioned smart grid project would install thousands of new digital transformers and grid sensors in homes and utility substations to enable a grid-smart data system. Section 12.4 shows that that the data response system is designed to strengthen the grid system. Sections 12.2 and 12.3 show examples of solar electricity and wind power that are sustainable, abundant, and affordable, based on excellent use of natural resources. President Obama's announcement reflects the U.S. adoption of clean technologies and the public's acceptance of a "going green" philosophy and way of life today.

The resolution to the energy and climate challenges has profound business impacts as well as a great societal effect; the effort to meet these challenges is sometimes referred as the *Third Industrial Revolution*. Furthermore, our nation faces the major challenge of upgrading the current electricity grid and the energy management infrastructure. The lack of new grid infrastructure forms a bottleneck to the investment in new renewable generation and in tapping the full power of renewable energy, such as a utility-scale solar electricity system.

The strong and rapidly growing consumer demand for clean energy promotes a new resolve to meet global needs with inexpensive electricity from clean energy sources. The advanced energy technology requires a development partnership among the elements of renewable energy, optimized energy efficiency, and the smart grid. The following text illustrates the challenges and shares our knowledge of employing several renewable energies and optimal grid infrastructure to reduce greenhouse gas emissions more than 25% by 2020—and more than 80% by 2050. Until the challenges are met, Americans cannot truly enjoy energy freedom, economic resilience to oil fluctuation, and national security.

12.2 SOLAR ELECTRICITY SYSTEMS AND THEIR RELATIONSHIP WITH THE GRID

Solar electricity systems are anticipated to most likely become commercially successful without government rebate in five more years. This anticipation is based on the current best knowledge of cost and adoption risks of the solar electricity that offers a superior future technology trajectory. Today, as the solar power business grows, we see the cost of solar photovoltaic (PV) panels declining rapidly. In the United States, the president's Solar America Initiative, in collaboration with the Department of Energy, has targeted grid parity, where the electricity cost based on renewable energy production is the same as the coal-fired traditional power cost.

More progress is needed in the balance-of-plant aspect of energy production, which is defined as the solar cost per installed system ready for use. Active research to accelerate the progress of the cost reduction in this area is underway. In many parts of the world, the solar PV system is appropriate for cost effective distributed generation. All the world's current and expected major electricity load centers are within practical transmission range of excellent solar radiation locations (Figure 12.1).

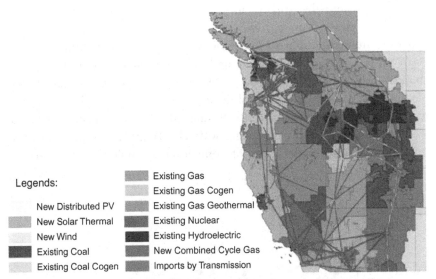

Legends:

	Existing Gas
	Existing Gas Cogen
New Distributed PV	Existing Gas Geothermal
New Solar Thermal	Existing Nuclear
New Wind	Existing Hydroelectric
Existing Coal	New Combined Cycle Gas
Existing Coal Cogen	Imports by Transmission

Figure 12.1 The map of West Energy illustrates the energy sources and electric transmission grid. The diversity of energy sources and smart grid should enable the "utility grade" power generation on a par with the current coal-based power plants.

As we show in this section, the solar electricity system can utilize the sun's natural energy to generate electricity. The solar electricity generation can be consumed or fed back into the utility grid. This results in less energy purchased from the utility company, so the consumer's monthly bill decreases (the monthly bill is then just for financing payment). During the solar system warranty period, the solar PV system is free for electricity generation and free from maintenance. Consumers pay for only the energy used from the grid (and the amortization of the system and installation costs). The solar PV systems connected to the grid can be very attractive, to reduce the more expensive peak-hour costs.

The term *cleantech* refers to technologies that produce and use energy and other raw materials more efficiently and that create significantly less waste or toxicity than prior commercial products. The cleantech energy production industry has developed clean alternative energy, utilizing the current benchmark technologies, such as solar and wind power. Examples of alternative energy sources include wind power, solar photovoltaic cells, solar thermal electric power, solar heating, and the like. Scientists and engineers continue to search for viable clean energy alternatives to our current traditional power production methods. Even though some renewable energy sources are variable in nature, several renewable energy sources can be integrated into the grid system quite well. For example, studies show the compatibility of sun and wind energy, which are complementary and may be quite manageable for integration into a grid system (Abbess 2009).

What is a solar electricity system? A solar electricity system utilizes the abundant sun's energy to produce electricity for consumers. For example, the solar PV cell is a physical device that converts light into electricity. Figure 12.2 is a schematic to illustrate the physical mechanism of a solar photovoltaic cell in producing solar electricity. The solar electricity industry comprises many types of competing technologies with various

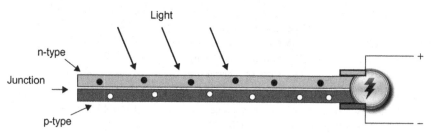

Figure 12.2 A solar PV cell that converts sunlight to electricity; see the text for details.

cost structure, efficiency, and scalability factors that are important to the renewable energy industry sector. The availability and future prospects are very promising at this time for the following three solar technologies: solar thermal power, solar PV panels, and solar heaters. For example, utility-scale solar thermal power plants have been constructed rapidly in the last two decades. Moreover, the solar PV panels offer scalable power that has been installed on thousands of rooftops in California.

Solar PV systems, which are made up of individual solar cells, are becoming more and more affordable and reliable all the time. The solar PV panels are made modular, scalable, and suitable for distributed generation. Moreover, the scalable solar panels can be utilized for utility-scale power plants.

Several types of devices may be required to connect solar PV systems so that they are suitable for individual consumer energy use and for supplying power to the electric grid. The most important unit is the inverter. The inverter unit is an electronic device that turns direct current (DC) from the solar electricity into an alternating current (AC) that is matched to the incoming main electric utilities standard and is used by almost all home appliances and electrical devices. The concern for safety also requires the solar electric system to be enabled by circuit breakers for safe maintenance. Circuit breakers are typically connected on both the dc and ac sides of circuitry path.

A solar thermal power plant (STPP) employs utility-scale steam turbine technology. As shown in Figure 12.3, an STPP collects the solar energy in a large real estate footprint for thermal energy to produce electricity. A circular array of solar light reflectors is used to concentrate the light on

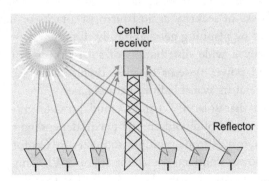

Figure 12.3 Solar thermal generator receiving a highly concentrated sunlight beam, heats up pressurized gas in the receiver and drives the hot gas through a turbine that produces electricity with combined hot gas and turbine operation.

a receiver located on the top of a tower. The light is absorbed as heat energy, which heats up the steam/gas or air to very high temperatures that produce pressurized hot gas or air to drive the turbine. An STPP has a typical footprint equivalent to the scale of a coal-fired utility power plant.

An STPP employs direct sunlight and, hence, requires its plant site to be in regions of high solar radiation. The thermal energy storage typically is achieved through liquid or solid media to extend the hours of the electric cycle. Figure 12.3 shows a portion of a typical solar thermal power design.

The United States is the world leader in installed concentrated solar power capacity, with 429 MW currently in commission. Three gigawatts of power is expected to be operational by 2011. Seven gigawatts total power is in development at this time. The United States alone anticipates powering 2 million homes by solar thermal power in the year 2020 (Environmental and Energy Study Institute 2009).

All solar power, including solar PV and solar thermal, generated about 0.1% or less of the total U.S. energy supply in 2008, but the installation of solar power is growing quickly. The U.S. Department of Energy's 2009 preliminary forecasts anticipate an annual growth rate in U.S. domestic solar PV generation of 21.3% through 2030 (and some analysts have even higher predictions for the growth rates) (U.S. EIA 2009). The solar PV technology is scalable. As Woodrow Clark mentioned previously, increasing demand will bring down the cost of PV modules and solar electricity systems when they are in volume production.

Variable but forecastable renewable energies (wind turbines and solar PV power sources) become more reliable in net output when integrated with each other than one source alone. This net output is suitable to meet the demand. Risks of security against terrorist attacks or natural disasters can be mitigated by planning geographically disperse energy sources. One plan is to employ a wide distribution of solar power from many sunny areas, smart grid power systems that are discussed in Section 12.5, and backup fuel-generators available from natural gases.

It is possible that solar PV panels could provide enough energy to power a family's total household energy needs, depending on area climate. In current average bids from multiple suppliers in California to utilities, the long-term power contracts are about $0.14/kWh. This bid estimate is based on a solar thermal power plant of 175 MW that starts construction in year 2009.

Finally, the distributed generation of solar PV electricity can be connected to the grid. Germany is noteworthy for its prior high-profile feed-in tariff in promoting solar electricity. California and several other states offer benefit incentives, such as net metering, which bills customers for a net electricity difference resulting from the electricity coming from the power grid and the electricity generated by the local solar electricity system. Several governments have successfully offered incentive packages to promote renewable energies and energy efficiency in the world.

12.3 WIND POWER

As power is generated by the new utility-scale renewable power plants, such as wind power plants, the power transmission grid needs to have the capacity to fully deliver the power to consumers without the distribution blockage. The goals of America's clean energy market can be empowered by addressing the advanced technology platform of a nationwide electric power system. The year 2009 ARPA-E grants of the United States heavily invested in projects with the capability to allow intermittent energy sources like wind and solar to provide a steady power flow to consumers.

We need an interstate power transmission superhighway for the electric power system. Immense solar power farms in America's deserts face this transmission challenge for moving through the power grid to the consumers. The congestion of the grid can create significant limitations that can reduce the potential advantage for large renewable power generation to pump power into the electric grid.

What limits the renewable energies (as a commercial bottleneck) is the outdated power transmission grid mentioned in the last section. For example, the current system cannot accommodate the present and future needs of delivering hundreds of megawatts of wind power to users. In one scenario, the total of 200,000 miles of power transmission needed to be rebuilt may provoke fights among 500 divided owners and numerous property owners. In another scenario, the large power generation usually requires extra storage system. For today's market, there is no commercially advantageous solution. Active development in the storage system area is underway.

The layout of the current power grid eventually should be accessible to a flexible change of total power (e.g., gigawatts) for power interconnection and transmission lines. The current transmission lines cannot increase their transmission capability by the hundreds of megawatts needed to meet the challenges. Such power pumping shows up as challenges at times even over a distance of a few hundred miles. The commercial pain is the severe congestion today for long-distance power transmission.

We have to achieve cleantech or renewable energy vision. The Kyoto Protocol set a target that the world needs to reduce greenhouse gas emissions by more than 25% by 2020 and more than 80% by 2050 (Center for American Progress 2009). One solution comes from an advanced energy technology, that is, wind power generation that it is very cost effective for today's commercial use.

Wind turbine manufacturing is cleaner than the volume production of solar PV cells. Today, America utilizes barely 1% of the power produced by wind energy. America's goal is to achieve 20% from wind power by year 2030 (U.S. Department of Energy 2008). Wind power turbines of the Maple Ridge Wind Farm near Lowville, New York, are capable of producing a total maximum of 320 MW. The farm has shut down at times due to the limitations on pumping capacity of the electric power system. Wind farms, too, face power transmission challenges. One cannot easily pump a large amount of power to the grid for various reasons that we discuss later. For users to utilize the full potential of wind power or other environmentally friendly energy, it is imperative for the nation to significantly improve or build a system of populated, optimized transmission lines.

Wind power is a type of solar induced energy. Wind is always present on our planet due to uneven heating of the earth's surface by the sun and the so-called Coriolis effect, which relates to the wind being dragged by the constant rotation of the earth on its axis.

The conversion of wind to electrical power is generated by a wind turbine. The modern wind power technology (such as the wind turbine) has been perfected over the last decade. A wind turbine power plant typically generates the electricity in much the same way (through electromagnetic induction) as the alternator in a car. As shown in Figure 12.4, a wind power station is usually positioned such that its rotor always faces the wind. The power engine has a drive train system that often includes a gearbox. There is a wealth of information about wind power. Interested readers are referred to the literature.[1]

The wind power depends on three variables: wind velocity, the radius of the generator, and temperature, which determines air density.

[1] Readers can refer to http://en.wikipedia.org/wiki/Wind_power, also the "Calculation of Wind Power" article from the Renewable Energy Website, REUK.co.uk, printed on November 5, 2009, www.reuk.co.uk/Calculation-of-Wind-Power.htm:
Power = $0.5 \times$ Swept area $\times$ Air density $\times$ Velocity^3 and
Air density = Constant $\times [T°C/(273.15 + T°C)]^1$.

Figure 12.4 A wind turbine.

The following is a simplified summary of this relationship to the operational state of the wind turbine:

1. The power increases with the cube of the velocity (e.g., a twofold increase in velocity leads to an eightfold increase of power output).
2. The power increases with the square of the radius (e.g., a twofold increase in velocity leads to a fourfold increase of power output).
3. The power increases with decreasing temperature (with about 3.3% of power for the change of every 10°C in air temperature).

Not only does wind power production make economic sense, there are also the greater social benefits of clean energy and a sense of personal freedom (by moving toward a zero-net energy residence, a type of energy independence). The current transmission lines cannot meet the goals of the advanced energy technologies. With a new grid system, the smart grid can be designed to meet the challenges of and suit ideally the demands of electricity production, distribution, and utilization.

The wind power makes good sense environmentally and economically. Turbine components are generally either recyclable or inert to the environment. The price of the wind turbine is a critical parameter for the return on investment. A typical payback period for the energy cost is about half a year. A residential wind turbine can be employed in homes or routed to storage, such as battery banks. Some farmers have utilized their land for wind farms, where the wind power generation does not affect how they farm or produce crops.

12.4 DATA RESPONSE AND POWER TRANSMISSION LINES

Power system management and optimization are really about data management, response, and efficiency. When the demand reaches a significantly high level or an energy reduction is needed, the smart demand response should help customers in energy conservation and reduction and thus in enhancing system reliability. The energy security is consistent with our nation's security concern so that the advanced energy technology supports the renewable power standard.

To address the transformation issue of the critical electrical power infrastructure, a major challenge is transmission lines that the government can adequately support in terms of policy and coordination.

The current grid protocol of the power infrastructure employs technology from about the turn of the 20th century (U.S. Energy Information Administration 2009). Initially 4000 individual electric utilities in the U.S. owned local grids and operated in isolation. Later, voluntary standards emerged through the electric utility industry to ensure coordination for linked interconnection operations. These voluntary standards were instituted after a major blackout in 1965 that affected New York, a large portion of the East Coast, and parts of Canada.

Due to the transmission lines' limitation across states, thousands of megawatts of wind projects are stalled or slowed down while many solar power deployments are experiencing similar challenges. In United States, for example, long distance power transmission has been the major barrier to the success of renewable power standard implementation in certain regions. The challenges in financing, permits, and pricing transmission systems have created nearly insurmountable obstacles in the past and present.

To address the need of a grid transformation, the vision of the energy industry is to employ an Internet Web model, as follows. As shown in Figure 12.5, the Internet Web of a smart grid takes the active system of a nerve network that determines, responds to, and controls the power needed for consumers. The network control system operates on a global scale to dispatch energy and manage the energy flow protocol but distributes control around the system. For example, data response management by the network control system recovers from a power block by circumventing it. This recovery is an attribute of a self-healing power network.

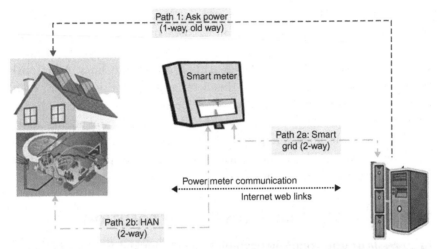

Figure 12.5 A smart grid system. A node with a smart meter enables a home automation network. This control system operates on a global scale to dispatch energy that determines, responds to, and controls the power needed by consumers.

The information exchange around the Internet Web uses the concept of distributed control, where the Web host computer or a designated computer server acts autonomously under a global protocol. Due to the information process capability in modern internet technology, consumers benefit from reduced cost by utilizing the internet to effectively manage the power grid.

The energy efficiency comes from the consumer choosing more efficient energy options over more costly ones. A smart grid can help utilities identify losses and support energy efficiency. The smart grid can manage its effective response to consumers. For energy consumers, power generation owners, buyers, and sellers, the nerve network in the Internet Web of a smart grid will be both flexible and economical to extend the services of the power purchase transaction. An electricity system would provide supply/demand coordination and would be interconnected in the grid to dispatch power.

Today, transmission and distribution lines have 500 owners and numerous property owners (see the sources cited for the U.S. Energy Information Administration). The coordination is mostly among the three regional interconnections (western interconnection, eastern interconnection, and Texas interconnection), and their grid systems are in turn coordinated by Federal Energy Regulatory Commission.

The diversity of energy sources and the smart grid have designed-in specifications that match the current standard of the existing coal-based power plants. The existing power grid needs to add additional transmission lines, shown in Figure 12.1 as dashed gray-white lines. Additional transmission lines are required for the electricity transmission within each interconnection region and among the three power interconnections. By getting renewable energy sources connected to the grid and adding additional transmission lines as required, sufficient power will flow to consumers. A new power grid can adopt cleaner, more efficient power plants than just the current coal-fired power plant.

12.5 THE SMART GRID AND A MARKET SOLUTION

In 2009, smart grid companies exhibited significant and fast growing spots alongside the cleantech market need in the United States. The need for a smart grid is fundamental to developing the modern energy networks[2] in the United States, the European Union, and every grid-connected nation worldwide.

Many tech giants, such as Cisco and Google, are planning to bring their products to the smart grid market. According to Cisco in its May 2009 announcement about its smart grid roadmap, the expanding smart grid market is one of its "new market priorities" valued at $100 billion in market size. Moreover, significant funds are infused into the smart grid projects by the government. The Obama administration has strongly propelled the smart grid development, and an electric smart power grid is regarded as an "urgent national priority." The Department of Energy awarded $47 million in funding in July 2009 to accelerate project timelines of the ongoing smart grid demonstration projects. Moreover, the collaborative approach of pubic and private organizations in the smart grid and cleantech will help an integrated and economics-driven technology of clean energy production. The advances in these products are most likely to address the challenges of our times in market demand, clean energy need, and greenhouse gas emission reduction.

[2]Major interests are dedicated to smart grid, such as "Smart Grids Europe 2010," accessed March 30–31, 2010. Please refer to the following work as well: "Smart Grid of European Platform in 2006," EUR 22040, http://ec.europa.eu/research/energy/pdf/smartgrids_en.pdf; "Interoperability of Energy Technology and Information Technology Operation with the Grid," IEEE, P2030/Draft 1.0 Skeletal Outline, 2009. Utilities play a lead role in defining their business process requirements, and specifying or testing the resulting standards in their operations.

What is a smart grid? The smart grid is a collection of energy control and monitoring devices, software, networking, and communications infrastructure that are installed in homes, businesses, and throughout the electricity distribution grid. This collective system generates a nerve system for the grid and for customers that provides the ability to monitor and control energy consumption comprehensively in real time.

The schematic of Figure 12.6, which is an expanded version of Figure 12.5, shows a smart integrated energy system that merges internet and grid features. This system has a smart grid with the power source(s), the data response, and a load center, such as a residential home. Figure 12.6 illustrates several concepts, including data collection, communication, control, and a smart grid system. A smart meter collects power usage data for the utilities and consumers, and it has internet communication capability, as mentioned previously.

Real-time data are fed back to a large distribution and transmission power grid. Moreover, the energy storage is extremely important, in that it assists in load leveling for transmitting any major power activities, comparable to a typical locally-rated load center. For example, a major power activity could be some solar electricity generation or a major charger for an electric car battery. The real-time data are useful for utilities to predict and hedge power usage.

The smart grid is ideally suited to meet the challenging demands in the production, distribution, and utilization of electricity. The innovation here is to take a century-old power grid infrastructure, turn it upside down, manage it as mentioned previously, and connect it to numerous renewable energy sources. Customers are interested in going green today. The cost of fossil-fuel-based energy is rising, due to both depleting resources and the cap and trade rules on greenhouse gas fuels. Although solar PV and wind power may have significant power output, they must be managed with load leveling suited to their output demand profiles. Moreover, much power may be wasted, since the power plants, such as nuclear power or coal fired, do not shut down when the consumer is asleep. Electricity analysis and management has to be directed toward energy management of existing infrastructure. As a result, energy efficiency, modeling, and data analysis are needed to be fed into the smart grid. The data are for private use, and a computerized electricity management has robust integrity with regard to network security, reliability, and consumer participation. A smart grid may respond to diverse conditions that are indiscriminate to storage, resources, and electricity reliability.

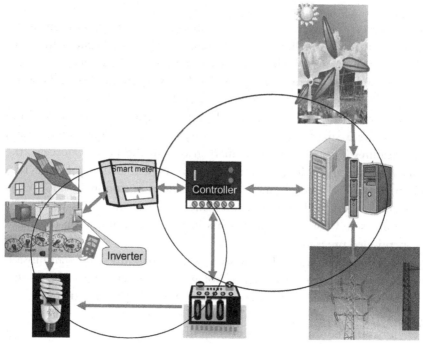

Figure 12.6 A smart integrated energy system that merges internet and grid features.

Let us recapture the important specifications of a smart grid discussed in this chapter. The smart grid should be developed and specified with features as follows:

1. Security to virus attacks in terms of the grid network control system.
2. Self-healing power for network and grid lines through effective data response.
3. Efficiency improvement, optimal demand response, and load leveling.
4. Response to a wide range conditions; diverse and indiscriminate as to storage or energy resource operating conditions.
5. A supply and demand chain to preserve the private competition that is compatible with a capitalist market without the government control.

There has been recent intense investor interest and corporate investment in the field of energy efficiency and carbon-free clean energy production. The smart grid is a bright and fast growing technology sector. Sustained and deep commitment by regulators, state law makers, utilities, and other stakeholders are needed to achieve the cost effective energy efficiency targets. For example, California utilities are recognized by their customers as energy efficiency and demand response experts.

Transformation of the mainstream energy market requires an advanced grid infrastructure with superior energy efficiency and green technology. A smart grid is becoming increasingly important for the wide use of solar, wind, and other renewable energy. This grid reflects excellent criteria for market growth. A smart grid is an investment field in a large market sector, and many recognize the opportunity to make a substantial social contribution while providing a good return on investment. Perhaps, a simple upgrade of the century-old power grids will form a rebuilt backbone to a smart grid system.

To begin building the transmission infrastructure provides great support in validating the U.S. commitment to a carbon-emission reduction and the Copenhagen Conference in December 2009.

ACKNOWLEDGMENTS

I have benefited from many discussions and editing from David Craven and Woodrow Clark. I am very grateful for help from colleagues and friends, who made this chapter a stimulating project.

REFERENCES

Associated Press News Reports., The President Has Set a Goal, (October 27, 2009), Arcadia, FL.
Abbess, J., *New Reports on Monday, August 17,* 2009. Wind Energy Variability and Intermittency in the UK, www.claverton-energy.com/wind-energy-variability-new-reports.html.
Environmental and Energy Study Institute, www.eesi.org/files/csp_factsheet_083109.pdf.
U.S. Energy Information Administration (EIA), Annual Energy Outlook March, 2009. www.eia.doe.gov/oiaf/aeo/index.html
www.americanprogress.org/issues/2009/01/pdf/romm_emissions_paper.pdf.
U.S. Department of Energy news, updated July, 2008, www1.eere.energy.gov/windandhydro/pdfs/41869.pdf, and wind power news, www.energy.gov/news/6253.htm.
U.S. Energy Information Administration, http://www.eia.doe.gov/www.eia.doe.gov/, http://www.tonto.eia.doe.gov/energy_in_brief/power_grid.cfmhttp://tonto.eia.doe.gov/energy_in_brief/power_grid.cfm, last updated: October 20, 2009; Home>Energy in Brief on "What Is the Electric Power Grid, and What Are Some Challenges It Faces?"

CHAPTER 13

Clarifying American Values through Sustainable Agriculture

Sierra Flanigan

Contents

13.1	The Murphy Apple Orchard	233
	How It All Began	236
	A Living Legacy	238
	Multiple Applications and Benefits	239
13.2	The SAD Truth	242
	Malnourished Values	243
13.3	A New Beginning	243
13.4	Intercampus Produce Exchange	244
13.5	Appreciating the Value of Food	245

The environmental and social challenges that face the current generation of college graduates are, without a doubt, daunting. How do we make sense, much less fix, decades of collective environmental abuse? Did something go terribly wrong? These problems affect us on a collective level; they are the fault of no one cause or people, but they are *our* responsibility, and thankfully are within our power to resolve.

Our generation is faced with an imperative to act. What I have learned is that nothing is more powerful than the conviction of "I can." Taking action is contagious and inspires further actions. It is about getting started, and it is up to all of us. It is about believing in ourselves, grasping the opportunity, and getting started.

13.1 THE MURPHY APPLE ORCHARD

On Earth Day 2009, my long-standing dream of sustainability became a reality. A group of students at Wheaton College in Norton, Massachusetts, broke ground and planted the first trees of what would become a living

Sustainable Communities Design Handbook
ISBN: 978-1-85617-804-4, DOI: 10.1016/B978-1-85617-804-4.00013-6
233

apple orchard and a flourishing symbol of sustainability for the college (Figures 13.1 and 13.2). Behind the president's house, adjacent to the tennis courts, and close to a neighboring preschool, we planted 15 baby honey crisp apple trees. In celebration, the Van Buren Boys band jammed to "Captain Planet" and "Johnny Apple Seed." Students and faculty sipped on cider, nibbled on apple treats, and boogied down to the apple-earth beats.

The day marked a marriage between education and environmental sustainability. The land had been an unused asset. In early April it was a matrix, an expansive hole-punched terrain ready for trees thanks to the giant auger

Figure 13.1 Murphy Orchard cofounders Sierra Flanigan and Chad Mirmelli honor late Professor Jeremiah Murphy at Orchard Dedication Ceremony.

Figure 13.2 This commissioned poster, animated by Wheaton alumnus, Nick Johansen, raised awareness across campus about the apple orchard groundbreaking celebration on Earth Day, 2009.

provided by campus services. After some entertainment, the crowd gathered as the orchard's cofounder, Chad Mirmelli, and I took the stage. Key administrators of the project, Chad and I shared our sentiments. The President of the college, Dr. Ronald Crutcher (Figure 13.3), reflected on the power of one community to come together as it did for this "shared purpose," despite economic distress.

Chad and I touched on the essence of sustainability, its versatility, and its potential to engage the entire campus. College Provost Pastril-Landis linked higher education with the apple orchard: "Remember Newton's apple and gravity," she said. She also pointed to the front of the podium where the college's seal glimmered in the sun (Figure 13.4). At its very

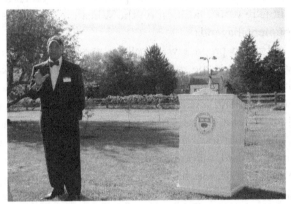

Figure 13.3 Wheaton College President, Dr. Ronald Crutcher, standing in front of baby trees, hails the Murphy Orchard in the worst economic climate since the Great Depression.

Figure 13.4 As the centerpiece of the Wheaton College seal, an apple tree symbolizes knowledge and prosperity.

core (excuse the pun), an apple tree shone as the focal point of the Wheaton College crest. Pastril–Landis spoke of the apple's deep-rooted symbolism in Wheaton's history. Sometimes win–win solutions in life just need a pair of hands to be held.

How It All Began

The orchard really stemmed from a need on campus. It was a familiar story: administrative commitments toward sustainability were cursory, there was little enthusiasm for sustainability among faculty and students alike, and even less of an expendable budget for it. I decided to devote myself to bolder environmental standards and accreditation on campus: I launched an eco-insight column in the school paper, pioneered several clubs and initiatives, and eventually won a position on the Wheaton student Senate with a strong environmental agenda.

The smell of green progress was finally in the air. For the first time, the president, the provost, and the student government association were in consensus that the issue of sustainability deserved to be formally recognized and addressed head-on. At last, sustainability was taking the forefront. Chad and I gave a special presentation to the board of trustees about the need for Wheaton to go green. This would involve a number of investments—and on that day, the stock market crashed, suffocating the issue. We were back to square one. The worst recession in decades had taken the wind out of our sails.

So the question now became, How could Wheaton demonstrate its new-found commitment to sustainability without the hefty upfront costs of many measures like wind and solar demonstrations? Was there some way we could better utilize the resources at our disposal to take action, to take on a leadership role in sustainability, all with very little workable capital?

On a cool fall day Chad and I looked out at our campus and noticed one significant but overlooked asset that had been right in front of us: land. It hit me like a ton of bricks. "What about a self-sustaining apple orchard," I asked. Seeing the sparkle in Chad's eye, I knew this was it: Plant an apple orchard on campus as a symbol of sustainability and a uniting force at Wheaton. It would be low in cost and high in impact.

The economic downturn, ironically, turned to be a blessing in disguise for us. Budget shortfalls forced us to think outside of the box. People were thirsty for progress and needed, more than ever, a source of civic engagement and community pride. With the perfect climate, open land, and glum campus demeanor, an orchard made perfect sense. It would be a highly visible demonstration of the college's commitment to liberal arts and the earth. It would also produce thousands and thousands of apples.

In less than two weeks, the puzzle pieces began to fall into place as the idea of an apple orchard caught the attention of entities across campus. Administrators, campus officers, and students who had been unenthusiastic in the past were excited and wanted to chip in one way or another. Chad and I were shocked by how the orchard captured so much interest. We were quickly granted an exceptional location on campus, money came in from multiple sources (the college venture fund, alumni, class gifts, and individual donors), and the student-run apple orchard was on a fast track.

We formed an Orchard Oversight Committee (Figure 13.5). This provided the support and cross-disciplinary insight for the success of the project. Representing members came from every major academic, student, and administrative hub on campus. The committee became a testament to the liberal arts experience at its finest. Chad and I also conducted weekly information sessions to educate community members. We used *green* advertising strategies to spread the word, like advertising on the backdrops of the library computers.

We developed a comprehensive budget, established a bank account, and spoke to as many students and publications as possible. We made a concerted effort to be included on everyone's agenda. We briefed and were guided by the Student Government Association, the faculty committee, the president's council, president's commission, alumni office, board of trustees, communications office, and buildings and grounds crew. Our goal was to get everyone involved, and we did.

Figure 13.5 The Orchard Oversight Committee discussing logistics in the soon-to-be orchard.

A Living Legacy

Jeremiah Murphy was a beloved professor who inspired countless students to change the world during his long tenure at Wheaton. In December 2009, Professor Murphy's unexpected death took the Wheaton community by surprise. Chad and I dedicated the orchard in Murphy's honor, (Figure 13.6).

Each Wheaton class traditionally donates a tree and plaque to the college. The class of 2009 broke records with 99% class participation—sending President Crutcher for a leap into Wheaton's Peacock Pond as he promised if we topped 93% participation—because of the gift's link to the orchard and its dedication to Professor Murphy. This gave students greater incentive to contribute, and as a result, the orchard had money and campuswide support. The orchard became ingrained in people's hearts, building student pride and accelerating sustainability on campus.

Phase I of the orchard was made up of the first 15 honey crisp dwarf trees (Figure 13.7). Within months, Phase II was accomplished, with another 35 myra fuji apple trees, including semi-dwarfs. Now the orchard will spread to a third location, adjacent to the old science observatory. Phase III will round out the orchard with 500 trees when fully established, producing more than a quarter of a million apples each year. This harvest will be more than sufficient to feed students on campus and for graduates and community members to join together in the fall for apple festivals, homecomings, cider pressing, bobbing, and pie baking.

Figure 13.6 Partners-in-sustainability Sierra Flanigan and Chad Mirmelli plant Phase I.

Figure 13.7 Baby apple tree blossom; in the background is a senior apple tree.

Multiple Applications and Benefits

An unexpected result of the orchard initiative is the many benefits it continues to create on campus. It encourages students to think outside of the classroom. Students are embracing experiential learning through the orchard; they are inventing their own pieces of the puzzle and spearheading independent endeavors. For instance, a commissioned apple orchard landscape painting now hangs in the once bland Student Government Association office (see Figure 13.8). Economic simulation models, computer graphic designs, and biological experimentation are in their early stages. Professors of anthropology, English, and psychology courses are using the orchard to highlight interdisciplinary learning and help undergrads "connect the dots" across campus. Art students are joining talents to customize an engraved tree stump they will design and lacquer, which will constitute the official plaque for the orchard.

The orchard also provides jobs on campus: an orchard manager, shadowed by credit-bearing and paid interns, manage the trees. These positions give students meaningful work experience and the chance to work with nature (Figure 13.9). Federal work-study jobs provide additional funds to the campus economy. Schuyler Horn (Figure 13.10), now manages the orchard and trains first-year seminar students to plant and care for trees. As manager, he is responsible for nurturing the baby trees, corresponding with faculty advisors, and monitoring growth.

Figure 13.8 Hanging in office of the Student Government Association, a portrait of the baby orchard was commissioned and painted by Wheaton student Johanna Rois-Beck '11.

Figure 13.9 Wheaton students goofing around during orchard planting Phase I.

Intern Anastasia Foresman, a Russian student, coordinates administrative duties with campus leaders. She is accountable for documenting the orchard's progression and maintaining communications with the Sustainability Committee. Both Anastasia and Schuyler report primarily to the head of grounds and the designated sustainability staff coordinator on campus, along with Chad and myself. We remain unofficial supervisors, clingy new parents, if you will.

Figure 13.10 In order to strategically yield the most fruit, orchard manager Schuyler Horn takes the lead during the construction of orchard structural scaffolding.

The Sustainability Education for Environmental Development Sessions (SEEDS) program is closely linked with the orchard. Through SEEDS, Wheaton students earn academic credit for designing and conducting lessons for local preschool children. SEEDS is available at no cost for parents and their children in the community, thus strengthening Wheaton's community relations and profile. Classes have already been held in the orchard to teach the importance of sustainability. Once the program is fully developed, the construction of an outdoor classroom and sensory garden will provide an oasis for young students. The first lesson for young kids is about how plants grow and sustain our healthy bodies, and the anatomy of apples.

Prior to the orchard, my mission to promote sustainability on campus felt diffused. With the inception of the Murphy Orchard, the project provided a focus for the college and galvanized community action on campus. The orchard has somehow struck a chord with the Wheaton community. It is profoundly linked with higher education. Sustainable agriculture enriches curricular activity and classroom diversity. The orchard has received a surprising amount of press coverage. It has even boosted Wheaton's college ranking and retention.

Still, the most fundamental aspect of the orchard is that it will actually bear fruit. When fully planted and mature, the orchard will yield thousands of bushels of honey crisp, fuji, heirloom, and many other varieties of apples. Fruit will feed students and the community and will be traded with a broader network of schools. The Wheaton campus became excited

(a)   (b)

Figure 13.11 Aerial shots of Phases one and two provide a blueprint for the Murphy Orchard's expansion.

about growing food locally, linking our education to our community and our food. I soon realized that food is a universal teacher and deserves more attention in the world of academia.

13.2 THE SAD TRUTH

As I explored the prevailing food system in America, I became both disheartened and enlightened. In America, sadly, consumer values take precedence over environmental justice and certainly community agriculture. The standard American diet (SAD) has caused us to overlook the very basis of our existence: the food we eat, how it nurtures our bodies, and how it enriches our communities. Our food system has been designed to profit agribusinesses, at the expense of our health and the health of our planet.

Needless to say, the impoverished suffer from the worst consequences of the food system. Without access to healthy food, areas of entrenched poverty are stuck with poor food choices, like fast food, items from "the dollar menu," all fueling diseases such as diabetes. I find it hard to believe that we can morally justify throwing away half our food supply while more than 2 billion people go to sleep hungry every night.

Food is both the primary source of our vitality and also a direct intermediary between the human and the natural worlds. Food connects us to the land, our bodies, and each other. However, most people know neither where their food comes from nor where their waste ends up. As the climate teeters off balance and development sprawls, narrow perspectives—exemplified by American agriculture—become increasingly problematic.

Malnourished Values

Today, the average piece of food travels more than 1500 miles from its origin of growth to the grocery store shelf, according to the documentary film *Food Inc*. Harsh environmental, social, and economic externalities associated with the handling of food are mostly unseen and therefore dismissed by the general public. This is a relatively new phenomenon.

The cultivation of food was once a much more integral and valued part of Western culture. Institutions and communities lived in tune with agriculture for more than 10,000 years. Villages, schools—even prisons and asylums—tended their own crops and livestock. Life was aligned with the health of the earth and its inhabitants as a single ecosystem. Eating in season was not a choice but a necessity. The industrial and technological revolutions changed all this, transforming the face of horticulture and the planet forever.

Mechanization, bioengineering, and massive federal subsidies commoditized and overwhelmed the American pallet, demoralizing food and undercutting the small family farm. According to the *U.S. Congressional Quarterly*, 63% of North America's native crop varieties have gone extinct since the European settlers arrived. Densely concentrated food production, fueled by artificial additives and specialization, sharpened this divide between food and people, people and land, and the shared community. The "brute force" manner in which food is manufactured has promulgated a food system defined by synthetic processing and driven by capitalistic motives.

13.3 A NEW BEGINNING

Sustainable agriculture counteracts the industrialization of food and rekindles a broken relationship. It reclarifies the bond that people have with the earth, forming a symbiotic relationship that can be carried on forever rather than a system that draws down precious resources and harms the soil, water, and people. If a student-run apple orchard can create such sweeping impacts on a single campus, the potential is limitless.

Like Wheaton, all institutions have notoriously bulky eco footprints. But the glass is half full. They also possess great human and resource potentials just ready to be harnessed. By providing a framework for community involvement, institutions can foster all sorts of student action and bolster food production. Equipped with the necessary attributes, sustainable agriculture on campus can rejuvenate community relations. Campus food generation diminishes the harmful consequences of conventional agribusiness and, at the same time, uplifts cultural attitudes.

So how realistic is this? Colleges like Bowdoin in Maine are producing the majority of their produce on campus. Community-sponsored agriculture (CSA) is expanding rapidly, as are farmers markets coast to coast. The "slow foods" movement has an 80 million person membership. These are highly successful operating models of food cultivation and distribution. Author Michael Polan reports that the local, organic, and ecologically sound agriculture is now a growing $3 billion dollar industry.

13.4 INTERCAMPUS PRODUCE EXCHANGE

If Wheaton could develop an orchard with such great success, other colleges and universities in the region could produce fruits and vegetables to share. Here is the plan that we are pursuing: The Intercampus Produce Exchange (ICPE) would consist of local networks of food yielding campuses acting in synchrony. We are starting in New England. The Wheaton orchard model would be applied to participating schools as a blueprint, in tandem with other sustainable agriculture and green practices. Together schools provide the opportune environment to grow and exchange food in a sustainable, economically feasible way.

The ICPE would be managed by an intercampus committee, a "bouquet" of different schools' representatives. Each campus would focus on a general area of agriculture, taking advantage of its assets—climate, land, greenhouses, rooftops, student interest, volunteer labor, –and the like—collaborating with one another to exchange produce. This will give students across many campuses the benefits of wholesome food produced in a sustainable way.

The ICPE would foster economic development through the promotion of federal work-study job positions and credit-bearing internships, creating hundreds of new student jobs (Figure 13.12). Campus demand for produce would shift from corporate to local sources, reviving regional economies. The environmental costs and the social inequities of our commercial agricultural system would be reduced. The life-cycle cost of food, which is rarely considered but includes the health of the earth and our bodies, would also be reduced.

The exchange would be linked by a website, an extensive monitoring system. The site would have a twofold purpose: to maximize food efficiency between campuses and to teach students about food. The exchange operator would work with dining hall directors to place orders and bring ICPE produce to campuses. The educational part of the site would allow

Figure 13.12 Mirmelli measures distance between apple trees.

professors and students to monitor the exchange and produce trends, costs, and benefits. Imagine Dartmouth pears, Tuft's lettuce, and Wheaton apples in dining halls and cafeterias throughout New England. The ICPE technical framework could change the way students think and value the food they eat.

This monitoring system would serve as a pilot that communities can replicate on a larger scale in the future. It would show consumers where food comes from, when it was tilled, and how. Furthermore, this tracking device would calculate the impact and externalities of each crop. Data generated would be made available on the internet, college home pages, and used in the classroom for comparison and analysis. By making "Food Space" as easy as "MySpace," students would be encouraged to consider the source, treatment, and environmental impact of their appetites.

The ICPE would transform the food paradigm on campuses. Healthy food, produced on neighboring campuses, coupled with education would result in heightened awareness about all resource uses. People would become reacquainted with the true value of food. Through experiential learning, the ICPE would rectify the fundamental disconnection between food and society, no less than reclarifying cultural values.

13.5 APPRECIATING THE VALUE OF FOOD

Food connects us to the planet. Our relationship with food is indicative of the way we value life. Shifting to a more sustainable food system—inspired by local actions such as the Intercampus Produce Exchange— would guide us to a more sustainable society.

Figure 13.13 As a harbinger for many a bloom to come, a baby apple tree buds for its first time in the Murphy Orchard.

In a culture that equates value with monetary gain, neither the value of food nor the quality of life is properly accounted for in America. Temporary gains fill a few pockets, but American society as a whole remains unfulfilled, unhappy, and unhealthy. When we look at the real costs of corporate agriculture, we understand that alternatives make sense. When we look at the multiple benefits of local actions, like the Murphy Apple Orchard, we understand how invaluable such initiatives really are.

Campuses are catalysts for change. Fundamentally, they are a setting based on the kind of collective intelligence required to shift from unsustainable to sustainable practices. My experience at Wheaton showed that colleges have tremendous untapped resources. They have huge potentials for all things green. Student-driven projects can have results that far exceed their highest expectations.

Food is a teacher and it forces us to really think about what matters the most, because without food, we cannot exist. The predominant food system in America is unsustainable and the true value of food has sadly depreciated. For this reason, the health of the food we eat, our bodies, and the environment we live in have been compromised.

The adoption of sustainable agriculture—spurred by campus action— plants a deeper seed of collective values within society, redefining wealth, reintegrating us within the global ecosystem, and reclarifying our integral connection with the earth.

For more information about the progression of the Murphy Apple Orchard, the ICPE, and EcoMotion Campus Services, visit us on the web at: www.ecomotion.us.

CHAPTER 14

Climate Change Mitigation from a Bottom-up Community Approach

Poul Alberg Østergaard and Henrik Lund

Contents

14.1	Introduction	247
14.2	Energy Demand in Frederikshavn	250
14.3	Current Energy System in Frederikshavn	253
14.4	Energy Resources and Energy Scenario for Frederikshavn	256
14.5	Energy System Integration in Frederikshavn	259
14.6	Public Involvement	263
References		275

14.1 INTRODUCTION

Throughout the world, increasing attention is being paid to climate change mitigation, but while ambitious national targets are hard to come by, several regions, cities, towns, institutions, and individuals have taken matters into their own hands. Rather than awaiting international agreement or national targets, these established their own ambitious targets for reducing carbon dioxide emissions and are in the process of findings ways and means to meet these targets.

This approach is in line with a Danish history of public involvement in energy policy making, as also exemplified by an ongoing campaign against a proposed carbon capture and storage system to be coupled to a coal-fired power plant in Aalborg, Denmark. The photo in Figure 14.1 shows one of many signs against this facility, with a reference to a similar case in Germany "In Schleswig Holstein 80,000 say no. What do you say?" Decisions within energy supply is not just left in the hands of policy makers or companies. It is very much within the interest of the general public.

While ambitious targets on a subnational level cannot replace ambitious national and international targets, in as much as the reductions will be limited and there will still be a plenty of free riders, they may serve as impetuous for more ambitious targets by demonstrating a will to policy makers as well as finding and demonstrating feasible options. It helps confront the

Sustainable Communities Design Handbook
ISBN: 978-1-85617-804-4, DOI: 10.1016/B978-1-85617-804-4.00014-8
247

Figure 14.1 Sign protesting against a proposed carbon capture and storage facility near Aalborg, Denmark.

previous paradigm that fossil-fueled economies are the only viable options and renewable energy systems cannot supply the required energy services at a competitive cost.

While researchers working from a bottom-up economic perspective commonly establish that ambitious targets may in fact be met at limited costs as exemplified by Lund (2010), Lund (2007), and Lund & Mathiesen (2009), researchers from a macroeconomic top-down approach commonly establish that any deviation from the current situation is in fact a deviation away from a "natural" pareto optimum—and therefore bears a cost of some magnitude.

The top-down approach is often found within policy making at an macroeconomic level, giving cause to hesitation in regard to commitment to serious changes. The bottom-up approach, however, is typically found within people of a microeconomic, institutional economic, or engineering background, the types of people likely to seek to promote ambitious targets.

This chapter describes one such case, where a town decided to establish a goal far more ambitious than that of the national government. The plan was made by university researchers in collaboration with local politicians, local stakeholders, and municipal engineers, thereby drawing on a substantial body of knowledge.

At an energy workshop in 2006, a number of Danish energy experts suggested that Denmark should convert the supply of a specific town to 100% renewable energy by 2015. The aim would be to investigate what innovative solutions would be required in such a specific setting as well as to utilize the town as a showcase for other similar projects nationally and worldwide.

Figure 14.2 Frederikshavn in the northern part of Denmark.

The experts suggested the 25,000-inhabitant town of Frederikshavn in the northern part of Denmark (Figure 14.2) for a number of reasons. The area is well defined and has a good mixture of urban landscape with dwellings and industry in addition to a smaller share of rural surroundings. It is also a town with a high level of local support and possibilities for renewable energy, of which near-shore wind power is already being exploited. Subsequently, in February 2007, the city council unanimously decided to proceed with the idea and established a project organization involving utilities and municipality administrators. Moreover, the local industry and Aalborg University are involved in the project.

This chapter provides an example of how a town like Frederikshavn, within a short space of time, could be converted into a town supplied 100% by renewable energy. It introduces a proposal for a 100% renewable energy system (RES) and deliberates the dynamics of the proposed system in light of the changes that need to be implemented if the overall Danish system is to be transformed into a 100% renewable energy supply. The proposal naturally serves as only an inspiration, as a true bottom-up community approach requires it to be community driven rather than expert driven.

For a more in-depth description of the various measures and the implementation over time, please see Clark (2009) and Østergaard & Lund (2010).

The project covers the town of Frederikshavn, the three suburbs of Strandby, Elling, and Kilden, as well as a limited number of isolated houses as indicated on the map.

Frederikshavn is an old industrial seaport with a ship wharf, diesel engine factory, a naval station, and ferries to Sweden and Norway. It is also a town in transition, with more and more emphasis on knowledge-based industry, such as Martin, supplying light effects for as diverse purposes as U2 rock tours and the Sheikh Zayed bin Sultan Al Nahyan Mosque in Abu Dhabi. The delimitation of the project area is in large part established to correspond with the boundary of the local electricity distribution company, Frederikshavn Elnet A/S.

The town of Frederikshavn should not be confused with the municipality of Frederikshavn, which encompasses a much larger area, extending to the northern tip of Denmark.

The entire area is indicated on the map, where the line from Nielstrup to Haldbjerg shows the delimitation. The darker highlighted areas are district heating areas. The lighted highlighted areas hatched are supplied with natural gas.

14.2 ENERGY DEMAND IN FREDERIKSHAVN

Designing sustainable energy systems involves a number of steps: mapping the energy demand, mapping the potential energy resources, and determining the optimal mix of various energy resources and potential energy savings. Many restrictions and objectives may have to be kept in mind including, for example, carbon dioxide emissions, self-sufficiency, and organizational, institutional, and economic factors as well as possibilities of implementation of the designed energy plan. In this case, focus was on supply-side options, although this by no means indicates that potential energy savings or changes in behavior should be neglected.

In Frederikshavn, the supply in 2007 had a renewable energy share of approximately 20% and a total final energy consumption of approximately 644 GWh (Figures 14.3 and 14.4). In terms of final energy demand, district heating accounts for the largest share as shown in Figure 14.1, although in terms of primary energy supply, the share is far smaller.

In fact, a main element of the plan is to increase the use of district heating even further. Evidently from the map in Figure 14.2, even within the

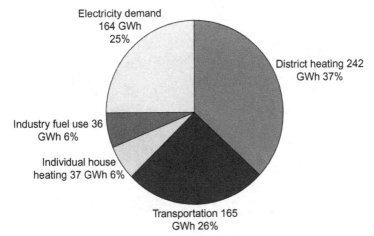

Figure 14.3 Final energy demand in Frederikshavn 2007.

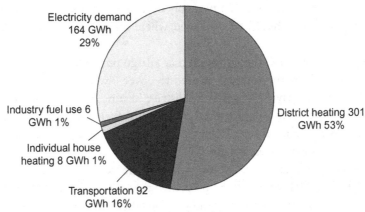

Figure 14.4 Final energy demand in Frederikshavn 2015 after implementation of changes.

contiguously built-up area of central Frederikshavn, there are potentials for an expansion of district heating, as some areas are currently supplied by natural gas for heating purposes. Using natural gas for heating purposes is less energy efficient than using district heating, as electricity generation in a thermal power plant produces a great deal of heat irrespective of whether the heat is utilized for heating—or discarded into nature. In addition, district heating offers other possibilities, which are very relevant in the transition to 100% RES. It introduces a flexibility in the choice of fuel; removing the decision from consumers, it enables certain large-scale options that would not otherwise be relevant, such as geothermal energy, and finally enables the use of waste heat from thermal power generation. District heating has played

Figure 14.5 Installation of district heating pipes.

a pivotal role in the Danish energy supply over the last three decades and for the same reasons it should be expanded where feasible, as demonstrated in figure 14.5.

An analysis of the industry found that a large proportion of the energy used (fuel used) by the industry is used to cover low-temperature heating demands, which may be covered by district heating, leaving only a small proportion that needs to be covered by an actual fuel use. The rationale for industries is the same as for residential buildings: Any fuel use that may be substituted by district heating should be.

Transport is among the sectors where there is a particular need for change. If fossil fuels are not to be used, then two options remain: Propulsion based on electricity (electricity, hydrogen, or other synthetic fuels based on hydrogen or electricity) or propulsion based on biomass. With biomass competing against other land uses—notably food production—electric vehicles should be favored, particularly in areas with a land-use constraint.

In the case or Frederikshavn, transportation accounted for 26% of the final energy consumption, as shown in Figure 14.3, but the actual demand is assumed to be lowered through efficiency improvements and fuel substitutions. Due to the very low efficiency of the internal combustion engine, introduction of electric vehicles would lower the final energy consumption radically. It is often estimated that electric vehicles would lower the demand by a factor of 3. If electricity is produced in a fuel-based power plant, some of this gain is lost in the conversion process there, alhough the heat produced may be used for heating purposes rather than vented away, as in a typical car.

Figure 14.6 Natural-gas-fired cogeneration heat and power plant in Strandby, Denmark.

Figure 14.7 Waste incineration plant in Frederikshavn.

14.3 CURRENT ENERGY SYSTEM IN FREDERIKSHAVN

Electricity and district heating is currently produced in two natural-gas–fired CHP (cogeneration of heat and power) plants in Frederikshavn and the northern suburb Strandby (see Figure 14.6) as well as in a waste incineration plant (see Figure 14.7).

Strandby has been very active in converting its energy system due to some fiery souls working for local district heating plant. This already resulted in an 8000 m² array of solar collectors (see Figure 14.8) producing an annual production of 4 GWh in combination with an additional heat

Figure 14.8 Array of solar collectors in Strandby.

Figure 14.9 Absorption heat pump in Strandby.

storage of $1500\,m^3$ of water. All Danish district heating supplying CHP plants are supplied with such storage to enable load shifting, as production of heat and electricity would otherwise have to follow one another. Solar collectors are utilized better if combined with a storage, hence the added storage capacity.

As a novelty, Strandby CHP plant has also invested in an absorption heat pump (see Figure 14.9) which draws heat from the exhaustion gas and thereby increases total efficiency from 94 to 98%. As opposed to conventional compression heat pumps, absorption heat pumps do not use electric power but rather use high-temperature heat to drive the thermodynamic cycle.

Another heat pump has been installed in Frederikshavn and connected to the district heating grid there. This is a semi-conventional compression heat

Figure 14.10 Near-shore wind turbines at and near the harbor.

pump of $1\,MW_{th}$ drawing heat from the wastewater at the town's wastewater treatment plant. This is expected to utilize $2\,GWh$ of electricity annually to extract $4\,GWh$ of heat from the wastewater and produce $6\,GWh$ of heat for the district heating supply. This corresponds to the heating demand in 400 houses. The water returning from the district heating network is increased from a temperature of approximately 40°C to 85°C before being returned to the district heating system, and this is a special characteristics of the particular heat pump. Most heats pumps are not well-adapted for such high temperatures, but this heat pump from the Danish company Advansor is particularly designed to be appropriate for a high-temperature heat supply, as needed in traditional district heating systems, where heat losses and user installations require a high water temperature. The heat pump went online in September 2009 and is controlled dynamically, based on electricity spot market prices. It is thus a technology that can assist in keeping the balance between electricity production and demand—a flexibility also required in systems with high penetrations of wind power.

Frederikshavn has four near-shore wind turbines at and near the harbor (see Figure 14.10), so wind already supplies a significant amount of electricity in the area.

The houses that are not connected to the district heating network use oil- and natural-gas-fired boilers and a small amount of wood.

Frederikshavn already entered an agreement with the electric vehicle project, Project Better Place, regarding the introduction of electric vehicles in Frederikshavn; however, actual activities have not started yet.

14.4 ENERGY RESOURCES AND ENERGY SCENARIO FOR FREDERIKSHAVN

Renewable energy sources are resources that are replenished continuously or are at least within a time span of few years. Frederikshavn only has limited access to energy resources. There are some biomass resources, a certain low-temperature geothermal potential, a limited solar potential, and a reasonably large potential for wind power. In addition to this comes a large waste resource from citizens within and beyond the project area. The biomass fraction of waste is usually regarded as renewable energy sources, while, for example, plastic products (made of oil derivatives) are a nonrenewable energy source. For practical reasons, the entire waste resource fraction is included as forming part of the renewable energy sources. It may furthermore be argued that, from marginal perspective, utilising the nonrenewable fraction merely speeds up a process that would otherwise take place in nature. This, however, is a discussion worthy of a doctoral dissertation.

While the area itself mainly is built-up, the municipality includes large areas of farm land. The project therefore includes a biogas plant able to produce 225 GWh of biogas annually, based on manure from livestock and other farm animals (Figure 14.11). This biogas may either be used directly in vehicles or CHP plants or may be upgraded to methane or converted into methanol.

It was originally planned to build an upgrading facility to convert some biogas to methane by removing the 40% carbon dioxide fraction of the biogas. Plans were well underway with elements including the upgrading facility, installation of a fueling station in Frederikshavn for natural

Figure 14.11 Manure is one of the potential renewable energy sources as it may be converted to biogas.

gas vehicles, investment in natural gas vehicles, and arrangements for the upgraded biogas to be transported through the existing natural gas grid; however the Danish government was unwilling to exempt the upgraded biogas from natural gas taxes. This made the project infeasible. While the community approach did work in terms of generating the idea and doing the groundwork for establishing this solution, national policy and regulations thus hindered its development. Framework conditions may change though, thereby making this and other similar projects feasible.

Other biomass resources such as straw and wood are also available in the area, but the amounts are limited. In the project, the use is limited to the area's fair share of the overall biomass resource in Denmark. Figure 14.12 shows export of biomass from the harbor in Aalborg, 60 km to the south of Frederikshavn.

Being a coastal town, Frederikshavn has reasonable wind resources, though west coast locations are preferred to east coast locations in Denmark. Wind power therefore forms an important element in Frederikshavn's ambitions. The first 12 MW of expansion is expected to be implemented during 2009 (see Figure 14.13); however, the final permissions were still not in place by the end of 2009. It is expected that the first two turbines will be installed with approximately one year's delay. The permit, however, allows for very large wind turbines, with a total height of up to 200 m, making them farther above sea level than any natural feature in Denmark.

Frederikshavn has good potential for geothermal resources (see Østergaard & Lund (2010)), but the water temperature is only around 58°C at a depth of 2000 m, which is insufficient for district heating purposes. An absorption heat pump may thus be applied to draw the heat

Figure 14.12 Export of biomass from the harbor in Aalborg.

Figure 14.13 Visualisation of future turbines taking advantage of Frederikshavn's wind resources.

Figure 14.14 The geothermal plant in Thisted, Denmark.

from this resource using steam from a waste incineration plant to drive the thermodynamic cycle.

Figure 14.14 shows the geothermal plant in Thisted, Denmark, which produces 15.4 GWh of heat per year. The high-temperature heat source for this plant is steam from a straw-fired boiler.

A geothermal plant supplying around 20 MW has an initial cost of around $15 million under Danish conditions, of which the main cost lies in the establishment of the well. Running costs however are fairly low though, giving reasonable heat costs for consumers. This demonstrates why district heating systems open up options that are not relevant for individual users.

As exemplified by the already existing heat pumps, there are also other options. One of the obvious options is the use of compression heat pumps

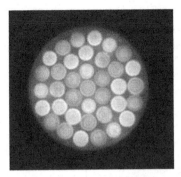

Figure 14.15 2 W LED lamps give the same light output as 25 W filament lightbulbs.

drawing heat from the ocean. While geothermal energy is a very realistic option in Frederikshavn, the scenario presented here includes only conventional heat pumps.

In terms of demand, the issue at hand is to combine the available resources with the available resources; hence, the previous changes toward district heating, electric vehicles, and methane vehicles.

As mentioned, the analyses here have not focused on energy savings, that is, improved insulation of buildings, power savings, and an increased efficiency in the industry as well as further transition to electric vehicles in transport. However, a large potential exists to be exploited. In Figure 14.15, a 2 W LED lamp giving the same light output as a 25 W filament lightbulb, indicates the large potential for electricity savings waiting to be realized.

14.5 ENERGY SYSTEM INTEGRATION IN FREDERIKSHAVN

Observing Figure 14.16, it is apparent, that the proposed system has many interdependencies and exploits many synergies. It is far from old-fashioned systems with one chain of energy ending up in the electricity supply and one chain in transport. An important lesson from Danish energy planning has been that exploitation of these interdependencies and synergies may form a solid a basis for energy-efficient energy systems; and to design future systems, even more interdependencies are introduced. Such interdependencies are even introduced purposely. For added system flexibility, the heat and electricity used in the biogas production, for instance, is drawn from district heating and electricity grids, respectively, rather than being produced on-site, which is more typical.

In 2015, the target is simply to design an energy system that, on an annual basis, is self-sufficient with renewable energy resources; however in the long term, the ambition is to create an energy system that does not

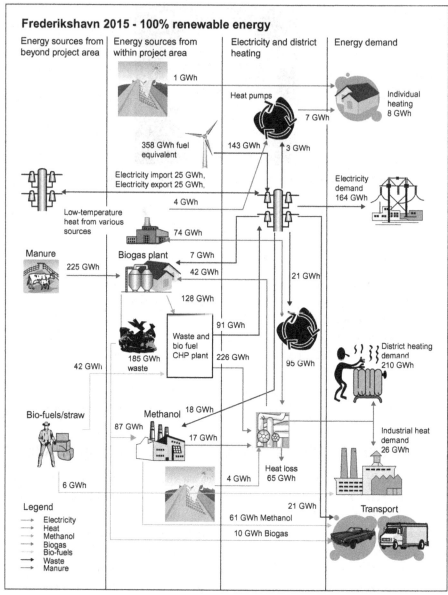

Figure 14.16 Flowchart of the proposed energy system in Frederikshavn 2015.

rely on neighboring areas to function as a buffer to balance the electricity system. Electricity systems are particularly sensitive, as production and demand needs to be balanced continuously, whereas heat systems are much less sensitive to short-term variations.

Figure 14.17 Exporting electricity imbalances using the transmission grid.

Relying on neighboring areas for buffer capacity is an option available only as long as the neighboring areas have not switched to energy systems similar to the proposed system for Frederikshavn. The target is not to entirely avoid exchange. However, it would not be acceptable for Frederikshavn to merely export any imbalances in electricity to the areas outside the town (Figure 14.17), as this would compromise the possibilities of a conversion to 100% renewable energy in these areas. The exchange should be limited to an appropriate level.

Figure 14.18 shows actual hourly data from western Denmark during a winter week in 2006 and demonstrates some of the problems that exist when it comes to balancing electricity production and electricity demand. The demand follows diurnal, weekly, and yearly cycles. The local CHP plants (relatively small-scale plants supplying district heating to towns and villages while also supplying electricity to the grid) operate at a nearly full load throughout the week, as shown in Figure 14.18, probably to supply heat to consumers. Only the large plants are centrally dispatchable and, therefore, under continuous control. This means that only they are used actively to balance supply and demand. In the given week, a certain amount is exported, although during most hours, this should probably be attributed to favorable electricity prices outside western Denmark.

Using the EnergyPLAN model, a number of analyses have been made to analyse particularly the hourly balance between electricity production and demand.

The EnergyPLAN model (Figure 14.19) is a deterministic input/output model based on hourly distribution data of demands and climate-given

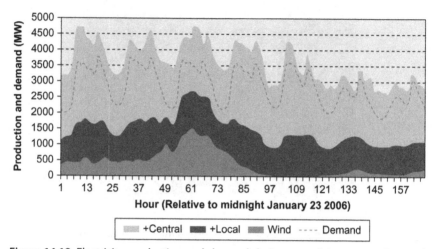

Figure 14.18 Electricity production and demand during a week in western Denmark (Data from Energinet.dk).

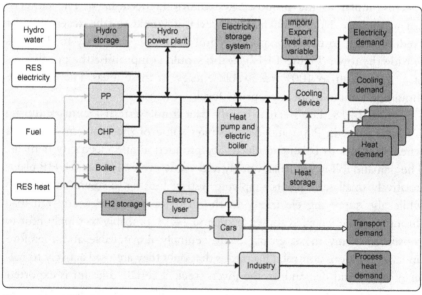

Figure 14.19 The EnergyPLAN model (Model from energyplan.aau.dk).

production, leaving the remaining production units under the control of the model to optimize the balance between power production and demand (if such a strategy is requested) as well as ensuring the most energy efficient or economically viable productions are employed. See also Lund (2010).

Some demands are also under the control of the model. This includes heat pumps, storage systems, hydrogen production, and to some extent, vehicles. Heat storage and heat pumps, for instance, are very useful in terms of rescheduling power production from CHP plants and, thus, accommodate fluctuating energy sources, such as wind power. The model also includes a number of alternative vehicles, for instance, sophisticated technologies like V2G (vehicle to grid), in which vehicles supply electricity to the electric grid. These are particularly relevant when analyzing future systems, where local load balancing is important.

Basically, the model distinguishes between technical regulation (identifying the least fuel consuming solution) and market economic regulation (identifying the consequences of operating each station on an electricity market with the aim of optimizing the business and economic profit). See also Østergaard (2009).

The analyses show that, on an annual basis, all energy demands in the energy town Frederikshavn can be met by 100% renewable energy through the energy system configuration in Figure 14.16; however, the production of electricity cannot meet the demands during all hours. The analyses indicate that the system needs to import approximately 25 GWh of electricity. On an annual basis, this import is compensated for by a similar export during other hours, but this needs further attention.

14.6 PUBLIC INVOLVEMENT

The described energy scenario has mainly been developed in a cooperation between the municipally in Frederikshavn and university researchers. The energy city Frederikshavn has also involved a number of professional stakeholders, like Project Better Place and the large Danish utility DONG Energy; however, the aim is also to involve the general public in Frederikshavn to anchor the transition better in the population.

Therefore, meetings, exhibitions, thematic weeks in the local highschool as well as in schools have taken place, and one of the largest initiatives has been the development of an online 3D model (Figure 14.20), which combines a geographical information system interface with an energy systems analyses tool, an adapted version of EnergyPLAN.

The user has the impression of flying over the project area and has a limited number of options for changing the energy system and seeing the effects visually in the urban landscape. Wind turbines appear off the coast, vehicles turn green, solar collectors appear on rooftops, and buildings change

Figure 14.20 The user interface of the 3D energy city model.

color if heat savings or electricity savings are implemented. The system is then analyzed based on the user-given parameters, and the results are shown both for aggregate numbers and hourly electricity values. The user may thus see how changes affect not only aggregate annual numbers, like carbon dioxide emissions, but also the balance between electricity production and demand.

Energy city Frederikshavn is a project conceived and implemented from a community approach rather than a project instigated at the government level through direct intervention or framework conditions. The project draws on a variety of actors, ranging from large corporations to fiery souls, but the local anchoring remains pivotal. Though the project was resistant to a recent swing in local government from the Social Democrats to the Liberal Party, it still requires a local anchoring for the project to succeed. Mobilizing the general public is therefore pertinent for the success of the project.

For further information on Frederikshavn and the energy city Frederikshavn project, see www.energycity.dk.

REFERENCES

Clark, W. (Ed.), 2009. Sustainable Communities. Springer, New York, Dordrecht, Heidelberg & London.
Lund, H., June 2007. Renewable energy strategies for sustainable development. Energy 32 (6), 912–919.
Lund, H., 2010. Renewable Energy Systems: The Choice and Modeling of 100% Renewable Solutions. Academic Press, Burlington, MA.

Lund, H., Mathiesen, B.V., May 2009. Energy system analysis of 100% renewable energy systems: the case of Denmark in years 2030 and 2050. Energy 34 (5), 524–531.

Østergaard, P.A., September 2009. Reviewing optimisation criteria for energy systems analyses of renewable energy integration. Energy 34 (9), 1236–1245.

Østergaard, P.A., Lund, H., 2010. A renewable energy system in Frederikshavn using low-temperature geo-thermal energy for district heating. Appl. Energy In press.

CHAPTER 15

Conclusions
Toward a Global Sustainable Future

Woodrow W. Clark, II, Ph.D

Contents

15.1 The Third Industrial Revolution in the United States: Second Decade of the
      21st Century   267
15.2 "Green" Careers and Businesses   269
15.3 Public Policy: Local, State, and National   270
15.4 Sustainable and Smart, Agile Communities   272
Reference   275

15.1 THE THIRD INDUSTRIAL REVOLUTION IN THE UNITED STATES: SECOND DECADE OF THE 21ST CENTURY

The publication of this book marks the end of the first decade of the 21st century, 2010. The Third Industrial Revolution (3IR) has already started. In that sense, it also marks the beginning of the next decade and the strong need for everyone to do something about stopping global warming and reversing it. The challenge is here today. The mechanisms and tools are available now, not 10-30 years from now, and advancing daily. This book provides some of the latest mechanisms and tools but recognizes that more are being discovered and used every day.

Hence, we present a challenge to the reader: use this book as a reference and guidebook. But even more significantly, use this book as a roadmap along the pathway to a sustainable future. The contributors provide a guideline and series of examples or mechanisms that require partnerships. The shift from the Second Industrial Revolution (2IR) to the Third Industrial Revolution (3IR) in the last few decades is a key concept, in that it provides a framework for action on all levels: from public policy to economics to legal contracts to technologies and future integrated infrastructure systems that fit within the local and larger environmental concerns.

Wind power from large to small systems to solar concentrators (for central plants located in distant places) then to solar panels or photovoltaics for smaller areas (homes and buildings) developed rapidly in the last decade and are now available in a variety of configurations. Today, national and international organizations set standards, codes, and policies for the technologies as well as for liability, insurance, and updated versions of the technologies.

Without getting into too much here, the integration of all these aspects and tools of sustainable development is *not* easy. There exist today conflicts at all levels. For example, the use of wind and sunshine for generating power are emerging technologies that stop and reverse global warming. Today much of the focus from the 3IR is on these renewable technologies that are, by definition, intermittent. That is, the sun does not always shine and the wind is never constantly blows, even at sea or offshore.

The key and one of the main aspects of the 3IR is to have technologies that can store energy. These technological devices would store the energy from the sun and wind for use when either sunshine or wind is not available. New technologies, ranging from current batteries to fuel cells and ultracapacitors, are becoming more viable and cost effective.

Nonetheless, many promoters of sustainable development push for solar and wind power generation, especially at the central plant level, which usually is miles away from the needed demand from consumers, businesses, and governments. The problem is that there are needs for solar and wind generation in local communities. In the European Union and Asia, this is being done more than in the United States.

Partnership mechanisms play another significant role as part of "sustainable development." What needs to be considered where sustainable communities are being promoted, developed, and implemented are the specific tools or mechanisms to accomplish such activities. Despite the political promises and delegations and especially with actual legislation and laws, making communities sustainable requires skill in many areas. For example, after policies are enacted, people, usually in teams, need to plan for these communities. That requires skill in design, planning, and a variety of areas, from dealing with contractors and construction companies to creating infrastructures with plumbing, electrical, air conditioning, and heating. Today, people who are experienced and skilled in these areas can be found. Only a few years ago, they were rare.

Today, often the specific skill areas come from union members or someone experienced in one area only. General contractors often had either a variety of skill sets or subcontracted for them. Rarely does one person or group (company, network, nonprofit organization, or government) have all the skills.

15.2 "GREEN" CAREERS AND BUSINESSES

Even more significant for the 3IR, there will be a need for technological experts. For example, as solar improves in efficiency, it becomes smaller and requires support in maintenance and improvements. Related to technologies are the legal contracts needed for insurance, operations, liability, and warranty coverage. Above all, accounting and economics plays a significant roles in sustainable development. Thus, the use of partnership mechanisms is just that: the need for each field to communicate and talk regularly. Collaboration in terms of constant interaction is a basic requirement of sustainable development. Legal staff members must talk to maintenance people and economists, as well as the other way around. Even more critical are the partnerships and interactions with the policy and decision makers, be they elected officials, board members, or local building owners.

Still more significant is that today and into the next decade or two, there will be a need for sustainable builders who are licensed in new technologies, including solar panels and storage devices. The labor unions have built an incredible amount of expertise through training all over the United States as well as globally. Some nonprofit organizations, such as the USGBC (U.S. Green Building Council) with its LEED (Leadership in Energy and Environmental Design), have begun giving points toward certification credits for buildings and soon for communities or clusters of buildings.

These individual contractors will soon be installing these systems as though they are part of the basic building infrastructures that evolved in the 2IR. So now solar systems will become part of the mortgage for all buildings in the 3IR, much like plumbing and electrical systems were in the 2IR. Now the issue is to integrate these systems and make them all efficient and cost effective and hence sustainable.

Today in the second decade of the 21st century, the costs for solar power, like that of wind power in the last decade, will be reduced dramatically. Solar technologies have already dramatically changed and their prices lowered, especially with the support of government incentive, tax, and rebate programs. The issue of lower costs for wind turbines was resolved in the late 1990s, after several decades, because their cost came down so that they are now cheaper to buy and install than natural gas. Still the intermittency of wind causes concern, but the move to the 3IR is apparent and here now. The same reduction in costs for new technologies such as solar power and photovatic cells are seen.

15.3 PUBLIC POLICY: LOCAL, STATE, AND NATIONAL

Consider another aspect of sustainable communities: requirements from local, state, and national agencies. One of the key requirements comes from local and state utilities: connect the renewable energy power generated from a home, office complex, or other community to the grid itself. This requirement is not to stop or interfere with the renewable power generated. It is not a law (reflected often in "net metering" requirements) but more of a method to account for the supply and demand of power generated.

The California energy crisis was the tip of a much larger problem in California as well as the United States and other industrialized nations. While private companies took over much of the state's energy genera- tion capacity, similar issues confront California and other nation-states, such as infrastructures for water, waste, and transportation, which are sepa- rate but interconnected sectors. These sectors remain in a crisis mode but are ignored and unattended to, since there have been few visible crises in them. However, the impact of hurricanes and storms as a result of cli- mate change on communities, in the southern part of the United States and globally, has begun to make the public more aware and ready to take constructive action, starting with the need to conserve and use energy efficiently.

Even though California is one of the states that uses the least amount of electricity per capita, the impact of its dominant use of fossil fuels (about 58% from coal and natural gas) negatively affects the environment and pollutes the atmosphere. It is far broader than energy deregulation; it is attitude, life style, ethics, policy, psychology, and so forth. The simple policy of deregulation is but a microtip of an incredibly complex series of issues about global warming, waste, and misuse of natural resources.

A part of the solution that came from the California energy crisis (2000–2003) is the creation of "agile energy systems," in which communi- ties have clusters of buildings, like colleges, local governments, residential divisions, shopping malls, and office buildings, that have their own "on-site power generation systems." The "central grid" still depends heavily on fos- sil fuels, like oil, gas, and coal as well as nuclear energy, to generate on-site power from renewable energy sources, such as solar, biomass, wind, and other sources. But the agile energy system model (Clark and Bradshaw, 2004), which is becoming a reality throughout California, is a combina- tion of local on-site energy generation (e.g., solar systems, combined heat and power, use of biomass, and other renewable sources for energy) along with the central grid power generation. For some clusters of buildings, like colleges, the use of solar for power is significant during the day, but on nonsunny days and at nighttime, the central grid becomes the power source for the community. A key component to buildings today is their design (such as LEED standards) so that they are environmentally sound. The design and construction of buildings and clusters must be addressed as a shift from centralized to decentralized or a combination of centralized and decentralized energy production.

Many renewable energy companies consider any kind of government or utility rules and regulations to be an interference with their business. However, there are important reasons for appropriate regulation, and the need for partnership mechanisms helps confront and resolve these issues.

Consider one local issue with regional implications: interconnections. Most power utilities have policies and programs about how local power generation will be connected to the power grid. There are both logical and financial reasons for this. One is that the use of renewable power generation has its restrictions. As mentioned previously, the key factor is intermittency. The grid acts like a battery or storage device for the facilities that need power when the sun is not shinning or the wind is not blowing.

The other key factor is financial. What should the cost be for the interconnection between the renewable power generation and the utility grid? There is no easy answer to that, but the issue is an important one. Some utilities simply use the time of day rule, which can mean that power provided to a customer can be one price at a certain time of day (high price at prime demand times, such as mid-afternoon to early evenings) and another price at another time (low demand, such as early morning).

No matter what the local policy, the key is to monitor or meter the power, if it is going to the power utility grid or coming from it. Such meters require both public policy rules and oversight, such as checks on the system or monitoring by "smart grid" technologies, such as the use of the Internet or wireless satellite systems. The technical issue is that there is a constant need to review, revise, and update power generation systems and how they are monitored. For communities to be sustainable, these other 3IR technologies are needed.

Sustainable development focuses on the kind of community that combines all the aspects of any society, from natural resources to human activities, including business development to technology innovation and commercialization. A consensus has grown where most organizations consider the concept of sustainable development as the interaction with the natural resources for economic concerns, infrastructures, and to protect and preserve the environment. The interaction of these sectors and others provide compelling business-minded reasons for pursuing sustainable development.

However, economic concerns (as one aspect for the definition of *development*) must always be secondary to protecting and promoting the environment. Protection of the environment must come first, since the environment is irreplaceable. This may concern some of the readers, but it is core to sustainability. If the environment is destroyed while developing any community, then the future of that area, region, or community is in jeopardy. Today, we realize that the damage to the environment anywhere in the world affects everyone throughout the world.

The concept of the 3IR takes sustainable development as the primary value. That is why Europe and Japan have set the pace by which local communities will be sustainable and secure through the use of their own renewable energy sources, storage devices, and emerging technologies, rather than the importing and use of fossil fuels.

The place to start is with small, relatively self-contained communities or villages within larger cities and regions. The issue is to get communities off their dependency on central-grid-connected energy, since most of these power generation sources come from fossil fuels, like coal, natural gas, and nuclear power. Local on-site power can be more efficiently used and based on the region's renewable energy resources, such as wind, solar power, biomass energy, among others. This model is now being accomplished in Denmark, where many communities are generating power with wind and biomass combined to provide the base load. Denmark has a goal for 50% renewable energy generation (primarily from on-site and local resources) by 2020. The country is well on its way to meeting and perhaps exceeding that national goal.

15.4 SUSTAINABLE AND SMART, AGILE COMMUNITIES

Sustainable and smart agile communities represent this new paradigm. Nations, states, and cities want to control and centralize power and authority. That has been the historical pattern. However, today, with the need to meet and address the global challenge of climate change, regional- and local-level solutions must develop, take action, and implement politics and programs now. This issue has never been more obvious than in the United States; given the conflict with the decision makers in Washington D.C., versus states like California. While that has changed with the new American federal government changes from the 2008 national elections, the problems are still there on a national level and exist within states for the need to take local actions for sustainable, agile, and smart communities.

One of the dangers confronting the United States today is that new federal laws protecting the environment or setting 3IR goals may be less than those already enacted by states. This national problem, which proposed to set 3IR standards, became obvious in areas for establishing renewable portfolio standards (RPS) at 17%, whereas at various state levels, the RPS was over 20%. The lower RPS would allow states to both fall behind and undermine the 3IR for them and most likely their region and the United States as a whole.

The starting point for public concern over global warming and climate change did not just happen in 2007 when former U.S. Vice President Al Gore won an Academy Award and the Nobel Peace Prize for his film, *An Inconvenient Truth*. Indeed, scholars and some political leaders have been concerned for decades. Gore himself has been involved with the subject since the early 1980s. However, most people make reference to former Norwegian Prime Minister Brundtland's *Report on Sustainable Development* for the United Nations as the starting point for modern-day concerns and actions.

While many would argue that the definition of *sustainable development* in the Intergovernmental Panel on Climate Change (IPCC) report was broad and vague, it nonetheless stimulated international concerns with four topics and how they are intertwined around the concept of sustainable development: environment, economics, natural resources, and human activities like transportation, building, and waste.

The international community took the lead. The first comprehensive indication of climate change came as the third IPCC report was developed. This report helped inform UN members for their discussions and decisions in Kyoto. From that time, scientists from around the world worked on the fourth IPCC report, which was issued in November 2007.

Because of global concerns, many nations and now regions, states, communities, and cities have developed their own policies to increase renewable energy power generation as part of the solution to respond to, and solve, climate change. Since the primary infrastructure sectors that affect global warming are energy and transportation, they must be examined to find ways to reverse the warming of the earth. A key element in achieving such goals is to consider how renewable energy can affect and change the transportation sector to be more environmentally sound and sustainable. Several technologies have been put forward, but in practice, no single technology can solve the problem on its own. Many contributions have to be combined and leveraged to coordinate with parallel activities in the energy sector.

The same issue, of lower standards for some nations while others have more aggressive standards and need goals, exists among nations and was reflected in the past decade and new UN meetings about the next steps to stop and reverse global warming.

On the local or regional level, sustainable and smart communities must have three components. First is the need for a master strategic plan for infra-structure that includes energy, transportation, water, waste, and telecommuni-cations along with the traditional dimensions of research, curricula, outreach, and assessments. Second is the array of issues pertaining to the design, architec-ture, and siting of buildings and overall facility master planning, which must be addressed from the perspective of "green" energy, efficient orientation, and be designed for multiple uses by the academic and local communities.

Developing dense, compact, walkable communities that enable a range of transportation choices leads to reduced energy consumption. *Communities* has a broad definition, because they can range from college campuses to cities, towns, and villages that are self-sustaining and provide for multiple uses ranging from housing, education, family events, and religion to busi-ness complexes, shopping streets, malls, and recreational activities. Third, a sustainable smart community is a vibrant, "experiential" applied model that should catalyze and stimulate entrepreneurial activities, education, and cre-ative learning along with research, commercialization, and new businesses.

Many communities, cities, and other organizations, such as academic institutions and private-sector businesses, recognize the need for policies that direct their facilities and infrastructures to be "green," based upon some cri-teria, such as the U.S. Green Building Council standards for achieving LEED certification. Individual buildings are to have "net-zero" carbon emissions. Many organizations are seeking to make their entire facilities "energy inde-pendent and carbon neutral." More recently (since June 2007), the USGBC created "community" or LEED neighborhood standards. This set of criteria reflects the broader concerns for clusters of buildings with designs integrated with basic infrastructure needs.

In the end, agile, sustainable communities must develop and imple-ment strategic plans for energy, waste, water, transportation, and telecom-munication. Each sustainable community must redesign, for example, the traditional central power plants and unsustainable infrastructure systems that exist today into ones that use on-site renewable energy, recycling, waste control, and water and land use as well as green building standards. Downsizing, providing backup and redundant power among other things, is a new and different role for "public" and now "private" companies. Today, agile, sustainable, and smart communities are necessary for a less polluted environment and providing a "green" world for tomorrow. The 3IR solutions to global warming and climate change exist now, we just need to design and implement them.

Most American cities have the potential to implement some, if not all, of these activities. With a little guidance, our communities, colleges, shopping areas, office buildings, homes, and retirement centers—our towns and cities—can have locally distributed renewable energy, clean water, recycled garbage and waste, and an efficient community transportation system that runs on renewable energy sources for power. America could be sustainable and free from the carbon-intensive, fossil-fueled inefficient centralized energy generation of the 2IR. Instead of lagging behind, the United States could become a leader in a world on the cusp of historic change and enter the 3IR with gusto.

But even better, communities and their leaders must figure out strategies to be energy independent, save millions of dollars, and save millions of tons of toxic greenhouse gases while obtaining energy independence from 2IR foreign oil and gas supplies, which come from sources around the world that are unsecurable, uncontrollable, and unsafe. In other words, the future for any community and nation-state means that it must be secure and not dependent on nonlocal fuels for energy. Local renewable energy power generation is the basic component to the 3IR and the future of humanity.

REFERENCE

Clark II, W.W., Bradshaw, T., 2004. Agile Energy Systems: Global lessons from the California Energy Crisis. Elsevier Press, Oxford, UK.

APPENDIX A

California Standard Practice Manual
Economic Analysis of Demand-Side Programs and Projects

Woodrow W. Clark, II, Ph.D, Arnie Sowell, and Don Schultz, Ph.D
For California Governor Gray Davis

Contents

	Foreword	278
A.1	Basic Methodology	278
	Background	278
	Demand-Side Management Categories and Program Definitions	280
	Basic Methods	282
	Balancing the Tests	285
	Limitations: Externality Values and Policy Rules	285
	Externality Values	285
	Policy Rules	285
A.2	Participant Test	286
	Definition	286
	Benefits and Costs	286
	How the Results Can Be Expressed	287
	Strengths of the Participant Test	287
	Weaknesses of the Participant Test	288
	Formulae	288
A.3	The Ratepayer Impact Measure Test	291
	Definition	291
	Benefits and Costs	291
	How the Results Can Be Expressed	291
	Strengths of the Ratepayer Impact Measure (RIM) Test	292
	Weaknesses of the Ratepayer Impact Measure (RIM) Test	293
	Formulae	294
A.4	Total Resource Cost Test	296
	Definition	296
	Benefits and Costs	296
	How the Results Can Be Expressed	297
	Strengths of the Total Resource Cost Test	300

Sustainable Communities Design Handbook
ISBN: 978-1-85617-804-4, DOI: 10.1016/B978-1-85617-804-4.00018-5

	Weakness of the Total Resource Cost Test	300
	Formulas	300
A.5	Program Administrator Cost Test	302
	Definition	302
	Benefits and Costs	302
	How the Results Can Be Expressed	302
	Strengths of the Program Administrator Cost Test	303
	Weaknesses of the Program Administrator Cost Test	303
	Formulas	303
Appendix a	Inputs to Equations and Documentation	304
Appendix b	Summary of Equations and Glossary of Symbols	306
	Basic Equations	306
	Benefits and Costs	307
	Glossary of Symbols	309
Appendix c	Derivation of Rim Life-Cycle Revenue Impact Formula	311
	Rate Impact Measure	311

FOREWORD

The Standard Practices Manual (SPM) was done and completed in 2001 by the Governor's Green Economic Team and released to the public. Over 40 people actively participated in its creation, which was lead by Dr. Don Schultz and Dr. Woodrow W. Clark, II. The other key author was Arnie Sowell, who led the California State Department of Consumer Services in Facilities and is now a policy advisor to the State Assembly. The SPM was initially used extensively by the California Public Utilities Commission and then State Agencies for over a decade. What it did was establish the need and process for finance and accounting for state funded projects through a life-cycle analysis economic process, rather than the conventional cost-benefit analysis. Today, a decade later, the SPM stands as a leading benchmark on how to calculate and monitor publicly funded projects from the state to the local levels.

A.1 BASIC METHODOLOGY
Background

Since the 1970s, conservation and load management programs have been promoted by the California Public Utilities Commission (CPUC) and the California Energy Commission (CEC) as alternatives to power plant construction and gas supply options. Conservation and load management (C&LM) programs have been implemented in California by the major utilities through the use of ratepayer money and by the CEC pursuant to the CEC legislative mandate to establish energy efficiency standards for new buildings and appliances.

While cost-effectiveness procedures for the CEC standards are outlined in the Public Resources Code, no such official guidelines existed for utility-sponsored programs. With the publication of the Standard Practice for Cost-Benefit Analysis of Conservation and Load Management Programs in February 1983, this void was substantially filled. With the informal "adoption" one year later of an appendix that identified cost-effectiveness procedures for an "All Ratepayers" test, C&LM program cost effectiveness consisted of the application of a series of tests representing a variety of perspectives-participants, non-participants, all ratepayers, society, and the utility.

The Standard Practice Manual was revised again in 1987–88. The primary changes (relative to the 1983 version), were: (1) the renaming of the "Non-Participant Test" to the "Ratepayer Impact Test"; (2) renaming the All-Ratepayer Test" to the "Total Resource Cost Test."; (3) treating the "Societal Test" as a variant of the "Total Resource Cost Test"; and, (4) an expanded explanation of "demand-side" activities that should be subjected to standard procedures of benefit-cost analysis.

Further changes to the manual captured in this (2001) version were prompted by the cumulative effects of changes in the electric and natural gas industries and a variety of changes in California statute related to these changes. As part of the major electric industry restructuring legislation of 1996 (AB1890), for example, a public goods charge was established that ensured minimum funding levels for "cost effective conservation and energy efficiency" for the 1998–2002 period, and then (in 2000) extended through the year 2011. Additional legislation in 2000 (AB1002) established a natural gas surcharge for similar purposes. Later in that year, the Energy Security and Reliability Act of 2000 (AB970) directed the California Public Utilities Commission to establish, by the Spring of 2001, a distribution charge to provide revenues for a self generation program and a directive to consider changes to cost-effectiveness methods to better account for reliability concerns.

In the Spring of 2001, a new state agency—the Consumer Power and Conservation Financing Authority—was created. This agency is expected to provide additional revenues in the form of state revenue bonds that could supplement the amount and type of public financial resources to finance energy efficiency and self generation activities.

The modifications to the Standard Practice Manual reflect these more recent developments in several ways. First, the "Utility Cost Test" is renamed the "Program Administrator Test" to include the assessment of programs managed by other agencies. Second, a definition of self generation as a type of "demand-side" activity is included. Third, the description of the various potential elements of "externalities" in the Societal version of the TRC test is expanded. Finally the limitations section outlines the scope of this manual and elaborates upon the processes traditionally instituted by implementing agencies to adopt values for these externalities and to adopt the the policy rules that accompany this manual.

Demand-Side Management Categories and Program Definitions

One important aspect of establishing standardized procedures for cost-effectiveness evaluations is the development and use of consistent definitions of categories, programs, and program elements.

This manual employs the use of general program categories that distinguish between different types of demand-side management programs, conservation, load management, fuel substitution, load building, and self-generation. Conservation programs reduce electricity and/or natural gas consumption during all or significant portions of the year. "Conservation" in this context includes all "energy efficiency improvements". An energy efficiency improvement can be defined as reduced energy use for a comparable level of service, resulting from the installation of an energy efficiency measure or the adoption of an energy efficiency practice. Level of service may be expressed in such ways as the volume of a refrigerator, temperature levels, production output of a manufacturing facility, or lighting level per square foot. Load management programs may either reduce electricity peak demand or shift demand from on-peak to non-peak periods.

Fuel substitution and load building programs share the common feature of increasing annual consumption of either electricity or natural gas relative to what would have happened in the absence of the program. This effect is accomplished in significantly different ways, by inducing the choice of one fuel over another (fuel substitution), or by increasing sales of electricity, gas, or electricity and gas (load building). Self generation refers to distributed generation (DG) installed on the customer's side of the electric utility meter, which serves some or all of the customer's electric load, that otherwise would have been provided by the central electric grid.

In some cases, self generation products are applied in a combined heat and power manner, in which case the heat produced by the self generation product is used on site to provide some or all of the customer's thermal needs. Self generation technologies include, but are not limited to, photovoltaics, wind turbines, fuel cells, microturbines, small gas-fired turbines, and gas-fired internal combustion engines.

Fuel substitution and load building programs were relatively new to demand-side management in California in the late 1980s, born out of the convergence of several factors that translated into average rates that substantially exceeded marginal costs. Proposals by utilities to implement programs that increase sales had prompted the need for additional procedures for estimating program cost effectiveness. These procedures may be applicable in a new context. AB 970 amended the Public Utilities Code and provided the motivation to develop a cost-effectiveness method that can be used on a common basis to evaluate all programs that will remove electric load from the centralized grid, including energy efficiency, load control/demand-responsiveness programs and self-generation. Hence, self-generation was also added to the list of demand side management programs for cost-effectiveness evaluation. In some cases, self-generation programs installed with incremental load are also included since the definition of self-generation is not necessarily confined to projects that reduce electric load on the grid. For example, suppose an industrial customer installs a new facility with a peak consumption of 1.5 MW, with an integrated on-site 1.0 MW gas fired DG unit. The combined impact of the new facility is *load building* since the new facility can draw up to 0.5 MW from the grid, even when the DG unit is running. The proper characterization of each type of demand-side management program is essential to ensure the proper treatment of inputs and the appropriate interpretation of cost-effectiveness results.

Categorizing programs is important because in many cases the same specific device can be and should be evaluated in more than one category. For example, the promotion of an electric heat pump can and should be treated as part of a conservation program if the device is installed in lieu of a less efficient electric resistance heater. If the incentive induces the installation of an electric heat pump instead of gas space heating, however, the program needs to be considered and evaluated as a fuel substitution program. Similarly, natural gas-fired self-generation, as well as self-generation units using other non-renewable fossil fuels, must be treated as fuel-substitution. In common with other types of fuel-substitution, any costs of gas transmission and distribution, and environmental externalities, must be accounted for. In addition, cost-effectiveness analyses of self-generation should account for utility interconnection costs. Similarly, a thermal energy storage device should be treated as a load management program when the predominant effect is to shift load. If the acceptance of a utility incentive by the customer to install the energy storage device is a decisive aspect of the customer's decision to remain an electric utility customer (i.e., to reject or defer the option of installing a gas-fired cogeneration system), then the predominant effect of the thermal energy storage device has been to substitute electricity service for the natural gas service that would have occurred in the absence of the program.

In addition to Fuel Substitution and Load Building Programs, recent utility program proposals have included reference to "load retention," "sales retention," "market retention," or "customer retention" programs. In most cases, the effect of such programs is identical to either a Fuel Substitution or a Load Building program—sales of one fuel are increased relative to sales without the program. A case may be made, however, for defining a separate category of program called "load retention." One unambiguous example of a load retention program is the situation where a program keeps a customer from relocating to another utility service area. However, computationally the equations and guidelines included in this manual to accommodate Fuel Substitution and Load Building programs can also handle this special situation as well.

Basic Methods

This manual identifies the cost and benefit components and cost-effectiveness calculation procedures from four major perspectives: Participant, Ratepayer Impact Measure (RIM), Program Administrator Cost (PAC), and Total Resource Cost (TRC). A fifth perspective, the Societal, is treated as a variation on the Total Resource Cost test. The results of each perspective can be expressed in a variety of ways, but in all cases it is necessary to calculate the net present value of program impacts over the life-cycle of those impacts.

Table A.1 summarizes the cost-effectiveness tests addressed in this manual. For each of the perspectives, the table shows the appropriate means of expressing test results. The primary unit of measurement refers to the way of expressing test results that are considered by the staffs of the two Commissions as the most useful for summarizing and comparing demand-side management (DSM) program cost-effectiveness. Secondary indicators of cost-effectiveness represent *supplemental* means of expressing test results that are likely to be of particular value for certain types of proceedings, reports, or programs.

This manual does not specify how the cost-effectiveness test results are to be displayed or the level at which cost-effectiveness is to be calculated

Table A.1 Cost-Effectiveness Tests

Primary	Secondary
Participant	
Net present value (all participants)	Discounted payback (years) Benefit-cost ratio Net present value (average participant)
Ratepayer Impact Measure	
Life-cycle revenue impact per Unit of energy (kWh or therm) or demand customer (kW) Net present value	Life-cycle revenue impact per unit Annual revenue impact (by year, per kWh, kW, therm, or customer) First-year revenue impact (per kWh, kW, therm, or customer) Benefit-cost ratio
Total Resource Cost	
Net present value (NPV)	Benefit-cost ratio (BCR) Levelized cost (cents or dollars per unit of energy or demand) Societal (NPV, BCR)
Program Administrator Cost	
Net present value	Benefit-cost ratio Levelized cost (cents or dollars per unit of energy or demand)

(e.g., groups of programs, individual programs, and program elements for all or some programs). It is reasonable to expect different levels and types of results for different regulatory proceedings or for different phases of the process used to establish proposed program-funding levels. For example, for summary tables in general rate case proceedings at the CPUC, the most appropriate tests may be the RIM life-cycle revenue impact, Total Resource Cost, and Program Administrator Cost test results for programs or groups of programs. The analysis and review of program proposals for the same proceeding may include Participant test results and various additional indicators of cost-effectiveness from all tests for each individual program element. In the case of cost-effectiveness evaluations conducted in the context of integrated long-term resource planning activities, such detailed examination of multiple indications of costs and benefits may be impractical.

Rather than identify the precise requirements for reporting cost-effectiveness results for all types of proceedings or reports, the approach taken in this manual is to (a) specify the components of benefits and costs for each of the major tests, (b) identify the equations to be used to express the results in acceptable ways; and (c) indicate the relative value of the different units of measurement by designating primary and secondary test results for each test.

It should be noted that, for some types of demand-side management programs, meaningful cost-effectiveness analyses cannot be performed using the tests in this manual. The following guidelines are offered to clarify the appropriated "match" of different types of programs and tests:

1. For generalized information programs (e.g., when customers are provided generic information on means of reducing utility bills without the benefit of on-site evaluations or customer billing data), cost-effectiveness tests are not expected because of the extreme difficulty in establishing meaningful estimates of load impacts.

2. For any program where more than one fuel is affected, the preferred unit of measurement for the RIM test is the life-cycle revenue impacts per customer, with gas and electric components reported separately for each fuel type and for combined fuels.

3. For load building programs, only the RIM tests are expected to be applied. The Total Resource Cost and Program Administrator Cost tests are intended to identify cost-effectiveness relative to other resource options. It is inappropriate to consider increased load as an alternative to other supply options.

4. Levelized costs may be appropriate as a supplementary indicator of cost per unit for electric conservation and load management programs relative to generation options and gas conservation programs relative to gas supply options, but the levelized cost test is not applicable to fuel substitution programs (since they combine gas and electric effects) or load building programs (which increase sales).

The delineation of the various means of expressing test results in Table A.1 is not meant to discourage the continued development of additional variations for expressing cost-effectiveness. Of particular interest is the development of indicators of program cost effectiveness that can be used to assess the appropriateness of program scope (i.e., level of funding) for General Rate Case proceedings. Additional tests, if constructed from the net present worth in conformance with the equations designated in this manual, could prove useful as a means of developing methodologies that will address issues such as the optimal timing and scope of demand-side management programs in the context of overall resource planning.

Balancing the Tests

The tests set forth in this manual are not intended to be used individually or in isolation. The results of tests that measure efficiency, such as the Total Resource Cost Test, the Societal Test, and the Program Administrator Cost Test, must be compared not only to each other but also to the Ratepayer Impact Measure Test. This multi-perspective approach will require program administrators and state agencies to consider trade-offs between the various tests. Issues related to the precise weighting of each test relative to other tests and to developing formulas for the definitive balancing of perspectives are outside the scope of this manual. The manual, however, does provide a brief description of the strengths and weaknesses of each test (Sections A.2, A.3, A.4, and A.5) to assist users in qualitatively weighing test results.

Limitations: Externality Values and Policy Rules

The list of externalities identified in Section A.4, in the discussion on the Societal version of the Total Resource Cost test is broad, illustrative and by no means exhaustive. Traditionally, implementing agencies have independently determined the details such as the components of the externalities, the externality values and the policy rules which specify the contexts in which the externalities and the tests are used.

Externality Values

The values for the externalities have not been provided in the manual. There are separate studies and methodologies to arrive at these values. There are also separate processes instituted by implementing agencies before such values can be adopted formally.

Policy Rules

The appropriate choice of inputs and input components vary by program area and project. For instance, low income programs are evaluated using a broader set of non-energy benefits that have not been provided in detail in this manual. Implementing agencies traditionally have had the discretion to use or to not use these inputs and/or benefits on a project- or program-specific basis. The policy rules that specify the contexts in which it is appropriate to use the externalities, their components, and tests mentioned in this manual are an integral part of any cost-effectiveness evaluation. These policy rules are not a part of this manual.

To summarize, the manual provides the methodology and the cost-benefit calculations only. The implementing agencies (such as the California Public Utilities Commission and the California Energy Commission) have traditionally utilized open public processes to incorporate the diverse views of stakeholders before adopting externality values and policy rules which are an integral part of the cost-effectiveness evaluation.

A.2 PARTICIPANT TEST

Definition

The Participants Test is the measure of the *quantifiable* benefits and costs to the customer due to participation in a program. Since many customers do not base their decision to participate in a program entirely on quantifiable variables, this test cannot be a complete measure of the benefits and costs of a program to a customer.

Benefits and Costs

The *benefits* of participation in a demand-side program include the reduction in the customer's utility bill(s), any incentive paid by the utility or other third parties, and any federal, state, or local tax credit received. The reductions to the utility bill(s) should be calculated using the actual retail rates that would have been charged for the energy service provided (electric demand or energy or gas). Savings estimates should be based on gross savings, as opposed to net energy savings.[1]

In the case of fuel substitution programs, benefits to the participant also include the avoided capital and operating costs of the equipment/appliance not chosen. For load building programs, participant benefits include an increase in productivity and/or service, which is presumably equal to or greater than the productivity/service without participating. The inclusion of these benefits is not required for this test, but if they are included then the societal test should also be performed.

The costs to a customer of program participation are all out-of-pocket expenses incurred as a result of participating in a program, plus any increases in the customer's utility bill(s). The out-of-pocket expenses include the cost

---

[1]Gross energy savings are considered to be the savings in energy and demand seen by the participant at the meter. These are the appropriate program impacts to calculate bill reductions for the Participant Test. Net savings are assumed to be the savings that are attributable to the program. That is, net savings are gross savings minus those changes in energy use and demand that would have happened even in the absence of the program. For fuel substitution and load building programs, gross-to-net considerations account for the impacts that would have occurred in the absence of the program.

of any equipment or materials purchased, including sales tax and installation; any ongoing operation and maintenance costs; any removal costs (less salvage value); and the value of the customer's time in arranging for the installation of the measure, if significant.

How the Results Can Be Expressed

The results of this test can be expressed in four ways: through a net present value per average participant, a net present value for the total program, a benefit-cost ratio or discounted payback. The primary means of expressing test results is net present value for the total program; discounted payback, benefit-cost ratio, and per participant net present value are secondary tests.

The discounted payback is the number of years it takes until the cumulative discounted benefits equal or exceed the cumulative discounted costs. The shorter the discounted payback, the more attractive or beneficial the program is to the participants. Although "payback period" is often defined as undiscounted in the textbooks, a discounted payback period is used here to approximate more closely the consumer's perception of future benefits and costs.[2]

Net present value (NPVp) gives the net dollar benefit of the program to an average participant or to all participants discounted over some specified time period. A net present value above zero indicates that the program is beneficial to the participants under this test.

The benefit-cost ratio (BCRp) is the ratio of the total benefits of a program to the total costs discounted over some specified time period. The benefit-cost ratio gives a measure of a rough rate of return for the program to the participants and is also an indication of risk. A benefit-cost ratio above one indicates a beneficial program.

Strengths of the Participant Test

The Participants Test gives a good "first cut" of the benefit or desirability of the program to customers. This information is especially useful for voluntary programs as an indication of potential participation rates.

For programs that involve a utility incentive, the Participant Test can be used for program design considerations such as the minimum incentive level, whether incentives are really needed to induce participation, and whether changes in incentive levels will induce the desired amount of participation.

---

[2]It should be noted that, if a demand-side program is beneficial to its participants (NPV$p \geq 0$ and BCR$p > 1.0$) using a particular discount rate, the program has an internal rate of return (IRR) of at least the value of the discount rate.

These test results can be useful for program penetration analyses and developing program participation goals, which will minimize adverse rate-payer impacts and maximize benefits.

For fuel substitution programs, the Participant Test can be used to determine whether program participation (i.e., choosing one fuel over another) will be in the long-run best interest of the customer. The primary means of establishing such assurances is the net present value, which looks at the costs and benefits of the fuel choice over the life of the equipment.

Weaknesses of the Participant Test

None of the Participant Test results (discounted payback, net present value, or benefit-cost ratio) accurately capture the complexities and diversity of customer decision-making processes for demand-side management investments. Until or unless more is known about customer attitudes and behavior, interpretations of Participant Test results continue to require considerable judgment. Participant Test results play only a supportive role in any assessment of conservation and load management programs as alternatives to supply projects.

Formulae

The following are the formulas for discounted payback, the net present value (NPV_p) and the benefit-cost ratio (BCR_p) for the Participant Test.

$$NPV_p = B_p - C_p$$

$$NPV_{avp} = \frac{B_p - C_p}{P}$$

$$BCR_p = \frac{B_p}{C_p}$$

$$DP_p = \text{Min } j \text{ such that } B_j \geq C_j$$

where:
NPV_p = Net present value to all participants
NPV_{avp} = Net present value to the average participant
BCR_p = Benefit-cost ratio to participants
DP_p = Discounted payback in years
B_p = NPV of benefit to participants
C_p = NPV of costs to participants

B_j = Cumulative benefits to participants in year j

C_j = Cumulative costs to participants in year j

p = Number of program participants

j = First year in which cumulative benefits are cumulative costs

d = Interest rate (discount)

The Benefit (B_p) and Cost (C_p) terms are further defined as follows:

$$B_p = \sum_{t=1}^{N} \frac{BR_t + TC_t + INC_t}{(1+d)^{t-1}} + \sum_{t=1}^{N} \frac{AB_{at} + PAC_{at}}{(1+d)^{t-1}}$$

$$C_p = \sum_{t=1}^{N} \frac{PC_t + BI_t}{(1+d)^{t-1}}$$

where:

BR_t = Bill reductions in year t

BI_t = Bill increases in year t

TC_t = Tax credits in year t

INC_t = Incentives paid to the participant by the sponsoring utility in year t[3]

PC_t = Participant costs in year t to include:

• Initial capital costs, including sales tax[4]

• Ongoing operation and maintenance costs include fuel cost

• Removal costs, less salvage value

• Value of the customer's time in arranging for installation, if significant

PAC_{at} = Participant avoided costs in year t for alternate fuel devices (costs of devices not chosen)

AB_{at} = Avoided bill from alternate fuel in year t

The first summation in the B_p equation should be used for conservation and load management programs. For fuel substitution programs, both the first and second summations should be used for B_p.

---

[3]Some difference of opinion exists as to what should be called an incentive. The term can be interpreted broadly to include almost anything. Direct rebates, interest payment subsidies, and even energy audits can be called incentives. Operationally, it is necessary to restrict the term to include only dollar benefits such as rebates or rate incentives (monthly bill credits). Information and services such as audits are not considered incentives for the purposes of these tests. If the incentive is to offset a specific participant cost, as in a rebate-type incentive, the full customer cost (before the rebate must be included in the PC_t term.

[4]If money is borrowed by the customer to cover this cost, it may not be necessary to calculate the annual mortgage and discount this amount if the present worth of the mortgage payments equals the initial cost. This occurs when the discount rate used is equal to the interest rate of the mortgage. If the two rates differ (e.g., a loan offered by the utility), then the stream of mortgage payments should be discounted by the discount rate chosen.

Note that in most cases, the customer bill impact terms (BR_t, BI_t, and AB_{at}) are further determined by costing period to reflect load impacts and/or rate schedules, which vary substantially by time of day and season. The formulas for these variables are as follows:

$$BR_t = \sum_{i=1}^{I}(\Delta EG_{it} \times AE : E_{it} \times K_{it})$$

$$+ \sum_{i=1}^{I}(\Delta DG_{it} \times AC : D_{it} \times K_{it}) + OBR_t$$

AB_{at} = (Use BR_t formula, but with rates and costing periods appropriate for the alternate fuel utility)

$$BI_t = \sum_{i=1}^{I}[\Delta EG_{it} \times AC : E_{it} \times (K_{it} - 1)]$$

$$+ \sum_{i=1}^{I}[\Delta DG_{it} \times AC : D_{it} \times (K_{it} - 1)] + OBI_t$$

where:

ΔEG_{it} = Reduction in gross energy use in costing period i in year t

ΔDG_{it} = Reduction in gross billing demand in costing period i in year t

$AC{:}E_{it}$ = Rate charged for energy in costing period i in year t

$AC{:}D_{it}$ = Rate charged for demand in costing period i in year t

K_{it} = 1 when ΔEG_{it} or ΔDG_{it} is positive (a reduction) in costing period i in year t, and zero otherwise

OBR_t = Other bill reductions or avoided bill payments (e.g., customer charges, standby rates).

OBI_t = Other bill increases (i.e., customer charges, standby rates).

I = Number of periods of participant's participation

In load management programs such as TOU rates and air-conditioning cycling, there are often no direct customer hardware costs. However, attempts should be made to quantify indirect costs customers may incur that enable them to take advantage of TOU rates and similar programs.

If no customer hardware costs are expected or estimates of indirect costs and value of service are unavailable, it may not be possible to calculate the benefit–cost ratio and discounted payback period.

A.3 THE RATEPAYER IMPACT MEASURE TEST[5]

Definition

The Ratepayer Impact Measure (RIM) test measures what happens to customer bills or rates due to changes in utility revenues and operating costs caused by the program. Rates will go down if the change in revenues from the program is greater than the change in utility costs. Conversely, rates or bills will go up if revenues collected after program implementation are less than the total costs incurred by the utility in implementing the program. This test indicates the direction and magnitude of the expected change in customer bills or rate levels.

Benefits and Costs

The benefits calculated in the RIM test are the savings from avoided supply costs. These avoided costs include the reduction in transmission, distribution, generation, and capacity costs for periods when load has been reduced and the increase in revenues for any periods in which load has been increased. The avoided supply costs are a reduction in total costs or revenue requirements and are included for both fuels for a fuel substitution program. The increase in revenues are also included for both fuels for fuel substitution programs. Both the reductions in supply costs and the revenue increases should be calculated using net energy savings.

The costs for this test are the program costs incurred by the utility and/or other entities incurring costs and creating or administering the program, the incentives paid to the participant, decreased revenues for any periods in which load has been decreased and increased supply costs for any periods when load has been increased. The utility program costs include initial and annual costs, such as the cost of equipment, operation and maintenance, installation, program administration, and customer dropout and removal of equipment (less salvage value). The decreases in revenues and the increases in the supply costs should be calculated for both fuels for fuel substitution programs using net savings.

How the Results Can Be Expressed

The results of this test can be presented in several forms: the life-cycle revenue impact (cents or dollars) per kWh, kW, therm, or customer; annual or first-year

---

[5]The Ratepayer Impact Measure Test has previously been described under what was called the "Non-Participant Test." The Non-Participant Test has also been called the "Impact on Rate Levels Test."

revenue impacts (cents or dollars per kWh, kW, therms, or customer); benefit-cost ratio; and net present value. The primary units of measurement are the life-cycle revenue impact, expressed as the change in rates (cents per kWh for electric energy, dollars per kW for electric capacity, cents per therm for natural gas) and the net present value. Secondary test results are the lifecycle revenue impact per customer, first-year and annual revenue impacts, and the benefit-cost ratio. LRI_{RIM} values for programs affecting electricity and gas should be calculated for each fuel individually (cents per kWh or dollars per kW and cents per therm) and on a combined gas and electric basis (cents per customer).

The life-cycle revenue impact (LRI) is the one-time change in rates or the bill change over the life of the program needed to bring total revenues in line with revenue requirements over the life of the program. The rate increase or decrease is expected to be put into effect in the first year of the program. Any successive rate changes such as for cost escalation are made from there. The first-year revenue impact (FRI) is the change in rates in the first year of the program or the bill change needed to get total revenues to match revenue requirements only for that year. The annual revenue impact (ARI) is the series of differences between revenues and revenue requirements in each year of the program. This series shows the cumulative rate change or bill change in a year needed to match revenues to revenue requirements. Thus, the ARI_{RIM} for year six per kWh is the estimate of the difference between present rates and the rate that would be in effect in year six due to the program. For results expressed as life-cycle, annual, or first-year revenue impacts, negative results indicate favorable effects on the bills of ratepayers or reductions in rates. Positive test result values indicate adverse bill impacts or rate increases.

Net present value (NPV_{RIM}) gives the discounted dollar net benefit of the program from the perspective of rate levels or bills over some specified time period. A net present value above zero indicates that the program will benefit (lower) rates and bills.

The benefit-cost ratio (BCR_{RIM}) is the ratio of the total benefits of a program to the total costs discounted over some specified time period. A benefit-cost ratio above one indicates that the program will lower rates and bills.

Strengths of the Ratepayer Impact Measure (RIM) Test

In contrast to most supply options, demand-side management programs cause a direct shift in revenues. Under many conditions, revenues lost from DSM programs have to be made up by ratepayers. The RIM test is the only test that reflects this revenue shift along with the other costs and benefits associated with the program.

An additional strength of the RIM test is that the test can be used for all demand-side management programs (conservation, load management, fuel substitution, and load building). This makes the RIM test particularly useful for comparing impacts among demand-side management options.

Some of the units of measurement for the RIM test are of greater value than others, depending upon the purpose or type of evaluation. The life-cycle revenue impact per customer is the most useful unit of measurement when comparing the merits of programs with highly variable scopes (e.g., funding levels) and when analyzing a wide range of programs that include both electric and natural gas impacts. Benefit-cost ratios can also be very useful for program design evaluations to identify the most attractive programs or program elements.

If comparisons are being made between a program or group of conservation/load management programs and a specific resource project, life-cycle cost per unit of energy, and annual and first-year net costs per unit of energy are the most useful way to express test results. Of course, this requires developing life-cycle, annual, and first-year revenue impact estimates for the supply-side project.

Weaknesses of the Ratepayer Impact Measure (RIM) Test

Results of the RIM test are probably less certain than those of other tests because the test is sensitive to the differences between long-term projections of marginal costs and long-term projections of rates, two cost streams that are difficult to quantify with certainty.

RIM test results are also sensitive to assumptions regarding the financing of program costs. Sensitivity analyses and interactive analyses that capture feedback effects between system changes, rate design options, and alternative means of financing generation and non-generation options can help overcome these limitations. However, these types of analyses may be difficult to implement.

An additional caution must be exercised in using the RIM test to evaluate a fuel substitution program with multiple end use efficiency options. For example, under conditions where marginal costs are less than average costs, a program that promotes an inefficient appliance may give a more favorable test result than a program that promotes an efficient appliance. Though the results of the RIM test accurately reflect rate impacts, the implications for long-term conservation efforts need to be considered.

Formulae

The formulae for the life-cycle revenue impact (LRI_{RIM})'s net present value (NPV_{RIM}), benefit-cost ratio (BCR_{RIM})'s first-year revenue impacts and annual revenue impacts are presented below:

$$LRI_{RIM} = \frac{C_{RIM} - B_{RIM}}{E}$$

$$FRI_{RIM} = \frac{C_{RIM} - B_{RIM}}{E}, \quad \text{for } t = 1$$

$$ARI_{RIM_t} = FRI_{RIM}, \quad \text{for } t = 1$$

$$= \frac{C_{RIM_t} - B_{RIM_t}}{E_t}, \quad \text{for } t = 2,\ldots, N$$

$$NPV_{RIM} = B_{RIM} - C_{RIM}$$

$$BCR_{RIM'} = \frac{B_{RIM}}{C_{RIM}}$$

where:

LRI_{RIM} = Life-cycle revenue impact of the program per unit of energy (kWh or therm) or demand (kW) (the one-time change in rates) or per customer (the change in customer bills over the life of the program). (Note: An appropriate choice of kWh, therm, kW, and customer should be made)

FRI_{RIM} = First-year revenue impact of the program per unit of energy, demand, or per customer.

ARI_{RIM} = Stream of cumulative annual revenue impacts of the program per unit of energy, demand, or per customer. (Note: The terms in the ARI formula are not discounted; thus they are the nominal cumulative revenue impacts. Discounted cumulative revenue impacts may be calculated and submitted if they are indicated as such. Note also that the sum of the discounted stream of cumulative + revenue impacts does not equal the LRI_{RIM})

NPV_{RIM} = Net present value levels

BCR_{RIM} = Benefit-cost ratio for rate levels

B_{RIM} = Benefits to rate levels or customer bills

C_{RIM} = Costs to rate levels or customer bills

E = Discounted stream of system energy sales (kWh or therms) or demand sales (kW) or first-year customers. (See Appendix c of this appendix for a description of the derivation and use of this term in the LRI_{RIM} test.)

The B_{RIM} and C_{RIM} terms are further defined as follows:

$$B_{\text{RIM}} = \sum_{t=1}^{N} \frac{\text{UAC}_t + \text{RG}_t}{(1+d)^{t-1}} + \sum_{t=1}^{N} \frac{\text{UAC}_{at}}{(1+d)^{t-1}}$$

$$C_{\text{RIM}} = \sum_{t=1}^{N} \frac{\text{UIC}_t + \text{RL}_t + \text{PRC}_t + \text{INC}_t}{(1+d)^{t-1}} + \sum_{t=1}^{N} \frac{\text{RL}_{at}}{(1+d)^{t-1}}$$

$$E = \sum_{t=1}^{N} \frac{\text{E}_t}{(1+d)^{t-1}}$$

where:

UAC_t = Utility avoided supply costs in year t

UIC_t = Utility increased supply costs in year t

RG_t = Revenue gain from increased sales in year t

RL_t = Revenue loss from reduced sales in year t

PRC_t = Program Administrator program costs in year t

E_t = System sales in kWh, kW, or therms in year t or first year customers

UAC_{at} = Utility avoided supply costs for the alternate fuel in year t

RL_{at} = Revenue loss from avoided bill payments for alternate fuel in year t (i.e., device not chosen in a fuel substitution program)

For fuel substitution programs, the first term in the B_{RIM} and C_{RIM} equations represents the sponsoring utility (electric or gas), and the second term represents the alternate utility. The RIM test should be calculated separately for electric and gas and combined electric and gas.

The utility avoided cost terms (UAC_t, UIC_t, and UAC_{at}) are further determined by costing period to reflect time-variant costs of supply:

$$\text{UAC}_t = \sum_{i=1}^{I} (\Delta\text{EN}_{it} \times \text{MC} : E_{it} \times K_{it}) + \sum_{i=1}^{I} (\Delta\text{DN}_{it} \times \text{MC} : D_{it} \times K_{it})$$

UAC_{at} = (Use UAC_t formula, but with marginal costs and costing periods appropriate for the alternate fuel utility.)

$$\text{UIC}_t = \sum_{i=1}^{I} [\Delta\text{EN}_{it} \times \text{MC} : E_{it} \times (K_{it} - 1)]$$
$$+ \sum_{i=1}^{I} [\Delta\text{DN}_{it} \times \text{MC} : D_{it} \times (K_{it} - 1)]$$

where:

(Only terms not previously defined are included here.)

ΔEN_{it} = Reduction in net energy use in costing period i in year t

ΔDN_{it} = Reduction in net demand in costing period i in year t
$MC{:}E_{it}$ = Marginal cost of energy in costing period i in year t
$MC{:}D_{it}$ = Marginal cost of demand in costing period i in year t

The revenue impact terms (RG_t, RL_t, and RL_{at}) are parallel to the bill impact terms in the Participant Test. The terms are calculated exactly the same way with the exception that the net impacts are used rather than gross impacts. If a net-to-gross ratio is used to differentiate gross savings from net savings, the revenue terms and the participant's bill terms will be related as follows:

$$RG_t = BI_t \times \text{(net-to-gross ratio)}$$

$$RL_t = BR_t \times \text{(net-to-gross ratio)}$$

$$RL_{at} = AB_{at} \times \text{(net-to-gross ratio)}$$

A.4 TOTAL RESOURCE COST TEST[6]

Definition

The Total Resource Cost Test measures the net costs of a demand-side management program as a resource option based on the total costs of the program, including both the participants' and the utility's costs.

The test is applicable to conservation, load management, and fuel substitution programs. For fuel substitution programs, the test measures the net effect of the impacts from the fuel not chosen versus the impacts from the fuel that is chosen as a result of the program. TRC test results for fuel substitution programs should be viewed as a measure of the economic efficiency implications of the total energy supply system (gas and electric).

A variant on the TRC test is the Societal Test. The Societal Test differs from the TRC test in that it includes the effects of externalities (e.g., environmental, national security), excludes tax credit benefits, and uses a different (societal) discount rate.

Benefits and Costs

This test represents the combination of the effects of a program on both the customers participating and those not participating in a program. In a sense, it is the summation of the benefit and cost terms in the Participant and the Ratepayer Impact Measure tests, where the revenue (bill) change and the incentive terms intuitively cancel (except for the differences in net and gross savings).

---

[6]This test was previously called the All Ratepayers Test.

The benefits calculated in the Total Resource Cost Test are the avoided supply costs, the reduction in transmission, distribution, generation, and capacity costs valued at marginal cost for the periods when there is a load reduction. The avoided supply costs should be calculated using net program savings, savings net of changes in energy use that would have happened in the absence of the program. For fuel substitution programs, benefits include the avoided device costs and avoided supply costs for the energy, using equipment not chosen by the program participant.

The costs in this test are the program costs paid by both the utility and the participants plus the increase in supply costs for the periods in which load is increased. Thus all equipment costs, installation, operation and maintenance, cost of removal (less salvage value), and administration costs, no matter who pays for them, are included in this test. Any tax credits are considered a reduction to costs in this test. For fuel substitution programs, the costs also include the increase in supply costs for the utility providing the fuel that is chosen as a result of the program.

How the Results Can Be Expressed

The results of the Total Resource Cost Test can be expressed in several forms: as a net present value, a benefit-cost ratio, or as a levelized cost. The net present value is the primary unit of measurement for this test. Secondary means of expressing TRC test results are a benefit-cost ratio and levelized costs. The Societal Test expressed in terms of net present value, a benefit-cost ratio, or levelized costs is also considered a secondary means of expressing results. Levelized costs as a unit of measurement are inapplicable for fuel substitution programs, since these programs represent the net change of alternative fuels which are measured in different physical units (e.g.,, kWh or therms). Levelized costs are also not applicable for load building programs.

Net present value (NPV_{TRC}) is the discounted value of the net benefits to this test over a specified period of time. NPV_{TRC} is a measure of the change in the total resource costs due to the program. A net present value above zero indicates that the program is a less expensive resource than the supply option upon which the marginal costs are based.

The benefit-cost ratio (BCR_{TRC}) is the ratio of the discounted total benefits of the program to the discounted total costs over some specified time period. It gives an indication of the rate of return of this program to the utility and its ratepayers. A benefit-cost ratio above one indicates that the program is beneficial to the utility and its ratepayers on a total resource cost basis.

The levelized cost is a measure of the total costs of the program in a form that is sometimes used to estimate costs of utility-owned supply additions. It presents the total costs of the program to the utility and its ratepayers on a per kilowatt, per kilowatt hour, or per therm basis levelized over the life of the program.

The Societal Test is structurally similar to the Total Resource Cost Test. It goes beyond the TRC test in that it attempts to quantify the change in the total resource costs to society as a whole rather than to only the service territory (the utility and its ratepayers). In taking society's perspective, the Societal Test utilizes essentially the same input variables as the TRC Test, but they are defined with a broader societal point of view. More specifically, the Societal Test differs from the TRC Test in at least one of five ways. First, the Societal Test may use higher marginal costs than the TRC test if a utility faces marginal costs that are lower than other utilities in the state or than its out-of-state suppliers. Marginal costs used in the Societal Test would reflect the cost to society of the more expensive alternative resources. Second, tax credits are treated as a transfer payment in the Societal Test, and thus are left out. Third, in the case of capital expenditures, interest payments are considered a transfer payment since society actually expends the resources in the first year. Therefore, capital costs enter the calculations in the year in which they occur. Fourth, a societal discount rate should be used.[7] Finally, marginal costs used in the Societal Test would also contain externality costs of power generation not captured by the market system. An illustrative and by no means exhaustive list of "externalities and their components" is given below (Refer to the Limitations section for elaboration.) These values are also referred to as "adders" designed to capture or internalize such externalities. The list of potential adders would include for example:

1. The benefit of avoided environmental damage. The CPUC policy specifies two "adders" to internalize environmental externalities, one for electricity use and one for natural gas use. Both are statewide average values. These adders are intended to help distinguish between cost-effective and non cost-effective energy-efficiency programs. They apply to an average supply mix and would not be useful in distinguishing among competing supply options. The CPUC electricity environmental adder is intended to account for the environmental damage from air pollutant emissions from power plants. The CPUC-adopted adder is intended to cover the human and material damage from sulfur oxides (SO_X), nitrogen oxides (NO_X), volatile organic compounds (VOCs, sometimes called reactive organic

[7]Many economists have pointed out that use of a market discount rate in social cost-benefit analysis undervalues the interests of future generations. Yet if a market discount rate is not used, comparisons with alternative investments are difficult to make.

gases or ROG), particulate matter at or below 10 micron diameter (PM_{10}), and carbon. The adder for natural gas is intended to account for air pollutant emissions from the direct combustion of the gas. In the CPUC policy guidance, the adders are included in the tabulation of the benefits of energy efficiency programs. They represent reduced environmental damage from displaced electricity generation and avoided gas combustion. The environmental damage is the result of the net change in pollutant emissions in the air basins, or regions, in which there is an impact. This change is the result of direct changes in power plant or natural gas combustion emission resulting from the efficiency measures, and changes in emissions from other sources, that result from those direct changes in emissions.

2. The benefit of avoided transmission and distribution costs—energy efficiency measures that reduce the growth in peak demand would decrease the required rate of expansion to the transmission and distribution network, eliminating costs of constructing and maintaining new or upgraded lines.

3. The benefit of avoided generation costs—energy efficiency measures reduce consumption and hence avoid the need for generation. This would include avoided energy costs, capacity costs and T&D line.

4. The benefit of increased system reliability: The reductions in demand and peak loads from customers opting for self generation provide reliability benefits to the distribution system in the forms of:

   a. Avoided costs of supply disruptions

   b. Benefits to the economy of damage and control costs avoided by customers and industries in the digital economy that need greater than 99.9 level of reliable electricity service from the central grid

   c. Marginally decreased System Operator's costs to maintain a percentage reserve of electricity supply above the instantaneous demand

   d. Benefits to customers and the public of avoiding blackouts.

5. Non-energy benefits: Non-energy benefits might include a range of program-specific benefits such as saved water in energy-efficient washing machines or self generation units, reduced waste streams from an energy-efficient industrial process, etc.

6. Non-energy benefits for low income programs: The low income programs are social programs which have a separate list of benefits included in what is known as the "low income public purpose test". This test and the specific benefits associated with this test are outside the scope of this manual.

7. Benefits of fuel diversity include considerations of the risks of supply disruption, the effects of price volatility, and the avoided costs of risk exposure and risk management.

Strengths of the Total Resource Cost Test

The primary strength of the Total Resource Cost (TRC) test is its scope. The test includes total costs (participant plus program administrator) and also has the potential for capturing total benefits (avoided supply costs plus, in the case of the societal test variation, externalities). To the extent supply-side project evaluations also include total costs of generation and/or transmission, the TRC test provides a useful basis for comparing demand- and supply-side options.

Since this test treats incentives paid to participants and revenue shifts as transfer payments (from all ratepayers to participants through increased revenue requirements), the test results are unaffected by the uncertainties of projected average rates, thus reducing the uncertainty of the test results. Average rates and assumptions associated with how other options are financed (analogous to the issue of incentives for DSM programs) are also excluded from most supply-side cost determinations, again making the TRC test useful for comparing demand-side and supply-side options.

Weakness of the Total Resource Cost Test

The treatment of revenue shifts and incentive payments as transfer payments, identified previously as a strength, can also be considered a weakness of the TRC test. While it is true that most supply-side cost analyses do not include such financial issues, it can be argued that DSM programs should include these effects since, in contrast to most supply options, DSM programs do result in lost revenues.

In addition, the costs of the DSM "resource" in the TRC test are based on the total costs of the program, including costs incurred by the participant. Supply-side resource options are typically based only on the costs incurred by the power suppliers.

Finally, the TRC test cannot be applied meaningfully to load building programs, thereby limiting the ability to use this test to compare the full range of demand-side management options.

Formulas

The formulas for the net present value (NPV_{TRC}), the benefit-cost ratio (BCR_{TRC}) and levelized costs are presented below:

$$NPV_{TRC} = B_{TRC} - C_{TRC}$$

$$\text{BCR}_{\text{TRC}} = \frac{B_{\text{TRC}}}{C_{\text{TRC}}}$$

$$\text{LC}_{\text{TRC}} = \frac{\text{LCRC}}{\text{IMP}}$$

where:
NPV_{TRC} = Net present value of total costs of the resource
BCR_{TRC} = Benefit-cost ratio of total costs of the resource
LC_{TRC} = Levelized cost per unit of the total cost of the resource (cents per kWh for conservation programs; dollars per kW for load management programs)
B_{TRC} = Benefits of the program
C_{TRC} = Costs of the program
LCRC = Total resource costs used for levelizing
IMP = Total discounted load impacts of the program
PCN = Net Participant Costs
The B_{TRC} C_{TRC}, LCRC, and IMP terms are further defined as follows:

$$B_{\text{TRC}} = \sum_{t=1}^{N} \frac{\text{UAC}_t + \text{TC}_t}{(1+d)^{t-1}} + \sum_{t=1}^{N} \frac{\text{UAC}_{at} + \text{PAC}_{at}}{(1+d)^{t-1}}$$

$$C_{\text{TRC}} = \sum_{t=1}^{N} \frac{\text{PRC}_t + \text{PCN}_t + \text{UIC}_t}{(1+d)^{t-1}}$$

$$\text{LCRC} = \sum_{t=1}^{N} \frac{\text{PRC}_t + \text{PCN}_t + \text{TC}_t}{(1+d)^{t-1}}$$

$$\text{IMP} = \sum_{t-1}^{N} \frac{\left[\left(\sum_{i=1}^{I} \Delta\text{EN}_{it}\right), \text{ or } (\Delta\text{DN}_{it}, \text{ where } i = \text{peak period})\right]}{(1+d)^{t-1}}$$

(All terms have been defined in previous sections.)
The first summation in the B_{TRC} equation should be used for conservation and load management programs. For fuel substitution programs, both the first and second summations should be used.

A.5 PROGRAM ADMINISTRATOR COST TEST

Definition

The Program Administrator Cost Test measures the net costs of a demand-side management program as a resource option based on the costs incurred by the program administrator (including incentive costs) and excluding any net costs incurred by the participant. The benefits are similar to the TRC benefits. Costs are defined more narrowly.

Benefits and Costs

The benefits for the Program Administrator Cost Test are the avoided supply costs of energy and demand, the reduction in transmission, distribution, generation, and capacity valued at marginal costs for the periods when there is a load reduction. The avoided supply costs should be calculated using net program savings, savings net of changes in energy use that would have happened in the absence of the program. For fuel substitution programs, benefits include the avoided supply costs for the energy-using equipment not chosen by the program participant only in the case of a combination utility where the utility provides both fuels.

The costs for the Program Administrator Cost Test are the program costs incurred by the administrator, the incentives paid to the customers, and the increased supply costs for the periods in which load is increased. Administrator program costs include initial and annual costs, such as the cost of utility equipment, operation and maintenance, installation, program administration, and customer dropout and removal of equipment (less salvage value). For fuel substitution programs, costs include the increased supply costs for the energy-using equipment chosen by the program participant only in the case of a combination utility, as above.

In this test, revenue shifts are viewed as a transfer payment between participants and all ratepayers. Though a shift in revenue affects rates, it does not affect revenue requirements, which are defined as the difference between the net marginal energy and capacity costs avoided and program costs. Thus, if $NPV_{pa} > 0$ and $NPV_{RIM} < 0$, the administrator's overall total costs will decrease, although rates may increase because the sales base over which revenue requirements are spread has decreased.

How the Results Can Be Expressed

The results of this test can be expressed either as a net present value, benefit–cost ratio, or levelized costs. The net present value is the primary test, and the benefit–cost ratio and levelized cost are the secondary tests.

Net present value (NPV_{pa}) is the benefit of the program minus the administrator's costs, discounted over some specified period of time. A net present value above zero indicates that this demand-side program would decrease costs to the administrator and the utility.

The benefit-cost ratio (BCR_{pa}) is the ratio of the total discounted benefits of a program to the total discounted costs for a specified time period. A benefit-cost ratio above one indicates that the program would benefit the combined administrator and utility's total cost situation.

The levelized cost is a measure of the costs of the program to the administrator in a form that is sometimes used to estimate costs of utility-owned supply additions. It presents the costs of the program to the administrator and the utility on per kilowatt, per kilowatt-hour, or per therm basis levelized over the life of the program.

Strengths of the Program Administrator Cost Test

As with the Total Resource Cost test, the Program Administrator Cost test treats revenue shifts as transfer payments, meaning that test results are not complicated by the uncertainties associated with long-term rate projections and associated rate design assumptions. In contrast to the Total Resource Cost test, the Program Administrator Test includes only the portion of the participant's equipment costs that is paid for by the administrator in the form of an incentive. Therefore, for purposes of comparison, costs in the Program Administrator Cost Test are defined similarly to those supply-side projects which also do not include direct customer costs.

Weaknesses of the Program Administrator Cost Test

By defining device costs exclusively in terms of costs incurred by the administrator, the Program Administrator Cost test results reflect only a portion of the full costs of the resource.

The Program Administrator Cost Test shares two limitations noted previously for the Total Resource Cost test: (1) by treating revenue shifts as transfer payments, the rate impacts are not captured, and (2) the test cannot be used to evaluate load building programs.

Formulas

The formulas for the net present value, the benefit-cost ratio and levelized cost are presented below:

$$NPV_{pa} = B_{pa} - C_{pa}$$

$$BCR_{pa} = \frac{B_{pa}}{C_{pa}}$$

$$LC_{pc} = \frac{LC_{pc}}{IMP}$$

where:

NPV_{pa} = Net present value of Program Administrator costs

BCR_{pa} = Benefit-cost ratio of Program Administrator costs

LC_{pa} = Levelized cost per unit of Program Administrator cost of the resource

B_{pa} = Benefits of the program

C_{pa} = Costs of the program

LC_{pc} = Total Program Administrator costs used for levelizing

$$B_{pa} = \sum_{t=1}^{N} \frac{UAC_t}{(1+d)^{t-1}} + \sum_{t+1}^{N} \frac{UAC_{at}}{(1+d)^{t-1}}$$

$$C_{pa} = \sum_{t-1}^{N} \frac{PRC_t + INC_t + UIC_t}{(1+d)^{t-1}}$$

$$LC_{pc} = \sum_{t=1}^{N} \frac{PRC_t + INC_t}{(1+d)^{t-1}}$$

(All variables are defined in previous sections.)

The first summation in the B_{pa} equation should be used for conservation and load management programs. For fuel substitution programs, both the first and second summations should be used.

APPENDIX a: INPUTS TO EQUATIONS AND DOCUMENTATION

A comprehensive review of procedures and sources for developing inputs is beyond the scope of this manual. It would also be inappropriate to attempt a complete standardization of techniques and procedures for developing inputs for such parameters as load impacts, marginal costs, or average rates. Nevertheless, a series of guidelines can help to establish acceptable procedures and improve the chances of obtaining reasonable levels of consistent and meaningful cost-effectiveness results. The following "rules" should be viewed as appropriate guidelines for developing the primary inputs for the cost-effectiveness equations contained in this manual:

1. In the past, Marginal costs for electricity were based on production cost model simulations that clearly identify key assumptions and characteristics of the existing generation system as well as the timing and nature of any generation additions and/or power purchase agreements in the future. With a deregulated market for wholesale electricity, marginal costs for electric generation energy should be based on forecast market prices, which are derived from recent transactions in California energy markets. Such transactions could include spot market purchases as well as longer term bilateral contracts and the marginal costs should be estimated based on components for energy as well as demand and/or capacity costs as is typical for these contracts.

2. In the case of submittals in conjunction with a utility rate proceeding, average rates used in DSM program cost-effectiveness evaluations should be based on proposed rates. Otherwise, average rates should be based on current rate schedules. Evaluations based on alternative rate designs are encouraged.

3. Time-differentiated inputs for electric marginal energy and capacity costs, average energy rates, and demand charges, and electric load impacts should be used for (a) load management programs, (b) any conservation program that involves a financial incentive to the customer, and (c) any Fuel Substitution or Load Building program. Costing periods used should include, at a minimum, summer and winter, on-, and off-peak; further disaggregation is encouraged.

4. When program participation includes customers with different rate schedules, the average rate inputs should represent an average weighted by the estimated mix of participation or impacts. For General Rate Case proceedings it is likely that each major rate class within each program will be considered as program elements requiring separate cost-effectiveness analyses for each measure and each rate class within each program.

5. Program administration cost estimates used in program cost-effectiveness analyses should exclude costs associated with the measurement and evaluation of program impacts unless the costs are a necessary component to administer the program.

6. For DSM programs or program elements that reduce electricity and natural gas consumption, costs and benefits from both fuels should be included.

7. The development and treatment of load impact estimates should distinguish between gross (i.e., impacts expected from the installation of a particular device, measure, appliance) and net (impacts adjusted to

account for what would have happened anyway, and therefore not attributable to the program). Load impacts for the Participants test should be based on gross, whereas for all other tests the use of net is appropriate. Gross and net program impact considerations should be applied to all types of demand-side management programs, although in some instances there may be no difference between gross and net.

8. The use of sensitivity analysis, i.e., the calculation of cost-effectiveness test results using alternative input assumptions, is encouraged, particularly for the following programs: new programs, programs for which authorization to substantially change direction is being sought (e.g., termination, significant expansion), major programs which show marginal cost-effectiveness and/or particular sensitivity to highly uncertain input(s).

The use of many of these guidelines is illustrated with examples of program cost effectiveness contained in Appendix b.

APPENDIX b: SUMMARY OF EQUATIONS AND GLOSSARY OF SYMBOLS

Basic Equations

Participant Test

$$NPV_p = B_p - C_p$$

$$NPV_{avp} = \frac{B_p - C_p}{P}$$

$$BCR_p = \frac{B_p}{C_p}$$

$$DP_p = \text{Min } j \text{ such that } B_j \geq C_j$$

Ratepayer Impact Measure Test

$$LRI_{RIM} = \frac{C_{RIM} - B_{RIM}}{E}$$

$$FRI_{RIM} = \frac{C_{RIM} - B_{RIM}}{E}, \quad \text{for } t = 1$$

$$ARI_{RIM_t} = FRI_{RIM}, \quad \text{for } t = 1$$

$$= \frac{C_{RIM_t} - B_{RIM_t}}{E_t}, \quad \text{for } t = 2,...,N$$

$$NPV_{RIM} = B_{RIM} - C_{RIM}$$

$$BCR_{RIM} = \frac{B_{RIM}}{C_{RIM}}$$

Total Resource Cost Test

$$NPV_{TRC} = B_{TRC} - C_{TRC}$$

$$BCR_{TRC} = \frac{B_{TRC}}{C_{TRC}}$$

$$LC_{TRC} = \frac{LCRC}{IMP}$$

Program Administrator Cost Test

$$NPV_{pa} = B_{pa} - C_{pa}$$

$$BCR_{pa} = \frac{B_{pa}}{C_{pa}}$$

$$LC_{pc} = \frac{LC_{pc}}{IMP}$$

Benefits and Costs
Participant Test

$$B_p = \sum_{t=1}^{N} \frac{BR_t + TC_t + INC_t}{(1+d)^{t-1}} + \sum_{t=1}^{N} \frac{AB_{at} + PAC_{at}}{(1+d)^{t-1}}$$

$$C_p = \sum_{t=1}^{N} \frac{PC_t + BI_t}{(1+d)^{t-1}}$$

Ratepayer Impact Measure Test

$$B_{RIM} = \sum_{t=1}^{N} \frac{UAC_t + RG_t}{(1+d)^{t-1}} + \sum_{t=1}^{N} \frac{UAC_{at}}{(1+d)^{t-1}}$$

$$C_{\text{RIM}} = \sum_{t=1}^{N} \frac{\text{UIC}_t + \text{RL}_t + \text{PRC}_t + \text{INC}_t}{(1 + d)^{t-1}} + \sum_{t=1}^{N} \frac{\text{RL}_{at}}{(1 + d)^{t-1}}$$

$$E = \sum_{t=1}^{N} \frac{E_t}{(1 + d)^{t-1}}$$

Total Resource Cost Test

$$B_{\text{TRC}} = \sum_{t=1}^{N} \frac{\text{UAC}_t + \text{TC}_t}{(1 + d)^{t-1}} + \sum_{t=1}^{N} \frac{\text{UAC}_{at} + \text{PAC}_{at}}{(1 + d)^{t-1}}$$

$$C_{\text{TRC}} = \sum_{t=1}^{N} \frac{\text{PRC}_t + \text{PCN}_t + \text{UIC}_t}{(1 + d)^{t-1}}$$

$$\text{LCRC} = \sum_{t=1}^{N} \frac{\text{PRC}_t + \text{PCN}_t + \text{TC}_t}{(1 + d)^{t-1}}$$

$$\text{IMP} = \sum_{t-1}^{N} \frac{\left[\left(\sum_{i=1}^{I} \Delta\text{EN}_{it} \right), \text{ or } (\Delta\text{DN}_{it}, \text{ where } i = \text{ peak period}) \right]}{(1 + d)^{t-1}}$$

Program Administrator Cost Test

$$B_{\text{pa}} = \sum_{t=1}^{N} \frac{\text{UAC}_t}{(1 + d)^{t-1}} + \sum_{t+1}^{N} \frac{\text{UAC}_{at}}{(1 + d)^{t-1}}$$

$$C_{\text{pa}} = \sum_{t=1}^{N} \frac{\text{PRC}_t + \text{INC}_t + \text{UIC}_t}{(1 + d)^{t-1}}$$

$$\text{LCPA} = \sum_{t=1}^{N} \frac{\text{PRC}_t + \text{INC}_t}{(1 + d)^{t-1}}$$

$$B_{\text{pa}} = \sum_{t=1}^{N} \frac{\text{UAC}_t}{(1 + d)^{t-1}} + \sum_{t+1}^{N} \frac{\text{UAC}_{at}}{(1 + d)^{t-1}}$$

$$C_{pa} = \sum_{t=1}^{N} \frac{PRC_t + INC_t + UIC_t}{(1 + d)^{t-1}}$$

$$LCPA = \sum_{t=1}^{N} \frac{PRC_t + INC_t}{(1 + d)^{t-1}}$$

Glossary of Symbols

Ab_{at} = Avoided bill reductions on bill from alternate fuel in year t

$AC{:}D_{it}$ = Rate charged for demand in costing period i in year t

$AC{:}E_{it}$ = Rate charged for energy in costing period i in year t

ARI_{RIM} = Stream of cumulative annual revenue impacts of the program per unit of energy, demand, or per customer. Note that the terms in the ARI formula are not discounted, thus they are the nominal cumulative revenue impacts. Discounted cumulative revenue impacts may be calculated and submitted if they are indicated as such. Note also that the sum of the discounted stream of cumulative revenue impacts does not equal the LRI_{RIM*}

BCR_p = Benefit-cost ratio to participants

BCR_{RIM} = Benefit-cost ratio for rate levels

BCR_{TRC} = Benefit-cost ratio of total costs of the resource

BCR_{pa} = Benefit-cost ratio of program administrator and utility costs

BI_t = Bill increases in year t

B_j = Cumulative benefits to participants in year j

B_p = Benefit to participants

B_{RIM} = Benefits to rate levels or customer bills

BR_t = Bill reductions in year t

B_{TRC} = Benefits of the program

B_{pa} = Benefits of the program

C_j = Cumulative costs to participants in year i

C_p = Costs to participants

C_{RIM} = Costs to rate levels or customer bills

C_{TRC} = Costs of the program

C_{pa} = Costs of the program

d = Discount rate

ΔDg_{it} = Reduction in gross billing demand in costing period i in year t

ΔDn_{it} = Reduction in net demand in costing period i in year t

DP_p = Discounted payback in years

E = Discounted stream of system energy sales (kWh or therms) or demand sales (kW) or first-year customers

ΔEG_{it} = Reduction in gross energy use in costing period i in year t

ΔEN_{it} = Reduction in net energy use in costing period i in year t

E_t = System sales in kWh, kW, or therms in year t or first year customers

FRI_{RIM} = First-year revenue impact of the program per unit of energy, demand, or per customer.

IMP = Total discounted load impacts of the program

INC_t = Incentives paid to the participant by the sponsoring utility in year t, first year in which cumulative benefits are > cumulative costs.

K_{it} = 1 when ΔEG_{it} or ΔDG_{it} is positive (a reduction) in costing period i in year t, and zero otherwise

$LCRC$ = Total resource costs used for levelizing

LC_{TRC} = Levelized cost per unit of the total cost of the resource

LC_{pa} = Total Program Administrator costs used for levelizing

LC_{pa} = Levelized cost per unit of program administrator cost of the resource

LRI_{RIM} = Life-cycle revenue impact of the program per unit of energy (kWh or therm) or demand (kW)—the one-time change in rates—or per customer—the change in customer bills over the life of the program.

$MC{:}D_{it}$ = Marginal cost of demand in costing period i in year t

$MC{:}E_{it}$ = Marginal cost of energy in costing period i in year t

NPV_{avp} = Net present value to the average participant

NPV_P = Net present value to all participants

NPV_{RIM} = Net present value levels

NPV_{TRC} = Net present value of total costs of the resource

NPV_{pa} = Net present value of program administrator costs

OBI_t = Other bill increases (i.e., customer charges, standby rates)

OBR_t = Other bill reductions or avoided bill payments (e.g., customer charges, standby rates).

P = Number of program participants

PAC_{at} = Participant avoided costs in year t for alternate fuel devices

PC_t = Participant costs in year t to include:
- Initial capital costs, including sales tax
- Ongoing operation and maintenance costs
- Removal costs, less salvage value
- Value of the customer's time in arranging for installation, if significant

PRC_t = Program Administrator program costs in year t

PCN = Net Participant Costs

RG_t = Revenue gain from increased sales in year t

RL_{at} = Revenue loss from avoided bill payments for alternate fuel in year
     t (i.e., device not chosen in a fuel substitution program)

RL_t = Revenue loss from reduced sales in year t

TC_t = Tax credits in year t

UAC_{at} = Utility avoided supply costs for the alternate fuel in year t

UAC_t = Utility avoided supply costs in year t

PA_t = Program Administrator costs in year t

UIC_t = Utility increased supply costs in year t

APPENDIX c: DERIVATION OF RIM LIFE-CYCLE REVENUE IMPACT FORMULA

Most of the formulas in the manual are either self-explanatory or are explained in the text. This appendix provides additional explanation for a few specific areas where the algebra was considered to be too cumbersome to include in the text.

Rate Impact Measure

The Ratepayer Impact Measure life-cycle revenue impact test (LRI_{RIM}) is assumed to be the one-time increase or decrease in rates that will re-equate the present valued stream of revenues and stream of revenue requirements over the life of the program.

Rates are designed to equate long-term revenues with long-term costs or revenue requirements. The implementation of a demand-side program can disrupt this equality by changing one of the assumptions upon which it is based: the sales forecast. Demand-side programs by definition change sales. This expected difference between the long-term revenues and revenue requirements is calculated in the NPV_{RIM}. The amount which present valued revenues are below present valued revenue requirements equals $-NPV_{RIM}$.

The LRI_{RIM} is the change in rates that creates a change in the revenue stream that, when present valued, equals the $-NPV_{RIM}$. If the utility raises (or lowers) its rates in the base year by the amount of the LRI_{RIM}, revenues over the term of the program will again equal revenue requirements. (The other assumed changes in rates, implied in the escalation of the rate values, are considered to remain in effect.)

Thus, the formula for the LRI_{RIM} is derived from the following equality where the present value change in revenues due to the rate increase or decrease is set equal to the $-NPV_{RIM}$ or the revenue change caused by the program.

$$-NPV_{RIM} = \sum_{t=1}^{N} \frac{LRI_{RIM} \times E_t}{(1 + d)^{t-1}}$$

Since the LRI_{RIM} term does not have a time subscript, it can be removed from the summation, and the formula is then:

$$-NPV_{RIM} = LRI_{RIM} \times \sum_{t=1}^{N} \frac{E_t}{(1 + d)^{t-1}}$$

Rearranging terms, we then get:

$$LRI_{RIM} = \frac{-NPV_{RIM}}{\sum_{t=1}^{N} \frac{E_t}{(1 + d)^{t-1}}}$$

Thus,

$$F = LRI_{RIM} - \sum_{t=1}^{N} \frac{E_t}{(1 + d)^{t-1}}$$

APPENDIX B

Request for Qualifications

Los Angeles Community College District
with lead author Woodrow W. Clark, II, Ph.D

Contents

B.1	Purpose and Scope of Request for Qualificatons	315
	Background	315
	Los Angeles Community College District Energy Systems Overview	315
	Energy Efficiency and Conservation Goal	316
	Purpose and Scope of Request For Qualifications (RFQ)	316
	Professional Services	317
	Performance Based Energy Savings Agreement	317
	RFQ Response	317
	Evaluation and Selection	318
	Notice of Short List	318
	LACCD Performance Based Energy Savings Agreement and Other Required Contract Documents	318
	LACCD General Conditions of the Construction Contract	319
	Future Solicitation for Short List	319
	Mandatory Pre-Response Conference	319
B.2	Schedule of Events	319
B.3	RFQ Instructions and General Provisions	320
	Questions Regarding RFQ and Point Of Contact	320
	Errors and Omissions	320
	Addenda	321
	Cancellation of Solicitation	321
	Compliance with RFQ	321
	Completion of RFQ Response	321
	Delivery of RFQ Response	321
	Exceptions	322
	Alternative RFQ Responses	322
	Withdrawal of RFQ Response	322
	RFQ Responses Become the Property of LACCD	322
	Confidential Material	323
	Reservation of Rights	323
	Non-Endorsement	323
	Disputes/Protests	324
	Award of Contract	324

Sustainable Communities Design Handbook
ISBN: 978-1-85617-804-4, DOI: 10.1016/B978-1-85617-804-4.00022-7
313

Execution of the Agreement   324
Failure to Execute the Agreement   325
Conflict of Interest   325
Business Outreach Policies and Procedures   325
Achieving the Small Business Participation Goal   325
Program Manager's Local, Small and Emerging Business Database   327
Construction Bond Assistance Program   328
Mentoring Program   329
Additional Provisions and Submittals   329
B.4   Scope of Services, Work, and Deliverables   329
Introduction   329
Service Provider   329
Performance of Services and Work   330
Preliminary Audit (PA)   330
PA Report Requirements   331
Investment Grade Audit (IGA)   333
IGA Submittal Requirements   335
Project Delivery   338
Required Design Reviews and Approvals   339
Protocol for Design, Specification and Construction   340
Resources Provided by LACCD   341
B.5   Qualification Requirements   341
Introduction   341
Comprehensive Energy Projects Solicited   342
Energy Projects Experience   342
Specific Project Information   343
Financial And Administrative Stability   344
Management and Personnel Qualification   344
Sample Project Response   345
B.6   Evaluation and Selection Criteria   346
Introduction   346
Evaluation Criteria   346
Point scoring schedule   346
Summary of Evaluation Categories   346
B.7   Content and Format of Response to RFQ   352
Introduction   352
Delivery of RFQ Responses   352
Preparation   352
RFQ Response Format   353

B.1 PURPOSE AND SCOPE OF REQUEST FOR QUALIFICATONS

Background

The Los Angeles Community College District serves over 100 cities and communities in an area encompassing 882 square miles. The District extends from Agoura Hills in the west San Fernando Valley to the City of San Fernando in the north and Monterey Park to the east. The service area includes Culver City on the west side of the greater Los Angeles basin, Monterey Park and San Gabriel on the east side as well as Palos Verdes Estates and San Pedro to the south. The LACCD colleges educate more than 120,000 students a year. The mission of the District is "to provide comprehensive lower-division general education, occupational education, transfer education, counseling and guidance, community services, and continuing education programs which are appropriate to the communities served and which meet the changing needs of students for academic and occupational, preparation, citizenship, and cultural understanding." The Western Association of Schools and Colleges accredits each of the nine colleges. A seven-member Board of Trustees, elected at large for four-year terms, governs the District.

Geographically, the colleges range in size from 22 to over 450 acres. Facilities include newly constructed classroom buildings as well as out-dated structures older than 50 years. On April 10, 2001 the voters authorized the District to issue $1.245 billion of general obligation bonds under Proposition A. In May 2003 the voters again authorized the District to issue approximately $980 million of general obligation bonds under Proposition AA. The bond proceeds are being used for construction, repair, improvement, and upgrade of District buildings, classrooms, and other facilities. Further information about the District and the ballot measures can be found on these web sites: www.laccd.edu and www.propositiona.org.

Los Angeles Community College District Energy Systems Overview

During fiscal year 2005/2006 LACCD campuses expended approximately $1 million for natural gas and approximately $12 million for electricity purchases serving over 5,460,000 gross square feet in 440 buildings. Given the size and complexity of building systems, there remains many opportunities to enhance building systems' efficiency and reduce overall utility costs and environmental impact. Opportunities for energy projects exist at all campuses in lighting systems, HVAC systems, building controls, automation, and energy infrastructure.

Energy Efficiency and Conservation Goal

It is the goal of the Los Angeles Community College District to improve campus-building performance and achieve the lowest environmental impact feasible, by continuing to practice responsible stewardship using available resources. Energy use intensity (EUI) measured in British thermal units per gross square foot per year shall be used to benchmark building performance for both state and non-state supported areas of the campuses.

Purpose and Scope of Request For Qualifications (RFQ)

The purpose of this RFQ is to identify through competitive means a list of qualified firms to provide cost effective and reliable energy related services to the various LACCD campuses. This RFQ is the qualifying process to satisfy competitive means pursuant to the Energy Conservation Contract Authority, California Government Code Section 4217.10–4217.18. This competitive process will establish a list of the best-qualified firms that have the size, resources, financial ability, expertise, and necessary experience to provide the services required for this program. LACCD envisions a process that will flow as follows:

1. A campus must select no more than three and no less than two firms from the approved list to do a preliminary review of potential energy efficiency projects on campus and write a report of the findings. This report and phase is called Preliminary Audit (PA).
2. The results of the PA from the selected two to three firms are reviewed by a third party Independent Peer Reviewer to verify feasibility and energy saving. After the peer review, the campus selects one firm to move forward to the next phase.
3. The selected firm does a more detailed and specific review and audit of the proposed energy efficiency projects and prepares a detailed written investment grade audit report of the findings and cost proposal. The Investment Grade Audit (IGA) is the second phase of a project.
4. The IGA is reviewed by the campus and a third party Independent Peer Reviewer to verify energy savings, costs, feasibility and that the project meets the required criteria to obtain financing.
5. Following a successful peer review of the IGA, the campus will decide whether to move forward and if so will then negotiate the price and enter into a Design/Build Agreement to construct the project.
6. The construction of the energy efficiency project is the Project Delivery (PD) phase. There is no minimum dollar amount for projects.

It is anticipated that provision of these services shall begin in the third quarter of 2008. LACCD does not guarantee any minimum level of business arising from this RFQ.

Professional Services

"Professional Services" includes any contract for services in connection with a project and includes architectural, engineering, planning, testing, general studies, or feasibility services. The selection of firms to provide professional services in connection with a project is on the basis of demonstrated competence and on the professional qualifications necessary for the satisfactory performance of the services required. Although the LACCD does not procure professional services on the basis of competitive bids, it is obligated to obtain the best services at fair and reasonable costs. Professional Services Agreements shall be issued to firms to provide Preliminary and Investment Grade Audits. This Professional Services agreement will be superseded by a Performance Based Energy Savings Agreement to construct the project and develop terms of financing.

Performance Based Energy Savings Agreement

The LACCD may enter into a "Performance Based Energy Savings Agreement" pursuant to the Energy Conservation Contract Authority, California Government Code Section 4217.10–4217.18. Such projects require appropriate due diligence measures to ensure the project is in the best interest of the Los Angeles Community College District. The anticipated cost for the alteration effected by the Performance Based Energy Savings Agreement must be less than the anticipated marginal cost to LACCD of energy that would have been consumed in the absence of those alterations. *Section 4217.*

RFQ Response

Firms wishing to participate in this solicitation shall submit a complete response to the RFQ by the date and time specified in Section B.2, Schedule of Events. Responses shall be reviewed and evaluated by an evaluation team comprised of representatives from LACCD. *RFQ responses shall be submitted in accordance with the instructions contained in Section B.7, Content and Format of Response to RFQ.* Responses or partial responses and modifications thereof received after closure time specified will not be considered.

Evaluation and Selection

All RFQ responses received by the date and time indicated in Section B.2, Schedule of Events, of this RFQ shall be evaluated by LACCD. The evaluation and selection of qualified finalists is described in Section B.6, Evaluation and Selection Criteria, of this RFQ. LACCD reserves the right to have confidential discussions if necessary to further evaluate responding firms and obtain additional information. LACCD reserves the right to obtain and utilize information from any source deemed appropriate in the evaluation and selection of responding firms.

Notice of Short List

LACCD reserves the right to reject any and all submittals and following the process, if pursued, to award none, one, or more contracts. A "Notice of Short List" will be publicly posted for five calendar days. Section 4, Scope of Services, Work, and Deliverables, of this RFQ identifies what the short listed firms are required to perform. Written notification will be made to all firms who have submitted a response to this solicitation.

LACCD Performance Based Energy Savings Agreement and Other Required Contract Documents

Pubic works involves the erection, construction, alteration, painting, repair, or improvement of any state structure, building, road, or other state improvement of any kind. In compliance with Contract Law the LACCD Performance Based Energy Savings Agreement is the instrument that shall be used for improvement projects that require construction and/or trades labor. *The Performance Based Agreement is the document that collectively represents the entire agreement between the Trustees and the Performance Based Agreement, and which supersedes any prior negotiations, representations, or agreements, either written or oral.* The LACCD General Conditions of the Construction Contract govern the LACCD Performance Based Energy Savings Agreement. The Performance Based Energy Savings Agreement accompanied by General Conditions of the Construction Contract, Payment Bond, Performance Bond, and a Certification form to identify persons authorized to execute contracts; all are parts thereof. All reference to bidding procedures in the General Conditions of the Construction Contract do not apply to this RFQ.

Other contract documents that make up the Agreement are included by reference and are as follows: Request for Qualifications (RFQ), cost proposal forms, insurance certificates, plans, specifications, and addenda.

LACCD General Conditions of the Construction Contract

LACCD General Conditions of the Construction Contract include all the required provisions of the contract relating to bidding, award, performance of the work, changes, claims and damages, payment, and completion to be in compliance with various applicable codes. To maintain consistency throughout the LACCD system, when there is a question on the interpretation of General Conditions of the Construction Contract, the construction administrator shall consult with the Build LACCD Contract administrator.

Future Solicitation for Short List

LACCD reserves the right to add additional qualified firms to the short-list at any time. Such additions shall be based on the firms meeting the same criteria that were used to select the initial list of short-listed vendors. From time to time, LACCD may issue subsequent solicitations to seek additional qualified firms in any of the Technology areas.

Mandatory Pre-Response Conference

A mandatory Pre-Response Conference will be held to clarify requirements and answer any questions relative to this RFQ. The conference will be held as follows:

Date:
Time:
Location:

It is strongly recommended that firms intending to respond to this RFQ attend the conference. This may be the only time responding firms can discuss the RFQ program with the LACCD project team.

B.2 SCHEDULE OF EVENTS

Action	Date
Release of Request for Qualifications (RFQ):	Monday, December 17, 2007
Mandatory Pre-Response Conference:	Thursday, January 17, 2008

Location:
  Los Angeles Community College District
  Office of the Executive Director
  Facility Planning and Development 6th Floor
  770 Wilshire Boulevard
  Los Angeles, 90017

Last Day to Submit Written Questions: Thursday, January 24, 2008
(Questions received after this date and time may not receive a written response).
Release of LACCD Written Response
to Questions: Thursday, January 31, 2008
RFQ Response Due: *Thursday, March 13, 2008*
(RFQ Responses received after this date and time will not be accepted).
All dates following the RFQ Response Due date are provided for planning purposes only and are subject to change without notice.
Committee Review of RFQ Responses: Thursday, April 10, 2008
Notice of Firms Selected for Interview: Thursday, April 17, 2008
Interviews of Selected Firms: Thursday, April 24, 2008
Notice of Short Listed Firms and
Intent to Award:*   Thursday, May 1, 2008
Investment Grade Audit and Contract Award:* Thursday, July 3, 2008

B.3 RFQ INSTRUCTIONS AND GENERAL PROVISIONS

Questions Regarding RFQ and Point of Contact

Any questions, interpretations or clarifications, either administrative or technical, about this RFQ must be requested in writing no later than the date indicated in the Schedule of Events. All written questions, not considered proprietary, will be answered in writing and conveyed to all responding firms. Oral statements concerning the meaning or intent of the contents of this RFQ by any person are not considered binding. Questions regarding any aspect of this RFQ should be directed to:

Errors and Omissions

If, prior to the date fixed for submission of RFQ Response, a respondent discovers any ambiguity, conflict, discrepancy, omission or other error in the RFQ or any of its exhibits and/or appendices, respondent shall immediately notify LACCD of such error in writing and request modification or clarification of the document. Modifications may be made by addenda prior to the RFQ response deadline. Clarifications will be given by written notice to all active firms who have been furnished an RFQ for responding purposes, without divulging the source of the request for it.

---

*NOTE: The above dates through deadline for receipt of responses may be adjusted upon advance written notice. All dates subsequent to that time are estimated and subject to change without notice.

Addenda

LACCD may modify this RFQ, any of its key action dates, or any of its attachments, prior to the date fixed for submission by issuance of a written addendum to all firms who have been furnished the RFQ for bidding purposes. Addenda will be numbered consecutively as a suffix to the RFQ Reference Number.

Cancellation of Solicitation

This solicitation does not obligate LACCD to enter into an agreement. LACCD retains the right to cancel this RFQ at any time for any reason. LACCD also retains the right to obtain the services specified in this RFQ in any other way. No obligation, either expressed or implied, exists on the part of LACCD to make an award or to pay any cost incurred in the preparation or submission of response to the RFQ.

Compliance with RFQ

To be compliant with the administrative requirements of this RFQ, responding firm must complete and return the list of submittals in Section B.7, Content and Format of Response to RFQ.

Completion of RFQ Response

Responses to the RFQ shall be complete in all respects as required by this solicitation. A submission may be rejected if conditional or incomplete, or if it contains any alterations or other irregularities of any kind, and will be rejected if any such defect or irregularity could have materially affected the quality of the submission. Documents which contain false or misleading statements, or which provide references that do not support an attribute or condition claimed by the responding firm, may be rejected. Statements made by a responding firm shall also be without ambiguity, and with adequate elaboration, where necessary, for clear understanding. Costs for developing RFQ Responses are entirely the responsibility of the responding firms and shall not be chargeable to LACCD.

Delivery of RFQ Response

The RFQ Response must be received in the Contract Services and Procurement Office no later than the time indicated on the date specified in Section B.2, Schedule of Events. The responding firm is responsible for the means of delivering the RFQ Response to the appropriate office on time. Delays due to the instrumentalities used to transmit the RFQ Response, including delay occasioned by the internal mailing system in the LACCD Office of Facilities, Planning and Development, will be the responsibility of the responding firm. Likewise, delays due to inaccurate directions given, even if by The LACCD Office of Facilities, Planning and Development staff, shall be the responsibility of the responding firm. The RFQ Response must be completed and delivered by the specified time in order to avoid disqualification for lateness due to difficulties in delivery. In accordance with Section B.7, Content and Format of Response to RFQ, responding firm must provide a minimum of one original hardcopy (marked as such), four copies and two electronic copies. *LATE, FAXED OR E-MAILED PROPOSALS WILL NOT BE ACCEPTED.*

Exceptions

In the event a respondent believes that this RFQ is unfairly restrictive or has substantive errors or omissions in it, the matter must be promptly brought to the attention of LACCD's Contact, either by telephone, e-mail, letter or facsimile, immediately upon receipt of the RFQ, in order that the matter may be fully considered and appropriate action taken by the LACCD prior to the closing time set for submission.

Alternative RFQ Responses

Only one RFQ Response is to be submitted by each respondent. Multiple RFQ Responses shall result in rejection of all RFQ Responses submitted by the respondent.

Withdrawal of RFQ Response

A RFQ Response may be withdrawn after it is received by LACCD by written or facsimile request signed by the responding firm or authorized representative, prior to the time and date specified for RFQ Response submission. RFQ Response may be withdrawn and resubmitted in the same manner if done so prior to the appropriate deadline. Withdrawal or modification offered in any other manner will not be considered.

RFQ Responses Become the Property of LACCD

RFQ Responses become the property of LACCD and information contained therein shall become public documents subject to disclosure laws after Notice of Intent to Award. LACCD reserves the right to make use of any information or ideas contained in the RFQ Response. RFQ

Responses may be returned only at LACCD's option and at the responding firm's expense. One copy shall be retained for official files. Responses to this RFQ and any other information that is currently or may become available as an outcome of the RFQ process may be used by LACCD to structure an RFQ or other solicitation.

Confidential Material

Respondent must notify LACCD in advance of any proprietary or confidential materials contained in the RFQ Response and provide justification for not making such material public. LACCD shall have sole discretion to disclose or not disclose such material subject to any protective order that responding firm may obtain.

Reservation of Rights

LACCD may reject any or all RFQ Responses and may waive any immaterial deviation in an RFQ Response. LACCD's waiver of an immaterial defect shall in no way modify the RFQ documents or excuse the responding firm from full compliance with the specifications if the responding firm is awarded the contract. RFQ Responses that include terms and conditions other than LACCD's terms and conditions may be rejected as being non-responsive. In the event all RFQ Responses are rejected or LACCD determines alternative solutions are in its best interest, LACCD may cancel this solicitation and pursue alternative sourcing options.

LACCD may make such investigations as deemed necessary to determine the ability of the responding firm to perform the work, and the responding firm shall furnish all such information and data for this purpose. LACCD reserves the right to reject any submittal made pursuant to this RFQ or any subsequent RFQ Response or bid if the evidence submitted by, or investigation of, such responding firm fails to satisfy LACCD that such responding firm is properly qualified to carry out the obligations of the contract and to complete the work specified. Additionally, LACCD reserves the right to request additional performance guarantees should, in the sole opinion of LACCD, financial stability or capability cannot be established.

Non-Endorsement

If selected as a qualified responding firm, the responding firm shall not issue any news releases or other statements pertaining to selection, which state or imply LACCD endorsement of responding firm's services.

Disputes/Protests

LACCD encourages potential respondents to resolve issues regarding the requirements or the procurement process through written correspondence and discussions. LACCD wishes to foster cooperative relationships and to reach a fair agreement in a timely manner.

Respondents, who desire to file protest, must do so within five calendar days after Notice of Intent to Award. The protesting firm shall submit a full and complete written statement detailing the facts in support of the protest. Protest must be sent by certified or registered mail or delivered in person to the Executive Director, Facility Planning and Development; within a reasonable time after receipt of the written statement of protest, LACCD will provide a decision on the matter. The decision will be in writing and sent by certified or registered mail or delivered in person to the protesting respondent. The decision of LACCD is final.

Award of Contract

LACCD reserves the right to reject any and all RFQ Responses and to award one or more contracts. Award, if any, will be to the responding firms, whose RFQ Responses best complies with all of the requirements of the RFQ documents and any addenda. A "Notice of Intent to Award" will be posted publicly for five calendar days prior to the award. Written notification will be made to unsuccessful responding firms.

The selected responding firms and LACCD shall commit to negotiation for the final scope of services to be accepted and execution of an agreement, in substantial accordance with the terms and conditions herein, within 30 days of the Notice of Intent to Award. Should the parties be unable to reach final agreement within this time frame, the parties may mutually agree upon a time extension to complete negotiations and contract execution. If the parties are unable to agree upon a time extension, or if LACCD determines that a time extension would not be beneficial to the project, LACCD reserves the right to terminate negotiations and proceed with a secondary finalist.

Execution of the Agreement

The Agreement shall be signed by the Service Providers and returned, along with the required attachments to LACCD within *14* calendar days from receipt of contract. Contracts are not effective until approved by the appropriate LACCD officials. Any work performed prior to receipt of a fully executed contract shall be at Service Provider's own risk.

Failure to Execute the Agreement

Failure to execute the Agreement within the time frame identified above shall be sufficient cause for voiding the award. Failure to comply with other requirements within the set time shall constitute failure to execute the Agreement. If the successful responding firm refuses or fails to execute the Agreement, LACCD may award the Agreement to the next qualified responding firm.

Conflict of Interest

Potential Service Providers are advised that Service Providers' officers and employees shall comply with the disclosure, disqualification, and other provisions of California's Political Reform Act of 1974 (Government Code Section 81000 et seq.) if their responsibilities include the making or participation in the making of a LACCD decision.

Business Outreach Policies and Procedures

It is the policy of the Board of Trustees (BOT) of the Los Angeles Community College District (District) to promote community economic development through the Proposition A and AA Bond Program, and to seek the maximum possible participation by local, small, emerging and disabled veteran owned businesses as a means of increasing the bidder pool on District projects thereby lowering bid prices.

The District has established a goal of achieving 28% participation by small business in the total value of the work performed in the Bond Program. Bidders or proposers who do not achieve 28% participation by small businesses may be deemed non-responsive.

A business is considered a "small business" if its gross revenue or number of employees fall within the definitions established by the Small Business Administration for the type of activity conducted by the business.

Achieving the Small Business Participation Goal

The small business outreach component of the Bond Program has as its key objective bringing new businesses into the bidder pool in the interest of achieving the lowest possible cost of construction. It is intended to promote personal interaction between prime bidders/proposers and a small business in a way that is cost effective and that does not unduly burden any participant in the Bond Program.

Prime bidders/proposers' outreach efforts to small businesses should be active and aggressive, and must be reasonably calculated to meet or exceed the small business participation goal of 28%. Bidders/proposers are encouraged to achieve the maximum possible participation by small businesses, including participation by lower tier subcontractors and small business vendors and suppliers.

All prime bidders/proposers analyze the work of the proposed contract for the purpose of achieving the 28% small business participation goal. Prime bidders/proposers should break down the work into sufficiently small units, portions or quantities so that it is economically feasible for small businesses to successfully compete for at least 28% of the work.

To maximize the personal interaction between prime consultants and contractors and small businesses, pre-submittal and pre-proposal conferences generally are mandatory. However, small businesses may be excused from attending such conferences due to unavoidable conflict.

Prime bidders/proposers should contact as many small business subconsultants or subcontractors as possible. It is recommended that at least three small business subconsultants or subcontractors be contacted in disciplines or trades which have been identified as economically feasible for small businesses. Prime bidders/consultants should advise the small business of the portion of the work upon which it is invited to bid, the name of the person to whom it should direct questions and the last date that bids are received by the prime bidder/consultant.

To identify small businesses that are bidding on the work, prime bidders/consultants are encouraged to make contact with the small businesses who have attended the Pre-Proposal/Bid conference; the sign-in sheet is posted by the Program Manager (PM) on the Bond Program website under the posting for the project.

Prime bidders/consultants are also encouraged to use the database of Small, Local and Emerging businesses maintained by the PM on the Bond Program website located at www.propositiona.org. In addition, prime bidders/consultants may use databases of other California or local agencies such as Caltrans, the Los Angeles County Metropolitan Transit Authority (LACTMA), the Metropolitan Water District (MWD), the Los Angeles Unified School District (LAUSD) or any other source.

Bidders/proposers should document their outreach efforts through a log or other clear record showing the names of small businesses invited to participate in the bid and the results of the outreach effort.

For bids and proposals to be considered responsive, they must demonstrate participation of at least 28% of the work by small businesses. If it is not possible to achieve the 28% small business participation goal, then prime bidders/proposers must demonstrate a good faith effort to achieve the maximum possible participation level by small businesses.

For purposes of calculating the percentage of participation by a small business, the entire amount of the contract or subcontract held by the small business is counted provided that the small business performs a minimum of 28% of the work.

If the reason for rejecting the bid of a small business was price, the prime consultant/contractor is to furnish the price bid by the rejected local business and the price bid by the selected sub or supplier. Since utilization of available small businesses is expected, only significant price differences are considered as cause for rejecting such businesses.

After opening of construction bids or review of professional proposals, the apparent low bidder or the apparent successful consultant are requested to provide logs or documentation to the PM for review of the bidders/consultants small business outreach effort. Such documentation is to include a copy of the form certifying the small business status of the business issued by any agency which, in the course of its business makes such certifications. Certifications from the federal Small Business Administration, Caltrans or other agency of the State of California, or the County or City of Los Angeles or any agency thereof are accepted for purposes of the Bond Program.

Bidders/proposers requested to submit documentation of their small business outreach efforts are to submit their documents not later than 4:00 P.M. on the third business day following a request to do so by the PM. Such information is submitted to the PM at its offices located at 515 South Flower St., Suite 900, Los Angeles, California 90017. Failure to provide such documentation is grounds to render the bid non-responsive.

In the event that the District is considering awarding away from the lowest bidder or not awarding a contract to a bidder because the bidder is determined to be non-responsive for failure to make a good faith outreach effort to small businesses, then the District, if requested, affords the bidder the opportunity to meet with District representatives to present evidence to the District of the bidder's good faith efforts in making its outreach. In no case does the District award away pursuant to the outreach policy of the BOT if the bidder makes a good faith effort, but fails to meet the expected levels of participation.

Program Manager's Local, Small, and Emerging Business Database

The PM maintains on the public website for the Bond Program (www. propositiona.org) an easily accessible and sortable database of consultants, contractors and suppliers of goods and materials which fall within the Local, Small and Emerging categories as defined below. Other categories such as disabled veteran-owned businesses are also included.

The database is designed in such a manner that users with older computers are still able to make use of it. To this end the information contained thereon is to be limited in the interest of minimizing the computer capacity necessary to use the database. At minimum, then name, phone number and email address of listed firms is included.

- For purposes of the Bond Program, the below terms have the following meanings:
- "Local" businesses are those with a principal office located in Los Angeles County.
- "Small" businesses are those which fall within the limits to gross revenue established by the federal Small Business Administration for the type activity conducted by the business.
- "Emerging" businesses are those in operation for less than five years.

Bond Program Advertising Policies

Construction projects are advertised in such a manner as to maximize the awareness of Local, Small and Emerging contractors of the project. Particular effort has been made to inform contractors whose businesses are located within a zip code considered to be proximate to the college at which the construction occurs.

Formal legal advertising for Bond Program construction projects are to continue to be in the *Daily Journal*, as has been the practice of the District. In addition, construction projects are to be listed in the *Dodge Reports*.

It is the responsibility of the College Project Managers to request that the PM place such advertisements, and give the PM correct information for the advertisement. It is the responsibility of the PM to place these advertisements of construction projects.

In addition to these advertisements, the College Administration may elect additional advertisements on such newspapers or other publications as are proximate to their campus. The PM has conducted a study of publications proximate to each District college. This list is accessible to interested parties at the website.

Construction Bond Assistance Program

The BOT of the District are aware that small and emerging construction companies have difficulty in obtaining performance bonds, and that this fact significantly limits their ability to submit bids. As a result the BOT has directed staff and the PM to develop a bond assistance program for small and emerging firms.

At the present time District staff and the PM are negotiating with an insurance firm which has successfully operated a bond guarantee program. When such a program is instituted for Proposition A and AA work, College Projects Managers are advised thereof by way of a Program Bulletin, and this section of the PMP is then updated.

Mentoring Program

The BOT of the District are aware that small and emerging businesses often lack the business skills and experience necessary to successfully compete for work. As a result the BOT have directed staff and the PM to develop a mentoring program to assist very small and emerging firms.

Additional Provisions and Submittals

Responding firms are advised that any subsequent contract executed as a result of the solicitation shall be subject to all applicable statutory, regulatory and policy requirements of LACCD. The successful responding firm will be expected to complete and submit upon request additional documentation in compliance with LACCD policy and regulations; this shall include but not be limited to Drug Free Certification and Payee Data record (Form 204).

B.4 SCOPE OF SERVICES, WORK, AND DELIVERABLES

Introduction

This section of the RFQ identifies the scope of services, work, and deliverables. The objective of this procurement is to establish a list of the best-qualified firms that have the size, resources, financial ability, expertise, and necessary experience to provide the services required for this program. Firms qualifying for the list will be eligible to participate in a three-step process, beginning with a Preliminary Audit (PA), an Investment Grade Audit (IGA), and Project Delivery (PD).

Service Provider

The term "Service Provider" herein refers to the successful short-listed firm qualified via this RFQ process selected by a campus for developing a specific project at a given campus. Service Providers shall provide cost effective and reliable energy related services to various campuses through-out the multi-campus LACCD system.

Performance of Services and Work

The following sections present specific requirements that shall be expected of the Service Provider for projects at all campuses. *Firms not willing to accept these terms shall be automatically disqualified. Failing to honor these requirements at some future date, after the short list is established, may be grounds for removal from the list and cancellation of any agreements in force.*

Preliminary Audit (PA)

A campus must select no more than three, and no less than two Service Providers to perform a preliminary review of the potential for implement-ing an energy project at one or more buildings on a specified campus. Such preliminary investigations and the resulting report are termed as a Preliminary Audit (PA).

1. A project may include one or more energy conservation measures at one or more buildings on the campus. Projects may be in a single technology area or can encompass multiple technology areas in several buildings.

2. The campus may indicate in its request for PA at such a time the cost effectiveness criteria or other conditions under which it would pro-ceed with the project.

3. Upon completion of the PA, the written PA report submitted by the Service Providers shall be reviewed by the campus and a third party Independent Peer Reviewer to assess the feasibility of the proposed plan including but not limited to accuracy of calculations, methods of development, proposed schematic design, materials and products recommended.

4. A campus panel shall review all PA reports and select at least one Service Provider to provide an Investment Grade Audit for the project. The firms that were not selected to continue will not be compensated for their services

5. If the campus determines the Service Provider's PA is unlikely to meet the investment criteria established by the LACCD; the campus is under no obligation to pursue the project.

6. Whether or not a project moves forward to the IGA phase, all stud-ies, data, results, analyses and reports become the property of the cam-pus and Service Provider shall cooperate and provide all audit related information to the campus at the completion of the PA.

The PA shall identify proposed energy saving measures that represent sound energy engineering practices and new energy efficient equipment and/or modification(s) to existing equipment and/or systems to achieve the optimum utilization of energy and overall lowest life-cycle cost. The PA shall include but not be limited to the following:

1. Analysis of utility usage data and billing for the campus site including the individual facilities and/or projects proposed in the audit.
2. Site visits and interviews with maintenance engineering and development personnel to determine the operational characteristics of facilities included in the audit.
3. Establishing energy conservation measures and the calculations to determine the avoided operational cost for implementing measures recommended. LACCD recognizes and acknowledges these cost estimates are based on common practices and random samples of existing conditions and are to be consistent with ASHRAE (Association of Heating, Refrigeration and Air-Conditioning Engineers) and California title 24 energy calculation methodologies and may not necessarily reflect the actual cost savings that would otherwise be calculated in an IGA.

PA Report Requirements

The written PA report shall include the necessary background information and site observations and related energy consumption information specific to the audit. In addition, a general discussion of the findings and proposed recommendation for project implementation, including "turnkey" cost estimates including a list of assumptions used in the analysis. The report will include a simple detailed cash flow by measure for 20 years cash in accordance with U.S. Treasury 30 year bond rate as published in the *Wall Street Journal*. Appendices shall include manufacturer's equipment performance data and or recognized agency performance test information and/or certification(s) as supporting documentation. Additional information specific to the proposed recommendation(s) may be included at the Service Provider's discretion. The report shall include at minimum the following general sections.

Executive Summary

A written overview of the proposed project including existing operational background assumptions based on interviews with campus personnel, energy usage data and cost information obtained as part of the audit process, followed with the proposed energy measures and an estimation of the resultant avoided cost in energy operations expenses on an annualized basis.

Facility Audit Process

Describe how the PA was performed including the individuals who performed the audit, time frame, and evaluation of results and presentation of findings. Describe the methods and means by which you interviewed key campus personnel and acquired pertinent data with minimal impact to campus and individual schedules.

Utility Tariff Analysis

Examine utility bills for the past 36 months and establish base year consumption for electricity, gas, steam, water, etc. in terms of energy units (kWh, kW, ccf, therms, gallons, or other units used in bills) and in terms of dollars. Describe the process used to determine the base year (averaging, selecting most representative contiguous 12 months, etc.). Consult with facility personnel to account for any anomalous billings that could skew the base year representation.

Reconcile annual end-use estimated consumption with the annual base year consumption to within 5% for electricity (kWh), fuels, and water. Also reconcile Electric Peak demand (kW) for each end use within 5%. The miscellaneous category can be no greater than 5%. This reconciliation will place reasonable limits on potential savings

Cost Benefit Analysis

The basis for a project cost must be presented in sufficient detail in a PA report so as to enable the campus, LACCD Office of Facilities, Planning and Development and its consultants to make an independent assessment of the reasonableness of the Service Provider's cost. Cost details must show associated item quantities, equipment costs, installation costs, engineering costs, construction management costs, commissioning costs, contingencies, Service Provider's project development costs directly applicable to the project, overhead and profit assumptions, campus dictated due-diligence costs resulting from campus review, and other costs as applicable to the specific project being developed. Equipment and installation costs must also be broken down by major subsystems where applicable.

Savings Calculations

The basis for project savings must be provided in sufficient detail to enable the campus, LACCD Office of Facilities, Planning and Development and its consultants to make an independent evaluation on the reasonableness of the savings projections. *Inclusion of maintenance and labor savings in cash flow estimates is strictly prohibited, and should not be done unless explicitly specified by the campus in the project specific solicitation.* However, these may be presented as an information item for the campus' consideration. Specifically, the savings estimates must state the following:

a. Savings in natural gas or other fossil fuel resources.

b. Savings in electricity energy (kWh) and demand (kW) for the measure.

c. Savings in water usage.

d. Savings as a percentage of historical (most recent year) use.

e. If demand savings is included, a narration of why it is valid to include demand savings in the estimate.

f. Current utility rates and time of use rates used in estimates and their conformity with actual rate schedules applicable to the building(s) at the time the proposal is developed.

g. The method by which savings were estimated.

h. How the existing energy use assumptions were estimated.

i. Calculations and equipment data sheets to substantiate Service Provider's estimates.

j. Future Savings Projections.

Grant/Rebate Incentive Applications

Describe your experience with administrating incentive programs. Provide examples of how you have provided these services to other customers. Describe your monitoring and verification process after implementation to insure project goals.

Service Provider's Staff Experience

Provide resumes for each of the individuals who will be assigned to this project. Include name, current duties, specific relevant experience, and role this person will play on this project.

Service Provider's Project Experience

State the number of years your firm has provided services similar in size, scope and complexity. Provide a list of representative projects completed within the past five years. Include a description of the firm including size, organizational structure, and office locations.

Investment Grade Audit (IGA)

The second step in the process is to complete an Investment Grade Audit (IGA). Generally an IGA is requested after a PA suggests that there is likely to be strong potential for pursuing an energy efficiency project.

1. Following a PA a campus may seek a project specific proposal for design and construction to develop the energy project identified. The IGA phase confirms the economic potential of a project and the campus may, upon due diligence, accept the findings. The IGA is the basis for the overall scope and total costs of an energy improvement project and is a required submittal to obtain financing for the project.
2. The campus may indicate in its request for an IGA the cost effectiveness criteria or other conditions for the IGA, which it shall use to determine whether or not to construct the project. In addition, campus may list preferences for material, make, and construction preferences for components envisioned in the proposed project. The Service Provider shall consider these preferences in all cost estimates. A campus may also specify a maximum budget for the project.
3. The IGA requires review by a third party Independent Peer Reviewer to assess and verify the finding. Submittals shall be in strict compliance with LACCD's polices and procedures.
4. Neither the campus nor the LACCD can guarantee that the campus will decide to construct the project or if it does any maximum or minimum time interval between conclusion of an IGA and financing of a project.
5. Whether or not a campus determines to construct a project, all studies, data, results, analyses and reports become the property of the campus and Service Provider shall cooperate and provide all study related information to the campus at the completion of the IGA.

The following definition of terms applies.

Building Envelope

Building Envelope shall mean the entire structure including but not limited to walls, partitions, glazing, insulation, roofing, support structures, mechanical, electrical, plumbing, controls, and any other systems or equipment that directly or indirectly effect the energy consumption of a building or facility.

HVAC Systems

HVAC System shall mean the heating, ventilation, air-conditioning systems including but not limited to air supply and delivery ducts, shafts, airways, controls, dampers, fire-safety devices, valves, coils, regulating devices, air moving equipment, self-contained cooling equipment, heat exchangers, and other equipment directly or indirectly related to controlling space temperature and environmental conditioned space.

Lighting

Lighting shall mean electric illuminating devices and lighting and day-lighting control systems and task lighting, including but not limited to electric power fixtures of any type wattage and voltage or configuration that provide illumination in, around and for the buildings use. Additionally, Lighting shall allow for architectural design and other factors including reflective surfaces and other conditions that affect the measured light levels in compliance with IES and LACCD standards.

Power Generation

Power Generation shall mean electric and/or power equipment including but not limited to cogeneration, combined heat and power, heat recovery, steam generator, fuel cell, photovoltaic and solar system that delivers heat as measured in BTUs and/or electricity as measured in kilowatts to any part of the campuses utility systems.

Following review of the IGA the campus may decide not to pursue the project(s). However, if the campus decides to proceed with the proposed project(s), the campus and the Service Provider shall enter into a Performance Based Energy Savings Agreement.

Should the Service Provider determine any time during the IGA that savings will not meet campus' requirements as listed in its request for the IGA, the Service Provider shall inform the campus in writing and cease all work on the IGA.

IGA Submittal Requirements

The IGA written report will include the necessary background information and site observations and related energy consumption information specific to the scope of audit undertaken. In addition, a general discussion of the findings and proposed recommendation for project implementation, including "turnkey" cost estimates including a list of assumptions used in the analysis. The report will include a simple detailed cash flow by measure for 20 years cash in accordance with U.S. Treasury 30 year bond rate as published in the *Wall Street Journal*. Appendices will include manufacturer's equipment performance data and/or recognized agency performance test information and/or certification(s) as supporting documentation. Any additional information specific to the proposed recommendation(s) maybe included at the company's discretion. The report shall include at minimum the following general sections:

Executive Summary

Include a written description of the proposed project(s) overview including existing operational background assumptions based on interviews with campus personnel, energy usage data and cost information obtained as part of the audit process, followed with the proposed energy measures and an estimation of the resultant avoided cost in energy operations expenses on an annualized basis. At a minimum the following items shall be included in the IGA Executive Summary:

a. Summary table of recommended energy conservation measures, with each energy conservation measures estimated design and construction costs, the first year cost avoidance (in dollars and energy units), and simple payback.
b. Summary of annual energy use and costs of existing or base year condition.
c. Calculation of annual percentage savings expected if all recommended energy conservation measures were implemented.
d. Description of the existing facility, mechanical and electrical systems.
e. Summary description of energy conservation measures, including estimated costs and savings for each as detailed above.
f. Discussion of measures considered but not investigated in detail.
g. Conclusions and recommendations.

Facility Audit Process

Describe how the IGA was performed including the individuals who performed the audit, time frame, and evaluation of results and presentation of findings. Describe the methods and means by which you interviewed key campus personnel and acquired pertinent data with minimal impact to campus and individual schedules.

Utility Tariff Analysis

Examine utility bills for the past 36 months and establish base year consumption for electricity, gas, steam, water, etc. in terms of energy units (kWh, kW, ccf, therms, gallons, or other units used in bills) and in terms of dollars. Describe the process used to determine the base year (averaging, selecting most representative contiguous 12 months, etc.). Consult with facility personnel to account for any anomalous billings that could skew the base year representation.

Cost benefit analysis

Service Provider's future savings projections in the IGA report shall be based on Life-Cycle Cost Analysis assumptions that are customarily used by the State of California, California Energy Commission. This includes inflation rates, discount rates, and fuel and electricity escalation rates. All analysis shall be performed on a nominal cash flow basis, with nominal discount rates for Life-Cycle Cost analysis on a yearly basis. These assumptions will be provided by LACCD. Future savings shall reasonably consider such factors as equipment degradation, expected usage factor, incremental increase in maintenance costs, if applicable, overhaul reserves, etc. as required for a project.

Savings Calculations

The basis for project savings must be provided in sufficient detail to enable the campus, LACCD Office of the Chancellor and its consultants to make an independent evaluation on the reasonableness of the savings projections. *Inclusion of maintenance and labor savings in cash flow estimates is strictly prohibited, and should not be done unless explicitly specified by the campus in the project specific solicitation.* However, these may be presented as an information item for the campus' consideration. Specifically, the savings estimates must state the following:

a. Base year energy use and cost.
b. Post-retrofit energy use and cost.
c. Savings estimates including analysis methodology, supporting calculations and assumptions used.
d. Savings estimates must be limited to savings allowed by the campus as described above.
e. Percent cost-avoidance projected.
f. Description and calculations for any proposed rate changes.
g. Explanation of how savings duplication or interactions between retrofit options is avoided.
h. Operation and maintenance savings, including detailed calculations and description.
i. A computer simulation is required and shall include a short description and statement of key input data. If requested by campus, access shall be provided to the program and all assumptions and inputs used and/or printouts shall be provided of all input files and important output files and included in the Investment Grade Audit with documentation that explains how the final savings figures are derived from the simulation program output printouts.
j. Where manual calculations are employed, formulas, assumptions, and key data shall be stated.

Grant/Rebate Incentive Applications

Identify and list any and all rebate and incentive programs the proposed projects are eligible for including all related documentation necessary to successfully complete the application process. Service Provider is responsible for identifying and initiating the rebate and incentive process and shall provide the campus with a Ghant Chart (Microsoft Project) schedule identifying key milestones and responsible parties and task functions assigned.

Project Performance Measurement Criteria

Identify the measurement points and describe the calculation methods that will be used to determine the avoided energy costs that will occur after installing and/or implementing the energy conservation measures for the project. Provide a Measurement and Verification (M&V) Plan in accordance with the most up to date International Protocol for Measurement and Verification standards including a schedule of metered points, frequency of measurement recording and data acquisition and a sample of the report format and baseline comparison data that will be used to calculate the Schedule and Performance Risk Value. Clearly identify which IP M&V standards have been referenced.

Service Provider's Staff Experience

Provide resumes for each of the individuals who will be assigned to this IGA or to construct the project. Include name, current duties, specific relevant experience, and role this person will play on this IGA or to construct the project.

Service Provider's Project Experience

State the number of years your firm has provided services similar in size, scope, and complexity. Provide a list of representative projects completed within the past five years. Include a description of the firm including size, organizational structure, and office locations.

Project Delivery

The third step in the process is Project Delivery. This phase will begin after an IGA confirms the economic savings potential of the proposed project and the campus, upon due diligence, determines to proceed with construction of the project.

1. "Prior to and as a condition of entering into a Performance Based Energy Savings Agreement with any Service Provider, the Board of Trustees of

LACCD shall, pursuant to Government Code Section 4217.12, hold a regularly scheduled public hearing, at which the Board determines that entering into a Performance Based Energy Savings Agreement with the selected Service Provider(s) is in the best interests of the District, and finds that the anticipated cost to LACCD for the conservation services effected by the Performance Based Energy Savings Agreement will be less than the anticipated marginal cost to LACCD of energy that would have been consumed by LACCD in the absence of the conservation services."

2. In the Project Delivery phase the Service Provider becomes a Contractor and assumes the responsibility of the "Design/Builder."

3. In its request for Project Delivery the campus shall describe the project in terms of maximum cost, design objectives, and minimum acceptable standards of construction.

4. Design/Builder shall be required to use California State licensed architects and engineers to prepare the design and shall adhere to all laws and LACCD requirements.

Required Design Reviews and Approvals

General

a. Professional services undertaken on behalf of LACCD must be in compliance with a range of codes and regulations required by law and/or LACCD Trustee policy. The Design/Builder shall be responsible for adherence to the following requirements and for securing the following approvals.

b. The Design/Builder shall assume responsibility for risks, delay, and/or added costs in securing required reviews and approvals. Cost that may be incurred in revising a design to secure these approvals shall be borne by the entity.

c. The Design/Builder shall be responsible to pay the Division of the State Architect (DSA) plan review fees incurred.

d. The campus shall work with the Design/Builder to coordinate and facilitate required reviews.

e. Campus approval to proceed to construction
   (1) The Design/Builder shall secure a written approval in the form of a Public Board Action (PBA) to Authorize contract prior to the start of construction activities.
   (2) This approval confirms that required code reviews and DSA reviews have been obtained.

Protocol for Design, Specification and Construction

The campus preference in design approach and special amenities shall be indicated, and conditions, which limit the Design/Builder's options, shall be called out. Within these broad guidelines, the Design/Builder shall submit a technical proposal that will produce the best overall project for the intended purpose within the funds available.

Measurement and Verification Plan

Energy savings shall be determined by comparing energy use associated with a facility, or certain systems within a facility, before and after the energy conservation measure installation. The "before" case is called the baseline model. The "after" case is called the post-installation model. Baseline and post-installation models must be constructed using the methods associated with International Performance Measurement and Verification Protocol.

Choice of Equipment

Depending on the nature of the project, a campus, at its discretion, may specify a specific make of equipment or specify a certain quality criteria for the equipment to be installed at the campus. These choices may be required to ensure compatibility with systems already existing in the campus. An example would be a building automation system, where it is desirable to ensure the same make campus-wide for ease of operability and maintenance. Campus shall identify such preferences at the time an IGA is initiated. As part of the PD proposal, Design/Builder shall not substitute makes, unless such substitute is approved in writing by the campus.

Cost of Due Diligence

Campus and LACCD at their discretion may retain the services of independent consultants from time to time to review the PD proposals, studies, design and specifications for reasonableness. These due diligence checks may be made at various times during project development. In the PA and/or IGA written report, the campus shall furnish an estimate of the due diligence costs associated with this process as a percentage of the project cost. The Design/Builder shall include the same in the overall amount to be financed for the project, and shall reimburse the campus for the same if the project is financed and developed.

Schedule and Performance Risk Value

The *Schedule and Performance Risk Value* is a dollar amount valued at 10% of the project construction cost. The measurement for schedule and performance threshold is 90% to 100%. Performance less than 90% of what is proposed in the Design/Builder's PD Proposal is considered failure. Design/Builder has the option to correct failures at its own expense and the Service Provider will be responsible for reimbursing LACCD annually for any continued Performance shortfall.

The Schedule and Performance Risk Value is withheld and paid in arrears during the project closeout upon satisfactory completion and acceptance of the following:

a. Completion of the entire project scope:
   (1) Design and Preconstruction Services.
   (2) Energy Conservation and Capital Improvements.
b. Commissioning.
c. Successful performance test of systems and equipment to establish that the systems and equipment meet or improve on the performance standards set up in the Special Conditions and/or Performance Based Energy Savings Agreement.
d. Application for utility incentives, if any.

Resources Provided by LACCD

Campuses shall coordinate with Service Provider and/or Design/Builder to provide the following resources as needed:

1. Guidance to Service Provider and/or Design/Builder regarding future anticipated needs or changes to facilities or agency mission, which may impact Service Provider's and/or Design/Builder's contract or anticipated results.
2. Access to key facility staff in management and engineering.
3. Reasonable access to facilities relevant to stage of work in progress.
4. Access to relevant utility records.
5. Access to relevant facility plans and blueprints.
6. Access to maintenance records.
7. Access to staff for training.

B.5 QUALIFICATION REQUIREMENTS

Introduction

This section describes the compliance and qualification requirements that all responding firms must demonstrate to be responsive to this RFQ. Respondents *must* furnish verifiable evidence they meet the following requirements. Respondents who do not demonstrate compliance with these requirements shall be rejected as being non-responsive. A respondent seeking to be short-listed in several technology areas must meet the requirements and present the requested qualifications information for each of those technology areas.

A uniform system of prequalification based upon submitted documents is then applied by the LACCD Energy Team in rating the prospective bidder on the size of the project upon which each is qualified to bid. Respondents who fail to meet the prequalification requirement of LACCD will be eliminated from this RFQ solicitation process and deemed ineligible to be a short-listed firm for this program.

Comprehensive Energy Projects Solicited

Experience and qualifications presented under this section shall be technology specific. Respondent shall provide in ten pages or less a general discussion of the demonstrated experience, availability, and qualifications for the energy services listed below:

1. Preliminary Audits to assess extent of energy conservation and other utility cost reduction potential.
2. Investment Grade Audits to assess feasibility of proposed energy conservation and utility cost reduction measures.
3. Design and specifications on building systems and energy infrastructure modifications and upgrades related to all energy related projects.
4. Competitively bidding the installation work in such areas as electrical, mechanical, controls, civil and other building trades.
5. Construction of energy related improvements in existing buildings and campus energy infrastructure in all areas including lighting, HVAC, central plant systems and utilities, building control and automation systems.
6. Construction management and commissioning of energy related projects.
7. Operating, servicing, staffing, maintaining, evaluating ongoing performance, and trouble-shooting of energy related projects over their life-cycle.
8. Providing metering and energy information management and administrative services.
9. Direct experience in developing energy related improvements in educational institutions such as college and university campuses.
10. Liaison with government entities related to all environmental and permitting issues associated with such projects.

Energy Projects Experience

LACCD will review each RFQ response and contact selected references to determine whether respondent has the necessary technical and financial resources to successfully implement comprehensive energy services at various campuses. All firms responding to the RFQ must address the following topics as well as submit references and client list for projects where similar scopes of services were provided.

1. Three years of successful experience in developing energy conservation projects, encompassing work in all the areas identified under the preceding section.
2. Three years of successful experience in developing energy conservation projects in each technology area for which respondent seeks pre-qualification.
3. Successfully financed and constructed a total of $15 million worth of energy conservation projects in the United States during the last three years (2003–2006).
4. Successfully developed at least ten, or a total of $5 million worth of energy conservation projects during the last three years (2003–2006) in United States, for each of the technology areas for which respondent seeks pre-qualification.
5. Is respondent an active NAESCO member?
6. Is respondent an approved performance contractor for the United States Department of Defense and/or the Department of Energy?
7. Can respondent provide energy related services in multiple technology areas as may be required for a given project?

Specific Project Information

Respondent shall list and briefly describe ten recent energy conservation contracts entered into by your firm in the last three years that are relevant to this solicitation. Respondent may present the information in a matrix, or chose to provide a separate one page summary for each of the representative projects. For each project, provide information as follows:

1. Name of the project contact; their position, address, and telephone number; include customer reference name and contact if different from the project contact. LACCD at its option may contact selected references to verify this information.
2. Project, location, start and completion dates.
3. Briefly describe the energy service contract relationship, including function, number of buildings, and size in square feet.
   a. What was the total dollar amount of the contract?
   b. What was the dollar amount for the capital investment?
   c. How was the project financed and was it public tax exempt financing?

4. Was the project completed on original schedule? If not, explain.
5. Was the project completed within budget? If not, explain.
6. What was the actual annual energy savings and demand reduction achieved and actual annual energy cost savings?
7. Identify the specific energy related service that your firm provided (i.e., Section "Comprehensive Energy Projects Solicited") that were associated with this project. Identify the specific technology area applicable to this project.
8. Include letters of recommendation from past and present customers on recent projects.
9. Indicate key professionals and management personnel who were the primary lead on the referenced projects.

Financial and Administrative Stability

1. The financial and administrative ability of respondent shall be primarily determined by the review of the financial statement.
2. *Insurance:* Responding firms shall provide evidence of capacity to comply with the appropriate levels of insurance as required for future solicitation and during the course of any contract awarded. Evidence of such capacity in the form of a letter from the firm's insurance broker is required for the following:
   a. Insurance requirements as specified in Schedule K in the Performance Based Energy Savings Agreement.
3. Responding firms shall adhere to Hold Harmless Provision as specified in item 31 in the Performance Based Energy Savings Agreement.

Management and Personnel Qualification

Respondent must demonstrate their firm has the technical, management and staff capacity to provide appropriate levels of service to LACCD for the work anticipated. Customer references may be contacted and evaluated as part of this section. Demonstration of this capacity shall take the form of information identifying the management team, their qualifications, and their placement in the organization as follows:

1. Illustrate the management team in the form of an organization chart. Include any partnerships formed to respond to this RFQ.
2. Provide evidence each member of the team has at least five years experience providing energy management and energy conservation services. Response shall detail specific experience held by the management team members in the energy services sector.

3. Provide educational background for key personnel you intend to use on the project. No Felony Conviction—A signed affidavit must be submitted that no member of the vendor management team has been convicted of a felony within the last ten years.

4. Respondent shall identify and provide resumes of the experience and qualifications of key personnel who will be assigned to this project including designation of the project manager. Each key personnel (or lead consultant) assigned to various parts of this project must be qualified in the area to which they are assigned to work. Personnel with public agency experience are preferred and will be given preference in the point allocation process.

5. Respondent must demonstrate the corporate and/or personnel resources to be able to assume the entire project.

6. Indicate office locations and the specific offices that will be used to provide services under this RFQ. Indicate the approximate number of professionals at each office location.

7. Indicate geographical preference, if any, related to LACCD campuses where the firm wishes to provide energy conservation related services pursuant to this RFQ.

Sample Project Response

In order to evaluate the approach respondent proposes to use in providing services, LACCD requires all respondents to provide samples of similar work product as follows:

Provide a sample of similar work product for Preliminary Audit (PA), Investment Grade Audit (IGA) and Project Delivery (PD) as outlined in Section B.4, of this RFQ, Scope of Services, Work, and Deliverables. One sample for each is required.

Sample Project Approach

In addition to any IGA information that respondent may provide related to the approach, please address the following questions:

1. What additional information would you consider asking before you even begin to commence work?

2. Under what conditions would you rule out the economic potential of energy efficiency measures related to each technology area?

3. How would you strike a balance between project economics and long-term value of infrastructure improvements desired by the campus?

4. How would you interface work with ongoing studies or plans being developed by the Project Architect/Engineer?

5. How much time may be involved in providing a Preliminary Audit?

Respondent shall provide, in ten pages or less, a detailed example of a project developed, designed and constructed by their firm.

B.6 EVALUATION AND SELECTION CRITERIA

Introduction

Responses to this RFQ shall be reviewed and evaluated by an evaluation committee comprised of LACCD staff and campus representatives. The LACCD, at its sole discretion, may also enlist the services of paid consultants to aid in the evaluation process of this and any follow-on solicitations. All submittals shall be reviewed to verify that respondent has met the requirements set forth in Section B.5, Qualification Requirements. RFQ responses that, in LACCD's opinion, do not meet requirements will be rejected and removed from further evaluation.

Evaluation Criteria

The evaluation committee will assign points according to the point schedule for each evaluation criteria noted below. At the conclusion of this review, respondents scoring a Subtotal of 140 points or better in the categories below will advance to the semi-finals, which may consist of an interview by the evaluation committee. Following the interview, respondents who receive at least 160 points, as a Grand Total will be short-listed for the purposes of this RFQ.

Point Scoring Schedule

Evaluation categories	Maximum scoring possible
a. Comprehensive Energy Projects Solicited	25 Points
b. Energy Projects Experience	25 Points
c. Specific Project Information	25 Points
d. Financial and Administrative Stability	25 Points
e. Management and Personnel Qualification	25 Points
f. Sample Project Response	25 Points
g. Overall Quality of Response to RFQ	25 Points
h. *Subtotal (without interview)*	*175 Points*
i. Interview	25 Points
j. *Grand Total*	*200 Points*

Summary of Evaluation Categories

Comprehensive Energy Projects Solicited—Possible 25 Points

This category will consider the vendor's response to Section B.5, "Comprehensive Energy Projects Solicited" of the RFQ, Experience and qualifications presented under this section are technology specific. Respondent is evaluated on a general discussion of ten pages or less on the demonstrated experience, availability, and qualifications for the energy services listed below:

1. Preliminary energy audits to assess extent of energy conservation and other utility cost reduction potential.
2. Investment Grade and comprehensive energy feasibility studies on the energy conservation and utility cost reduction measures.
3. Design and specifications on building systems and energy infrastructure modifications and upgrades related to all energy related projects.
4. Competitively bidding the installation work in such areas as electrical, mechanical, controls, civil, and other building trades.
5. Construction of energy related improvements in existing buildings and campus energy infrastructure in all areas including lighting, HVAC, central plant systems and utilities, building control, and automation systems.
6. Construction management and commissioning of energy related projects.
7. Operating, servicing, staffing, maintaining, evaluating ongoing performance, and trouble-shooting of energy related projects over their life-cycle.
8. Providing metering and energy information management and administrative services.
9. Direct experience in developing energy related improvements in educational institutions such as college and university campuses.
10. Liaison with government entities related to all environmental and permitting issues associated with such projects.

The scoring range for the above category is 15 points for satisfactory, 20 points for above average and 25 points for excellent.

Energy Projects Experience—Possible 25 Points

This category will consider the respondent's response to Section B.5, "Energy Projects Experience" and "Specific Project Information" of the RFQ, particularly customer references, where similar energy services were provided. Points will be awarded based upon the firm's experience with comparable projects that demonstrate the respondent's experience, in-depth knowledge and background in providing the full spectrum of comprehensive energy services as follows:

1. Does respondent have three years of successful experience in developing energy conservation projects, encompassing work in all the areas identified under Section B.5, "Comprehensive Energy Projects Solicited"?
2. Does respondent have years of successful experience in developing energy conservation projects in each technology area for which respondent seeks pre-qualification?
3. Has respondent successfully financed and constructed a total of $15 million worth of energy conservation projects in the United States during the last three years (2003–2006).
4. Has respondent successfully developed at least ten, or a total of $5 million worth of energy conservation projects during the last three years (2003–2006) in United States, for each of the technology areas for which respondent seeks pre-qualification.
5. Is respondent an active NAESCO member?
6. Is respondent an approved performance contractor for the United States Department of Defense and/or the Department of Energy?
7. Can respondent provide energy related services in multiple technology areas as may be required for a given project?
8. Can respondent provide energy related services in all geographic regions in the State?
9. Can respondent provide energy related services at all campuses or at any geographical location concurrently.

The scoring range for the above category is 15 points for satisfactory, 20 points for above average and 25 points for excellent.

Specific Project Information—Possible 25 Points

This category will consider the ten most recent energy conservation contracts entered into by respondent in the last three years that are relevant to this solicitation. Points will be awarded based upon on respondent's demonstrated background, experience, and qualifications as outlined in Section B.5, "Specific Project Information" of the RFQ as follows:

1. Name of the project contact; their position, address, and telephone number; include customer reference name and contact if different from the project contact.
2. Project, location, start and completion dates.

3. Briefly describe the energy service contract relationship, including function, number of buildings, and size in square feet.

   a. What was the total dollar amount of the contract?

   b. What was the dollar amount for the capital investment?

   c. How was the project financed and was it public tax exempt financing?

4. Was the project completed on original schedule? If not, explain.

5. Was the project completed within budget? If not, explain.

6. What were the actual annual energy savings and demand reduction achieved and actual annual energy cost savings?

7. Identify the specific energy related service that your firm provided (i.e., Section B.5, "Comprehensive Energy Projects Solicited") that was associated with this project. Identify the specific technology area applicable to this project.

8. Include letters of recommendation from past and present customers on recent projects.

9. Indicate key professionals and management personnel who were the primary lead on the referenced projects.

   The scoring range for the above category is 15 points for satisfactory, 20 points for above average and 25 points for excellent.

Financial and Administrative Stability—Possible 25 Points

This category will consider the financial and administrative stability of potential respondents Points will be awarded based upon the respondent's capacity to comply with the criteria in the following areas:

1. Respondent's financial and administrative ability as determined by the review and results of respondent's financial statement

2. *Insurance:* Responding firms shall provide evidence of capacity to comply with the appropriate levels of insurance as required for future solicitation and during the course of any contract awarded as defined in Section B.5, "Financial and Administrative Stability," subsection "Insurtance" of this RFQ.

   The scoring range for the above category is 15 points for satisfactory, 20 points for above average and 25 points for excellent.

Management and Personnel Qualification—Possible 25 Points

This category will consider the professional, technical, and educational qualifications of personnel that will be assigned to the project as described in Section B.5, "Management and Personnel Qualification," of the RFQ. Respondent shall demonstrate that all key personnel have been successfully involved with projects of similar scope and magnitude. The evaluation shall be based upon experience of key personnel providing services described in Section B.5, "Comprehensive Energy Projects Solicited," "Energy Projects Experience," and "Specific Project Information" of the RFQ as follows:

1. Illustrate the management team in the form of an organization chart. Include any partnerships formed to respond to this RFQ.

2. Provide evidence each member of the team has at least five years experience providing energy management and energy conservation services. Response shall detail specific experience held by the management team members in the energy services sector.

3. Provide educational background for key personnel you intend to use on the project. No Felony Conviction—A signed affidavit must be submitted that no member of the vendor management team has been convicted of a felony within the last ten years.

4. Respondent shall identify and provide resumes of the experience and qualifications of key personnel who will be assigned to this project including designation of the project manager. Each key personnel (or lead consultant) assigned to various parts of this project must be qualified in the area to which they are assigned to work. Personnel with public agency experience are preferred and will be given preference in the point allocation process.

5. Respondent must demonstrate the corporate and/or personnel resources to be able to assume the entire project.

6. Indicate office locations and the specific offices that will be used to provide services under this RFQ. Indicate the approximate number of professionals at each office location.

The scoring range for the above category is 15 points for satisfactory, 20 points for above average and 25 points for excellent.

Sample Project Response—Possible 25 Points

In the evaluation of the response, it is presumed that the quality assurance standards employed in the preparation and delivery of the submittal is reflective of the respondent's overall quality assurance standards to be used in performance of actual campus projects. This category will consider responding firm's responses to the sample project in Section B.5, "Sample Project Response" of the RFQ as follows:

1. Respondent shall provide a sample of similar work product for Preliminary Audit, Investment Grade Audit and Project Delivery as

outlined in Section B.4, of this RFQ, Scope of Services, Work, and Deliverables. One sample for each is required.

2. In addition to any Investment Grade Audit information that respondent may provide related to the approach, respondent shall respond to the following questions:
   a. What additional information would you consider asking for before you even begin to commence work?
   b. Under what conditions would you rule out the economic potential of energy efficiency measures related to each technology area?
   c. How would you strike a balance between project economics and long-term value of infrastructure improvements desired by the campus?
   d. How would you interface work with ongoing studies or plans being developed by the Project Architect/Engineer?
   e. How much time may be involved in providing a Preliminary Audit?
3. Respondent shall provide, in ten pages or less, a detailed example of a project developed, designed and constructed by their firm.

The scoring range for the above category is 15 points for satisfactory, 20 points for above average and 25 points for excellent.

Overall Quality of Response to RFQ—Possible 25 Points

This category will consider the overall quality of the RFQ response and demonstrated understanding of the purpose, scope and objective of the project. In the evaluation of the overall quality of response to the RFQ, it is presumed that the quality assurance standards employed in the preparation and delivery of the submittal is reflective of the respondent's overall quality assurance standards to be used in performance of the contract. Respondent shall be judged on the overall quality of the response, completeness and clarity of content.

The scoring range for the above category is 15 points for satisfactory, 20 points for above average and 25 points for excellent.

Interview—Possible 25 Points

If the LACCD elects to conduct interviews, this category will consider the respondent's responses to questions asked during the oral interview, which elaborate on information provided in the RFQ. Emphasis will be placed on the LACCD's assessment of the respondent's experience in providing requested services as well as skills, quality, and depth of answers.

The scoring range for the above category is 15 points for satisfactory, 20 points for above average and 25 points for excellent.

B.7 CONTENT AND FORMAT OF RESPONSE TO RFQ

Introduction

To be considered responsive to this RFQ, respondent shall submit a response in the format identified in this section. All requirements and questions in the RFQ shall be addressed and all requested data should be supplied. LACCD reserves the right to request additional information which in LACCD's opinion is necessary to assure that the respondent's competence, number of qualified employees, business organization, and financial resources are adequate to perform according to contract.

Delivery of RFQ Responses

RFQ Response must be received in the Executive Director of Facility Planning and Development no later than the date and time indicated in Section B.2, Schedule of Events, of this RFQ. The respondent is responsible for the means of delivering the RFQ Response to the appropriate office on time. Delays due to the instrumentalities used to transmit the response including delay occasioned by the internal mailing system in LACCD will be the responsibility of the respondent. Likewise, delays due to inaccurate directions given, even if by LACCD staff shall be the responsibility of the respondent. The RFQ Response must be completed and delivered in sufficient time to avoid disqualification for lateness due to difficulties in delivery. *LATE RFQ RESPONSES WILL NOT BE ACCEPTED.*

RFQ Responses shall be sealed and addressed or delivered to:
RESPONSE TO RFQ 2007 002
LACCD Campus Wide Comprehensive Energy Services
Larry Eisenberg, Executive Director
Facilities, Planning and Development 6th Floor
770 Wilshire Boulevard
Los Angeles, CA 90017

Respondent must submit one original, four copies and two electronic copies.

Preparation

RFQ Response should be prepared in such a way as to provide a straightforward, concise delineation of capabilities to satisfy the requirements of this RFQ. Response should emphasize the respondent's demonstrated capability to perform work of this type. Expensive bindings, colored displays, promotional materials, etc., are not necessary or desired. However, technical literature describing the proposed services and extent of support included in the response should be forwarded as part of the response. Emphasis should be concentrated on completeness and clarity of content.

RFQ Response Format

RFQ Responses shall adhere to this required format for organization and content. The Response must be minimally divided into the individual sections listed below, indexed, and tabbed. Responses may contain additional sections or subsections as necessary to present response content in a concise and logical manner. Additional documentation or collateral material to substantiate the submittal claims may be appended to the response; however, *only content presented in the following sections will be evaluated:*

Section 1. Cover Letter and Responding Firm's Information

The cover letter shall include a brief statement of intent to perform the services and the signature of an authorized officer of the firm who has legal authority in such transactions. Unsigned RFQ Responses shall be rejected. In addition, respondent's cover letter shall expressly state that, should respondent's RFQ Response be accepted, the respondent agrees to enter into an Agreement under the terms and conditions as prescribed by LACCD.

Exhibit A, RFQ Response Certification form, shall be completed and included in Section 1. Also to be included in Section 1 is general information about the responding firm. The response in this section shall use the outline order and titles listed below:

0. Vendor General Information.
1. Name and address of firm.
2. Telephone number, facsimile number and e-mail address.
3. Name and titles of two people authorized to represent firm.
4. Federal Employer Identification Number.
5. Year firm was established.
6. Name and address of parent company (if applicable).
7. Indicate type of firm:

> Partnership
> Corporation (Indicate State incorporated in)
> Sole Proprietor
> Branch Office of:
> Joint Venture (List venture partners)
> Other (Explain)

8. If firm is not California based, discuss the mechanism, which will guarantee the local support services necessary for fulfilling an energy performance contract.

9. Indicate the number of years in business providing similar services. Indicate all other names for firm and the length of time firm had that name.

10. Provide a brief overview of your experience in providing energy performance contracting services directly to customers, particularly educational institutions.

Section 2. Exceptions

Any and all exceptions to the RFQ must be listed on an item-by-item basis and cross-referenced with the RFQ document. If there are no exceptions, respondent must expressly state that no exceptions are taken.

Section 3. Comprehensive Energy Projects Solicited

This section is a response to Section B.5, "Comprehensive Energy Projects Solicited," of the RFQ. Respondent shall provide in ten pages or less a general discussion of the demonstrated experience, availability, and qualifications for the energy services listed below:

1. Preliminary Audits (PA) to assess extent of energy conservation and other utility cost reduction potential.

2. Investment Grade Audits (IGA) to assess feasibility of proposed energy conservation and utility cost reduction measures.

3. Design and specifications on building systems and energy infrastructure modifications and upgrades related to all energy related projects.

4. Competitively bidding the installation work in such areas as electrical, mechanical, controls, civil and other building trades.

5. Construction of energy related improvements in existing buildings and campus energy infrastructure in all areas including lighting, HVAC, central plant systems and utilities, building control and automation systems, cogeneration systems and other renewable energy resources.

6. Construction management and commissioning of energy related projects.

7. Operating, servicing, staffing, maintaining, evaluating ongoing performance, and trouble-shooting of energy related projects over their life-cycle.

8. Providing metering and energy information management and administrative services.

9. Direct experience in developing energy related improvements in educational institutions such as college and university campuses.
10. Liaison with government entities related to all environmental and permitting issues associated with such projects.

Section 4. Energy Projects Experience

This section is a response to Section B.5, "Energy Projects Experience," of the RFQ and request customer references (past and present), as they relate to Section B.5, "Comprehensive Energy Projects Solicited," "Energy Projects Experience," and "Specific Project Information" of the RFQ. LACCD is very interested in respondent's background in providing similar energy services. Response should be organized, labeled and address each of, but not be limited to, the following nine items:

1. Does respondent have three years of successful experience in developing energy conservation projects, encompassing work in all the areas identified under Section B.5, "Comprehensive Energy Projects Solicited"?
2. Does respondent have years of successful experience in developing energy conservation projects in each technology area for which respondent seeks pre-qualification?
3. Has respondent successfully financed and constructed a total of $15 million worth of energy conservation projects in the United States during the last three years (2003–2006).
4. Has respondent successfully developed at least ten, or a total of $5 million worth of energy conservation projects during the last three years (2003–2006) in United States, for each of the technology areas for which respondent seeks pre-qualification.
5. Is respondent an active NAESCO member?
6. Is respondent an approved performance contractor for the United States Department of Defense and/or the Department of Energy?
7. Can respondent provide energy related services in multiple technology areas as may be required for a given project?
8. Can respondent provide energy related services in all geographic regions in the State?
9. Can respondent provide energy related services at all campuses or at any geographical location concurrently?

Section 5. Specific Project Information

This section is a response to Section B.5, "Specific Project Information," of the RFQ and will consider ten of the most recent energy conservation contracts entered into by respondent in the last three years that are relevant to this solicitation. Response should be organized, labeled and address each of, but not be limited to, the following nine items for each project:

1. Provide name of the project contact; their position, address, and telephone number; include customer reference name and contact if different from the project contact.
2. Provide project location, start, and completion dates.
3. Briefly describe the energy service contract relationship, including function, number of buildings, and size in square feet.
   a. What was the total dollar amount of the contract?
   b. What was the dollar amount for the capital investment?
   c. How was the project financed and was it public tax exempt financing?
4. Was the project completed on original schedule? If not, explain.
5. Was the project completed within budget? If not, explain.
6. What were the actual annual energy savings and demand reduction achieved and actual annual energy cost savings?
7. Identify the specific energy related service that your firm provided. Identify the specific technology area applicable to this project.
8. Include letters of recommendation from past and present customers on recent projects.
9. Indicate key professionals and management personnel who were the primary lead on the referenced projects.

Section 6. Financial and Administrative Stability

This section is a response to Section B.5, "Financial and Administrative Stability," of the RFQ and will consider the financial and administrative stability of potential respondents as described in Section B.5, "Comprehensive Energy Projects Solicited" and "Financial and Administrative Stability" of the RFQ. Response should be organized, labeled and address each of, but not be limited to, the following items:

1. Responding firms shall provide evidence of capacity to comply with the appropriate levels of insurance as required for future solicitation and during the course of any contract awarded. Evidence of such capacity in the form of a letter or certificate of insurance from the firm's insurance broker is required for the following:
   a. Insurance requirements as specified in Schedule K of the Performance Based Energy Savings Agreement.
2. A brief statement and the signature of an authorized officer of the firm who has legal authority in such transactions stating that responding firms

shall adhere to hold harmless provisions as specified section 31 of the Performance Based Energy Savings Agreement and LACCD's General Terms and Conditions.

3. A brief statement and the signature of an authorized officer of the firm who has legal authority in such transactions stating that responding firms shall adhere to the a Performance Based Energy Savings Agreement and LACCD's General Terms and Conditions.

Section 7. Management and Personnel Qualifications

This section is a response to Section B.5, "Management and Personnel Qualification," of the RFQ. Firms responding to the RFQ shall provide information regarding the professional, technical, and educational qualifications of personnel that will be assigned to the project. Response should be organized, labeled and address each of, but not be limited to, the following seven items:

1. Illustrate the management team in the form of an organization chart. Include any partnerships formed to respond to this RFQ.

2. Provide evidence each member of the team has at least five years experience providing energy management and energy conservation services. Response shall detail specific experience held by the management team members in the energy services sector.

3. Provide educational background for key personnel you intend to use on the project. No Felony Conviction—A signed affidavit must be submitted that no member of the vendor management team has been convicted of a felony within the last ten years.

4. Respondent shall identify and provide resumes of the experience and qualifications of key personnel who will be assigned to this project including designation of the project manager. Each key personnel (or lead consultant) assigned to various parts of this project must be qualified in the area to which they are assigned to work. Personnel with public agency experience are preferred and will be given preference in the point allocation process.

5. Respondent must demonstrate the corporate and/or personnel resources to be able to assume the entire project.

6. Indicate office locations and the specific offices that will be used to provide services under this RFQ. Indicate the approximate number of professionals at each office location.

7. Indicate geographical preference, if any, related to the LACCD campuses where the firm wishes to provide energy conservation related services pursuant to this RFQ.

Section 8. Sample Project Response

This section shall include responding firm's response to Section B.5 of the RFQ, Sample Project Response. Respondent shall be evaluated on their demonstrated understanding of the purpose, scope, and objective of the project. Response should be organized, labeled, and address each of, but not be limited to, the following areas:

1. Respondent shall provide one sample each of similar work products for a Preliminary Audit, an Investment Grade Audit and Project Delivery as outlined in Section B.4, of this RFQ, Scope of Services, Work, and Deliverables.
2. In addition to any Investment Grade information that respondent may provide related to the approach, respondent shall respond to the following questions:
   a. What additional information would you consider asking before you even begin to commence work?
   b. Under what conditions would you rule out economic potential of energy efficiency measures related to each technology area?
   c. How would you strike a balance between project economics and long-term value of infrastructure improvements desired by the campus?
   d. How would you interface work with ongoing studies or plans being developed by the Project Architect/Engineer?
   e. How much time may be involved in providing a Preliminary Audit?
3. Respondent shall provide, in ten pages or less, a detailed example of a project developed, designed and constructed by their firm.

EXHIBIT A
RFQ RESPONSE CERTIFICATION

The text of the following certification must be included in your RFQ Response:
I certify that I am authorized to represent the company named below and that the answers to the foregoing questions and all statements contained in this RFQ Response are true and correct.
Dated at _____ this _____ day of _____ 2004.
Name of company: _____
By: _____
Title/Position: _____

CLINTON FOUNDATION MEMORANDUM OF UNDERSTANDING
Please identify whether your response to this Request for Qualifications is in agreement with your executed Memorandum of Understanding with the Clinton Foundation.
I certify that I am authorized to represent the company named below and that submittals are in agreement with our executed Memorandum of Understanding with the Clinton Foundation.
Dated at _____ this _____ day of _____ 2004.
Name of company: _____
By: _____
Title/Position: _____

EXHIBIT B
NON-COLLUSION AFFIDAVIT
(Public Contract Code Section 7106)

_____ , being first duly sworn, deposes and says that he or
she is _____ of _____ the
_____ party making the foregoing bid:

- that the bid is not made in the interest of, or on behalf of, any undisclosed person, partnership, company, association, organization, or corporation;
- that the bid is genuine and not collusive or sham;
- that the bidder has not directly or indirectly induced or solicited any other bidder to put in a false or sham bid, and has not directly or indirectly colluded, conspired, connived, or agreed with any bidder or anyone else to put in a sham bid, or that anyone shall refrain from bidding;
- that the bidder has not in any manner, directly or indirectly, sought by agreement, communication, or conference with anyone to fix the bid price of the bidder or any other bidder, or to fix any overhead, profit, or cost element of the bid price, or of that of any other bidder, or to secure any advantage against the public body awarding the contract or anyone interested in the proposed contract;
- that all statements contained in the bid are true; and, further, that the bidder has not, directly or indirectly, submitted his or her bid price of any breakdown thereof, or the contents thereof, or divulged information or data relative thereto, or paid, and will not pay, any fee to any corporation, partnership, company, association, organization, bid depository, or to any member or agent thereof to effectuate a collusive or sham bid.

_____
[NAME OF BIDDER]

_____
[Signature of Bidder (if individual) or its Officer]

_____
[Typed Name of Person Signing]

_____
[Office or Title]

State of California   )
) ss
County of _____)

Subscribed and sworn to (or affirmed) before me this___day of,___ 200__by_____,
personally known to me or proved to me on the basis of satisfactory evidence to be the
person(s) who appeared before me._____

NP

My Commission Expires:_____

[NOTORIAL SEAL]

WARNING! PROPOSALS WILL NOT BE CONSIDERED UNLESS THIS AFFIDAVIT IS COMPLETED AND EXECUTED, INCLUDING THE AFFIDAVIT OF THE NOTARY AND THE NOTORIAL SEAL.

EXHIBIT C
PAYMENT BOND

Project No.

Bond No. _____

KNOW ALL PERSONS BY THESE PRESENTS:

THAT WHEREAS, the <u>Los Angeles Community College District</u> (hereinafter referred to as "District") by Board action on _____, 20___, has awarded Construction Contract Number____(hereinafter referred to as the "Contract") to _____as Principal (hereinafter referred to as "Principal") for the work described as follows:

AND WHEREAS, said Principal is required by the Contract and/or by Division 3, Part IV, Title XV, Chapter 7 (commencing at Section 3247) of the California Civil Code to furnish a payment bond in connection with the Contract;

NOW THEREFORE, we the Principal and _____ ("Surety"), an admitted surety insurer pursuant to Code of Civil Procedure, Section 995.120, are held and firmly bound unto the District in the penal sum of____ Dollars ($____) (this amount being not less than one hundred percent (100%) of the total sum payable by District under the Contract at the time the Contract is awarded by the District to the Principal) lawful money of the United States of America, for the payment of which sum well and truly to be made, we bind ourselves, our heirs, executors, administrators, successors and assigns, jointly and severally, firmly by these presents.

THE CONDITION OF THIS OBLIGATION IS SUCH that if Principal, its heirs, executors, administrators, successors, or assigns approved by District, or its subcontractors, of any contracting tier, shall fail to pay any person or persons named in Civil Code Section 3181, or amounts due under the Unemployment Insurance Code, with respect to work or labor performed under the Contract, or amounts required to be deducted, withheld, and paid over to the State of California Employment Development Department from the wages of employees of Principal and subcontractors pursuant to Section 13020 of the State of California Unemployment Insurance Code with respect to such work and labor, then Surety will pay for the same, in or to an amount not exceeding the penal amount hereinabove set forth, and also will pay in case suit is brought upon this bond, such reasonable attorney's fees as shall be fixed by the court, awarded and taxed as provided in Division 3, Part IV, Title XV, Chapter 7 (commencing at section 3247) of the California Civil Code.

This bond shall inure to the benefit of any of the persons named in section 3181 of the California Civil Code, so as to give a right of action to such persons or their assigns in any suit brought upon this bond.

No change, extension of time, alteration or modification of the Contract, or of the work to be performed thereunder, nor any rescission or attempted rescission of the Contract or this bond, nor any conditions precedent or subsequent in the bond or Contract attempting to limit the right of recovery of any claimant otherwise entitled to recover under the Contract or this bond, shall in any way affect Surety's obligations on this bond; and Surety does hereby waive notice of any change, extension of time, alteration or modification of the Contract or of work to be performed thereunder.

This bond shall be construed most strongly against the Surety and in favor of all persons for whose benefit such bond is given, and under no circumstances shall Surety be released from liability to those for whose benefit this bond has been given, by reason of any breach of the Contract by the District or Principal, but the sole conditions of recovery shall be that claimant is a person described in Section 3181 of the California Civil Code, and has not been paid the full amount of his or its claim.

Surety's obligations hereunder are independent of the obligations of any other surety for the performance of the Contract, and suit may be brought against Surety and such other sureties, joint and severally, or against any one or more of them or against less than all of them, without impairing the District rights against the others.

Correspondence or claims relating to this bond shall be sent to Surety at the address set forth below.

IN WITNESS WHEREOF, two (2) identical counterparts of this instrument, each of which shall for all purposes be deemed an original thereof, have been duly executed by the Principal and Surety named therein, on the _____ day of   , 200 the name and corporate seal of each corporate party being hereto affixed and these presents duly signed by its undersigned representative pursuant to authority of its governing body.

(SEAL AND NOTARIAL
ACKNOWLEDGMENT OF
SURETY)

Principal _____ **(Seal)**
BY _____
(Name and Title)

_____
_____
_____
(Mailing Address of Surety)

_____
Surety
BY _____
(Name and Title)

Note: Notary acknowledgment for Surety and Surety's Power of Attorney must be included or attached.

EXHIBIT E
PERFORMANCE BOND

Project No.: _____
Bond No.:   _____

KNOW ALL PERSONS BY THESE PRESENTS:

THAT WHEREAS, Los Angeles Community College District (hereinafter referred to as "District") by Board action on _____, 20__, has awarded Agreement Number _____ (hereinafter referred to as the "Contract") to____as Principal {hereinafter referred to as "Principal"), which Contract is by this reference made a part hereof, for the work described as follows:

_____

AND WHEREAS, said Principal is required under the terms of said Contract to furnish a bond for the faithful performance of said Contract.

NOW, THEREFORE, we the undersigned Principal and _____, an admitted Surety insurer pursuant to Code of Civil Procedure Section 995.120, as Surety, are held and firmly bound to the District, in the sum of _____ Dollars ($ _____) (this amount being not less than one hundred percent (100%) of the total price of the Contract awarded by the District to the Principal), lawful money of the United States of America, for payment of which sum well and truly to be made, we bind ourselves, our heirs, executors and administrators, successors or assigns, jointly and severally, firmly by these presents.

THE CONDITION OF THIS OBLIGATION IS SUCH THAT if Principal, its heirs, executors, administrators, successors or assigns approved by District, shall in all things stand to and abide by and well and truly keep and perform all the undertakings, terms, covenants, conditions and agreements in the Contract during the original term and any extensions thereof as may be granted by the District, with or without notice to Surety thereof, and during the period of any warranties and guarantees required under the Contract (including, but not limited to, the provisions of the Contract regarding liquidated damages payable to District), all within the time and in the manner therein designated in all respects according to their true intent and meaning, and shall indemnify, defend, protect, and hold harmless the District as stipulated in the Contract; then this obligation shall become null and void; otherwise, it shall be and remain in full force and effect.

Whenever Principal shall be, and is declared by the District to be, in default under the Contract, the Surety shall promptly either remedy the default, or shall promptly:

1.   Complete the Contract through its agents or independent contractors, subject to acceptance of such agents or independent contractors by District as hereinafter set forth, in accordance with its terms and conditions and to pay and perform all obligations of Principal under the Contract, including, without limitation, all obligations with respect to warranties, guarantees and the payment of liquidated damages; or, at Surety's election, or if required by the District,

2.   Obtain a bid or bids for completing the Contract in accordance with its terms and conditions, and upon determination by District of the lowest responsible Bidder, arrange for a contract between such Bidder and the District and make available as Work progresses (even though there should be a default or succession of defaults under the contract or contracts of completion arranged under this paragraph) sufficient funds to pay the cost of completion less the "balance of the Contract price" (as hereinafter defined), and to pay and perform all obligations of Principal under the Contract, including, without limitation, all obligations with respect to warranties, guarantees and the payment of liquidated damages; but, in any event, Surety's total obligations hereunder shall not exceed the amount set forth in the third paragraph hereof. The term "balance of the Contract price," as used in this paragraph, shall mean the total amount payable to Principal by the District under the Contract and any modifications thereto, less the amount previously paid by the District to the Principal.

Surety expressly agrees that the District may reject any agent or contractor which may be proposed by Surety in fulfillment of its obligations in the event of default by the Principal. Unless otherwise agreed by District, in its sole discretion, Surety shall not utilize Principal in completing the Contract nor shall Surety accept a bid from Principal for completion of the work in the event of default by the Principal.

No right of action shall accrue on this bond to or for the use of any person or corporation other than the District or its successors or assigns.

No change, extension of time, alteration or modification of the Contract, or of the work to be performed thereunder, shall in any way affect Surety's obligations on this bond; and Surety does hereby waive notice of any change, extension of time, alteration or modification of the Contract or of work to be performed thereunder.

Surety's obligations hereunder are independent of the obligations of any other surety for the performance of the Contract, and suit may be brought against Surety and such other sureties, joint and severally, or against any one or more of them or against less than all of them, without impairing the District rights against the others.

At the request of District, Surety shall join in and be a party to any arbitration proceedings brought under the Contract and any arbitration award entered against it in such proceedings shall be final and binding upon Surety.

Surety shall pay the reasonable attorney's fees and costs incurred by the District in any suit or arbitration brought or conducted upon this bond.

Correspondence or claims relating to this bond shall be sent to Surety at the address set forth below.

IN WITNESS WHEREOF, two (2) identical counterparts of this instrument, each of which shall for all purposes be deemed an original thereof, have been duly executed by the Principal and Surety named therein, on the ___day of ___, 200_ the name and corporate seal of each corporate party being hereto affixed and these presents duly signed by its undersigned representative pursuant to authority of its governing body.

(SEAL AND NOTARIAL ACKNOWLEDGMENT OF SURETY)

Principal   **(Seal)**

BY _____
(Name and Title)

_____
_____
_____
(Mailing Address of Surety)

_____
Surety

BY _____
(Name and Title)

Note: Notary acknowledgment for Surety and Surety's Power of Attorney must included or attached.

EXHIBIT F
LIST OF CAMPUSES

The Los Angeles Community College District is the largest community college district in the United States and is one of the largest in the world. The LACCD consists of nine colleges and covers an area of more than 882 square miles.

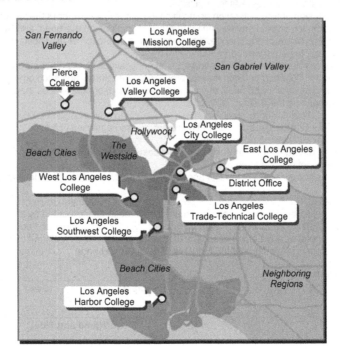

EXHIBIT G
REFERENCE ECM

ECM-1: Convert to Central Air Handling [Specific Objective].
ECM-2: Install Ultra-Violet 'C' Band on all air equipment.
ECM-3: Install CO_2 monitoring in all classrooms.
ECM-4: Install operable windows.
ECM-5: On/Off interlock on damper controls/windows/doors.
ECM-6: Dual-technology occupancy sensors.
ECM-7: Lighting upgrade to T5.
ECM-8: DDC open-architecture - BAS.
ECM-9: Real-Time Energy Monitoring on electrical circuits.
ECM-10: Air Quality Annunciators.
ECM-11: Install kW meters on all
ECM-12: Tower Free Cooling
ECM-13: Water Conservation [Domestic/Potable/Wastewater/Swimming Pool/Tower
         Drift/Blow-down]
ECM-14: Predictive Maintenance
ECM-14 Fume Hood Performance
ECM-15: Air Filtration
ECM-16: VFD Motor Control
ECM-17: Thermal Displacement Ventilation [TDV]
ECM-18: Dual-Fan-Dual-Duct Ventilation [DFDD]
ECM-19: Re-Heat Alternatives
ECM-20: Pulsed Output Water Meters
ECM-21: SeaHawk/Waterbug type Water Leakage Rate Detection
ECM-22: Install high efficiency transformers
ECM-23: 2-Phase Shift transformers (where applicable. High Computer density areas).
ECM-24: Upgrade facade (better glass and insulation)
ECM-25: Upgrade from T-12 to T-8 (T-5 may not be feasible for some locations)
ECM-26: Replace pneumatic controls with DDC controls
ECM-27: Install dual minimums on VAV terminal boxes
ECM-28: Supply temperature reset
ECM-29: Duct pressure reset
ECM-30: Convert constant flow hydronic systems to variable flow
ECM-31: Eliminate low Delta T from chilled water systems
ECM-32: Space temperature dead-band complying with T-24
ECM-33: Day lighting Integration
ECM-34: Condenser water temperature reset based on outdoor wet bulb
ECM-35: Condensing boilers
ECM-36: High efficiency extended surface filters
ECM-37: LED Lighting
ECM-38: Outdoor Lighting Upgrades – Photocells & other Photo-detection Devices
ECM-39: Operable Windows
ECM-40: Window Glazing Treatments
ECM-41: Energy Efficient Food Service Equipment
ECM-42: Premium Efficient Motors
ECM-43: Premium Efficient Pumps
ECM-44: Solar Powered Trash Compaction [e.g. Big-Belly]
ECM-45: Computer Virtualization

EXHIBIT H
CLINTON CLIMATE INITIATIVE

The following Energy Service Companies are currently signatories to the Clinton Climate Initiative Memorandum of Understanding:

Siemens
Johnson Controls
Honeywell
Trane
Noresco
Ameresco
Schneider Electric/TAC
Chevron Energy Services

APPENDIX C

Los Angeles Community College District Energy Program

Los Angeles Community College District with lead author
Woodrow W. Clark, II, Ph.D

Contents

Section One: Energy Strategy 370
Section Two: Central Plant 372
Section Three: Demand Side Management / Performance Contracting 374
Section Four: Renewable Energy Technologies 377
Section Five: Campus Plans 388
Section Six: Financial Structure 399
Section Seven: Power Purchase Agreement 401
Section Eight: Finance Partners 404
Section Nine: Feed-in Tariff Analysis / RECs 406
Section Ten: Education / Jobs 410

LACCD
Energy Program

TABLE OF CONTENTS

1. Energy Strategy
2. Central Plant
3. Demand Side Management / Performance Contracting
4. Planned Renewable Energy Technologies
    a. Solar
    b. Wind
    c. Solar thermal
    d. Geo thermal
    e. Thermal storage
    f. Electrolyzer and hydrogen fueling
    g. Solid state hydrogen storage
    h. Fuel cells
    i. Battery storage
5. Campus Plans
6. Financial Structure
7. Power Purchase Agreements
8. Finance Partners
9. Feed in Tariff Analysis / RECs
10. Education / Jobs

Section One:
Energy Strategy

LACCD COMPREHENSIVE ENERGY STRATEGIC PLAN

A Paradigm Change

1. Efficient renewable energy central plants
2. Demand management through performance contracts
3. One MW+ renewable energy power generation program (e.g. solar, wind, geothermal, etc.)
4. Sustainable education curriculum and jobs program

Section Two:
Central Plant

373

CENTRAL PLANT INTERIOR

Section Three:
Demand Side Management /
Performance Contracting

PERFORMANCE CONTRACTS

<u>Arithmetic</u>

Annual electric / gas bill before	$1,000,000
Annual electric / gas bill after	- $ 800,000
Annual difference	$200,000

376

PERFORMANCE CONTRACTS
Demand Side Management
Performance Contracting

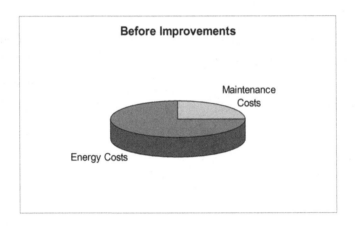

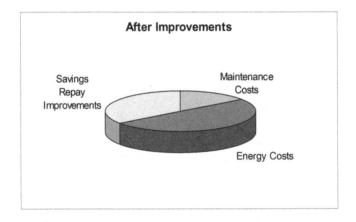

Section Four:
Renewable Energy Technologies

SOLAR

Rigid Panel

East Los Angeles College

Thin Film

Concentrator

URBAN WIND POWER

Architectural Windmills
- Designed for light wind
- Deploy multiple units
- 1 to 6kW output each
- Bird friendly

380

SOLAR THERMAL

GEO THERMAL

- Use the Earth as a heat sink for heat absorption/rejection.
- Maintains a relatively consistent temperature throughout the year compared to the air above it.

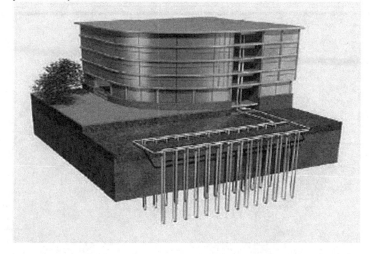

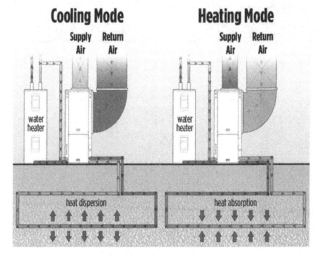

THERMAL STORAGE

ELECTROLYZER

The Hydrogenics Electrolyzer creates hydrogen by passing an electric current through water.

384

SOLID STATE HYDROGEN STORAGE

Ovonics solid state hydrogen storage

FUEL CELL

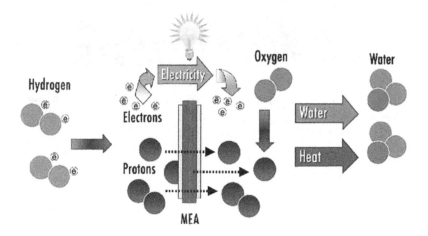

How a fuel cell works: A chemical reaction inside the fuel cell turns hydrogen into electricity and water.

LITHIUM ION BATTERY

iCeL is a unique, proven, scalable, intelligent, high density energy storage system. With many applications, most anything requiring energy can be powered by an iCeL.

FLOW BATTERY

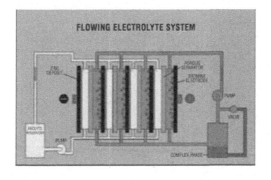

Section Five:
Campus Plans

389

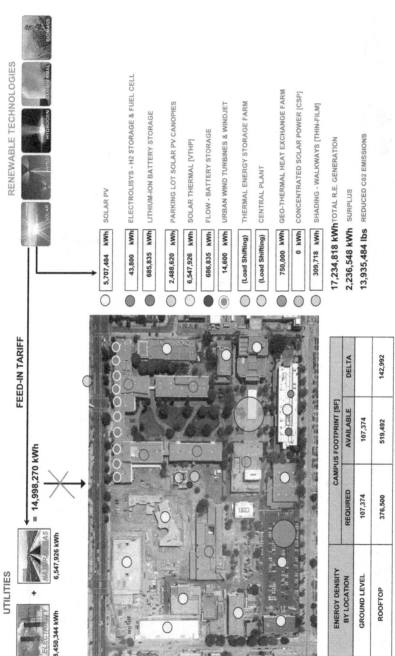

LOS ANGELES CITY COLLEGE

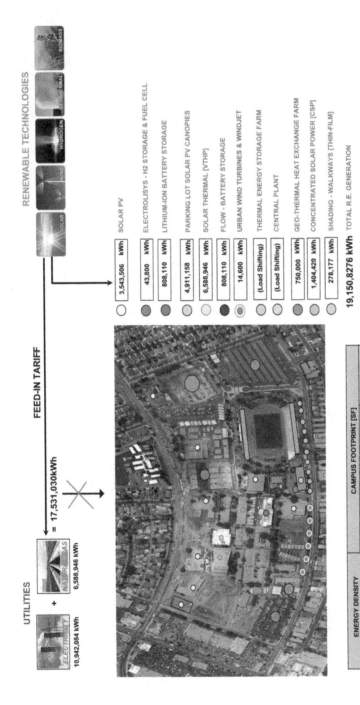

RENEWABLE TECHNOLOGIES

UTILITIES

FEED-IN TARIFF

= 17,531,030kWh

+

ELECTRICITY
10,942,084 kWh

NATURAL GAS
6,588,946 kWh

		kWh	SOLAR PV
	3,543,506	kWh	ELECTROLISYS - H2 STORAGE & FUEL CELL
	43,800	kWh	LITHIUM-ION BATTERY STORAGE
	808,110	kWh	PARKING LOT SOLAR PV CANOPIES
	4,911,158	kWh	SOLAR THERMAL [VTHP]
	6,588,946	kWh	FLOW - BATTERY STORAGE
	808,110	kWh	URBAN WIND TURBINES & WINDJET
	14,600	kWh	THERMAL ENERGY STORAGE FARM
	(Load Shifting)		CENTRAL PLANT
	(Load Shifting)		GEO-THERMAL HEAT EXCHANGE FARM
	750,000	kWh	CONCENTRATED SOLAR POWER [CSP]
	1,404,420	kWh	SHADING - WALKWAYS [THIN-FILM]
	278,177	kWh	

19,150,8276 kWh   TOTAL R.E. GENERATION

1,619,797 kWh   SURPLUS

16,741,578 lbs   REDUCED C02 EMISSIONS

ENERGY DENSITY BY LOCATION	CAMPUS FOOTPRINT [SF]		
	REQUIRED	AVAILABLE	DELTA
GROUND LEVEL	507,875	507,875	0
ROOFTOP	377,588	281,442	96,146

EAST LOS ANGELES COLLEGE

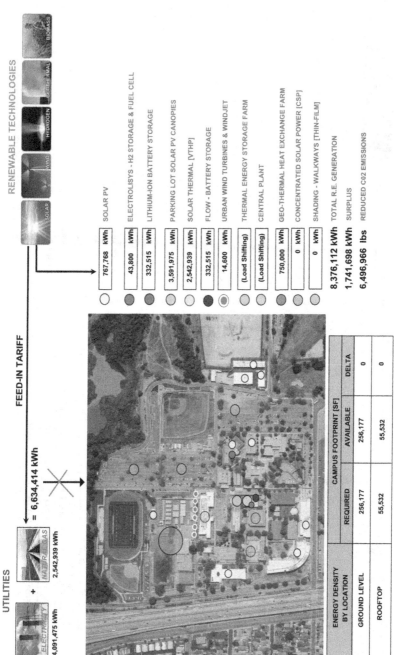

RENEWABLE TECHNOLOGIES

BIOMASS · GEOTHERMAL · HYDROGEN · WIND · SOLAR

UTILITIES

ELECTRICITY 4,091,475 kWh

+

NATURAL GAS 2,542,939 kWh

= 6,634,414 kWh

FEED-IN TARIFF

767,768	kWh	SOLAR PV
43,800	kWh	ELECTROLISYS - H2 STORAGE & FUEL CELL
332,515	kWh	LITHIUM-ION BATTERY STORAGE
3,591,975	kWh	PARKING LOT SOLAR PV CANOPIES
2,542,939	kWh	SOLAR THERMAL [VTHP]
332,515	kWh	FLOW - BATTERY STORAGE
14,600	kWh	URBAN WIND TURBINES & WINDJET
(Load Shifting)		THERMAL ENERGY STORAGE FARM
(Load Shifting)		CENTRAL PLANT
750,000	kWh	GEO-THERMAL HEAT EXCHANGE FARM
0	kWh	CONCENTRATED SOLAR POWER [CSP]
0	kWh	SHADING - WALKWAYS [THIN-FILM]

8,376,112 kWh   TOTAL R.E. GENERATION
1,741,698 kWh   SURPLUS
6,496,966 lbs   REDUCED C02 EMISSIONS

ENERGY DENSITY BY LOCATION	CAMPUS FOOTPRINT [SF]		
	REQUIRED	AVAILABLE	DELTA
GROUND LEVEL	256,177	256,177	0
ROOFTOP	55,532	55,532	0

LOS ANGELES HARBOR COLLEGE

RENEWABLE TECHNOLOGIES

SOLAR | WIND | HYDROGEN | GEOTHERMAL | BIOMASS

6,480,000	kWh	SOLAR PV
730,000	kWh	ELECTROLISYS - H2 STORAGE & FUEL CELL
412,450	kWh	LITHIUM-ION BATTERY STORAGE
2,109,000	kWh	PARKING LOT SOLAR PV CANOPIES
2,275,796	kWh	SOLAR THERMAL [VTHP]
412,450	kWh	FLOW - BATTERY STORAGE
54,750	kWh	URBAN WIND TURBINES & WINDJET
N/A		THERMAL ENERGY STORAGE FARM
N/A		CENTRAL PLANT
750,000	kWh	GEO-THERMAL HEAT EXCHANGE FARM
1,404,420	kWh	CONCENTRATED SOLAR POWER [CSP]
0	kWh	SHADING - WALKWAYS [THIN-FILM]

14,628,866 kWh TOTAL R.E. GENERATION
7,564,136 kWh SURPLUS
7,324,983 lbs REDUCED C02 EMISSIONS

UTILITIES

ELECTRICITY 4,788,934 kWh
+ NATURAL GAS 2,275,796 kWh
= 7,064,730 kWh

FEED-IN TARIFF

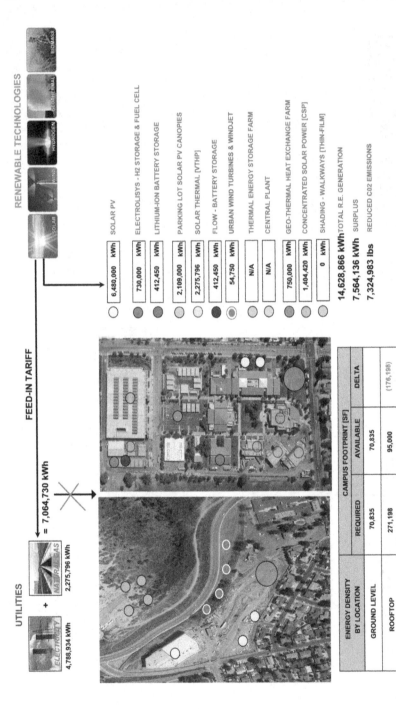

ENERGY DENSITY BY LOCATION	CAMPUS FOOTPRINT [SF]		
	REQUIRED	AVAILABLE	DELTA
GROUND LEVEL	70,835	70,835	
ROOFTOP	271,198	95,000	(176,198)

LOS ANGELES MISSION COLLEGE

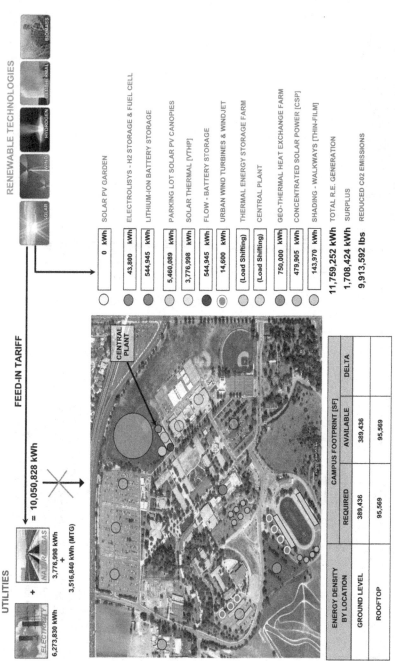

RENEWABLE TECHNOLOGIES

BIOMASS

GEOTHERMAL

HYDROGEN

WIND

SOLAR

0	kWh	SOLAR PV GARDEN
43,800	kWh	ELECTROLISYS - H2 STORAGE & FUEL CELL
544,945	kWh	LITHIUM-ION BATTERY STORAGE
5,460,089	kWh	PARKING LOT SOLAR PV CANOPIES
3,776,998	kWh	SOLAR THERMAL [VTHP]
544,945	kWh	FLOW - BATTERY STORAGE
14,600	kWh	URBAN WIND TURBINES & WINDJET
(Load Shifting)		THERMAL ENERGY STORAGE FARM
(Load Shifting)		CENTRAL PLANT
750,000	kWh	GEO-THERMAL HEAT EXCHANGE FARM
479,905	kWh	CONCENTRATED SOLAR POWER [CSP]
143,970	kWh	SHADING - WALKWAYS [THIN-FILM]

11,759,252 kWh TOTAL R.E. GENERATION

1,708,424 kWh SURPLUS

9,913,592 lbs REDUCED C02 EMISSIONS

FEED-IN TARIFF

UTILITIES

ELECTRICITY
6,273,830 kWh

+

NATURAL GAS
3,776,998 kWh
+
3,516,840 kWh (MTG)

= 10,050,828 kWh

CENTRAL PLANT

ENERGY DENSITY BY LOCATION	CAMPUS FOOTPRINT [SF]		
	REQUIRED	AVAILABLE	DELTA
GROUND LEVEL	389,436	389,436	
ROOFTOP	95,569	95,569	

LOS ANGELES PIERCE COLLEGE

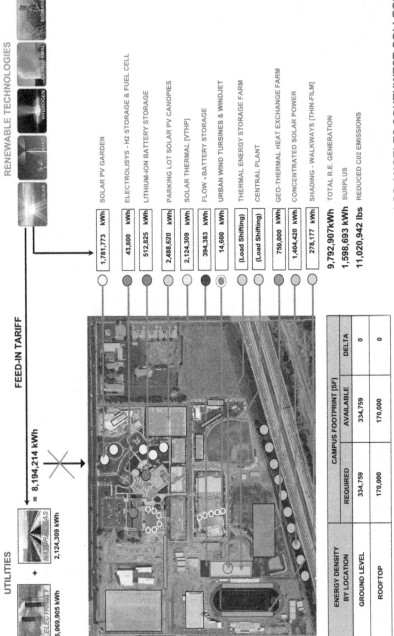

RENEWABLE TECHNOLOGIES

SOLAR · WIND · HYDROGEN · GEOTHERMAL · BIOMASS

UTILITIES

ELECTRICITY 6,069,905 kWh + NATURAL GAS 2,124,309 kWh = 8,194,214 kWh

FEED-IN TARIFF

1,781,773 kWh	SOLAR PV GARDEN	
43,800 kWh	ELECTROLISYS - H2 STORAGE & FUEL CELL	
512,825 kWh	LITHIUM-ION BATTERY STORAGE	
2,488,620 kWh	PARKING LOT SOLAR PV CANOPIES	
2,124,309 kWh	SOLAR THERMAL [VTHP]	
394,383 kWh	FLOW - BATTERY STORAGE	
14,600 kWh	URBAN WIND TURBINES & WINDJET	
(Load Shifting)	THERMAL ENERGY STORAGE FARM	
(Load Shifting)	CENTRAL PLANT	
750,000 kWh	GEO-THERMAL HEAT EXCHANGE FARM	
1,404,420 kWh	CONCENTRATED SOLAR POWER	
278,177 kWh	SHADING - WALKWAYS [THIN-FILM]	

9,792,907kWh TOTAL R.E. GENERATION
1,598,693 kWh SURPLUS
11,020,942 lbs REDUCED CO2 EMISSIONS

ENERGY DENSITY BY LOCATION	CAMPUS FOOTPRINT [SF]		
	REQUIRED	AVAILABLE	DELTA
GROUND LEVEL	334,759	334,759	0
ROOFTOP	170,000	170,000	0

LOS ANGELES SOUTHWEST COLLEGE

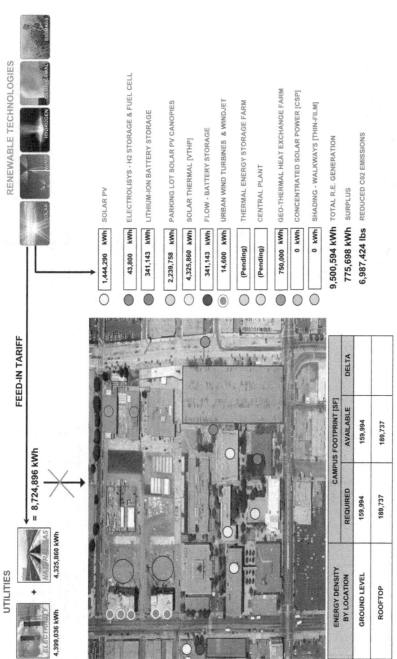

RENEWABLE TECHNOLOGIES

SOLAR	WIND	HYDROGEN	GEOTHERMAL	BIOMASS	

1,444,290	kWh	SOLAR PV
43,800	kWh	ELECTROLISYS - H2 STORAGE & FUEL CELL
341,143	kWh	LITHIUM-ION BATTERY STORAGE
2,239,758	kWh	PARKING LOT SOLAR PV CANOPIES
4,325,860	kWh	SOLAR THERMAL [VTHP]
341,143	kWh	FLOW - BATTERY STORAGE
14,600	kWh	URBAN WIND TURBINES & WINDJET
(Pending)		THERMAL ENERGY STORAGE FARM
(Pending)		CENTRAL PLANT
750,000	kWh	GEO-THERMAL HEAT EXCHANGE FARM
0	kWh	CONCENTRATED SOLAR POWER [CSP]
0	kWh	SHADING - WALKWAYS [THIN-FILM]
9,500,594 kWh		TOTAL R.E. GENERATION
775,698 kWh		SURPLUS
6,987,424 lbs		REDUCED C02 EMISSIONS

FEED-IN TARIFF

= 8,724,896 kWh

UTILITIES

ELECTRICITY
4,399,036 kWh

+

NATURAL GAS
4,325,860 kWh

ENERGY DENSITY BY LOCATION	CAMPUS FOOTPRINT [SF]		
	REQUIRED	AVAILABLE	DELTA
GROUND LEVEL	159,994	159,994	
ROOFTOP	189,737	189,737	

LOS ANGELES TRADE TECHNICALCOLLEGE

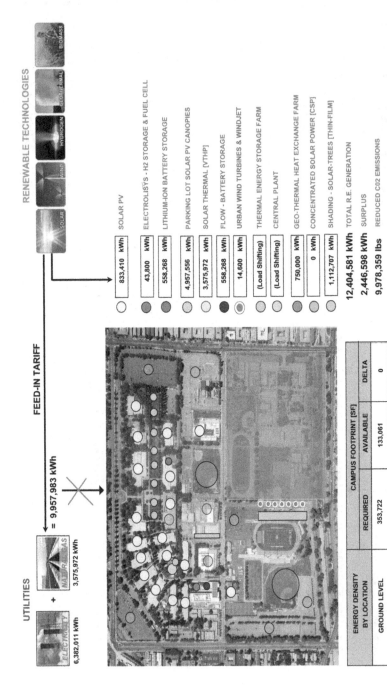

RENEWABLE TECHNOLOGIES

833,410	kWh	SOLAR PV
43,800	kWh	ELECTROLISYS - H2 STORAGE & FUEL CELL
558,268	kWh	LITHIUM-ION BATTERY STORAGE
4,957,556	kWh	PARKING LOT SOLAR PV CANOPIES
3,575,972	kWh	SOLAR THERMAL [VTHP]
558,268	kWh	FLOW - BATTERY STORAGE
14,600	kWh	URBAN WIND TURBINES & WINDJET
(Load Shifting)		THERMAL ENERGY STORAGE FARM
(Load Shifting)		CENTRAL PLANT
750,000	kWh	GEO-THERMAL HEAT EXCHANGE FARM
0	kWh	CONCENTRATED SOLAR POWER [CSP]
1,112,707	kWh	SHADING - SOLAR-TREES [THIN-FILM]
12,404,581 kWh		TOTAL R.E. GENERATION
2,446,598 kWh		SURPLUS
9,978,359 lbs		REDUCED C02 EMISSIONS

UTILITIES

ELECTRICITY
6,382,011 kWh

+

NATURAL GAS
3,575,972 kWh

= 9,957,983 kWh

FEED-IN TARIFF

ENERGY DENSITY BY LOCATION	CAMPUS FOOTPRINT [SF]		
	REQUIRED	AVAILABLE	DELTA
GROUND LEVEL	353,722	133,061	0
ROOFTOP	353,722	133,061	0

LOS ANGELES VALLEY COLLEGE

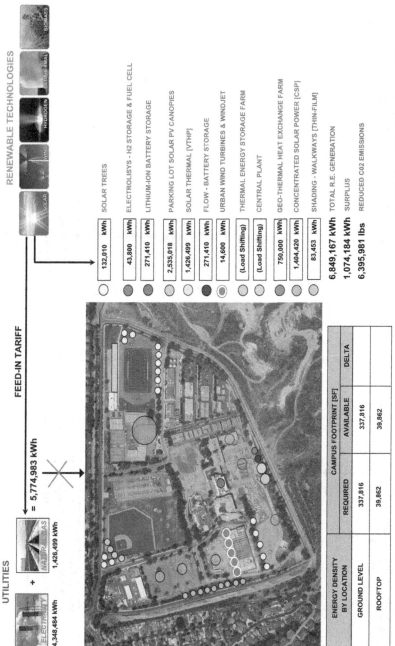

RENEWABLE TECHNOLOGIES

BIOMASS | GEOTHERMAL | HYDROGEN | WIND | SOLAR

UTILITIES

ELECTRICITY
4,348,484 kWh

+

NATURAL GAS
1,426,499 kWh

FEED-IN TARIFF

= 5,774,983 kWh

132,010	kWh	SOLAR TREES
43,800	kWh	ELECTROLISYS - H2 STORAGE & FUEL CELL
271,410	kWh	LITHIUM-ION BATTERY STORAGE
2,535,018	kWh	PARKING LOT SOLAR PV CANOPIES
1,426,499	kWh	SOLAR THERMAL [VTHP]
271,410	kWh	FLOW - BATTERY STORAGE
14,600	kWh	URBAN WIND TURBINES & WINDJET
(Load Shifting)		THERMAL ENERGY STORAGE FARM
(Load Shifting)		CENTRAL PLANT
750,000	kWh	GEO-THERMAL HEAT EXCHANGE FARM
1,404,420	kWh	CONCENTRATED SOLAR POWER [CSP]
83,453	kWh	SHADING - WALKWAYS [THIN-FILM]

6,849,167 kWh TOTAL R.E. GENERATION
1,074,184 kWh SURPLUS
6,395,981 lbs REDUCED C02 EMISSIONS

ENERGY DENSITY BY LOCATION	CAMPUS FOOTPRINT [SF]		
	REQUIRED	AVAILABLE	DELTA
GROUND LEVEL	337,816	337,816	
ROOFTOP	39,862	39,862	

WEST LOS ANGELES COLLEGE

RENEWABLE TECHNOLOGIES

1,359,732	kWh	SOLAR PV
43,800	kWh	ELECTROLISYS - H2 STORAGE & FUEL CELL
127,233	kWh	LITHIUM-ION BATTERY STORAGE
140,268	kWh	PARKING LOT SOLAR PV CANOPIES
737,132	kWh	SOLAR THERMAL [VTHP]
127,233	kWh	FLOW - BATTERY STORAGE
14,600	kWh	URBAN WIND TURBINES & WINDJET
(Load Shifting)		THERMAL ENERGY STORAGE FARM
(Load Shifting)		CENTRAL PLANT
750,000	kWh	GEO-THERMAL HEAT EXCHANGE FARM
0	kWh	CONCENTRATED SOLAR POWER [CSP]
0	kWh	SHADING - WALKWAYS [THIN-FILM]

3,299,998kWh		TOTAL R.E. GENERATION
1,062,866 kWh		SURPLUS
2,304,042 lbs		REDUCED C02 EMISSIONS

FEED-IN TARIFF

UTILITIES

= 2,237,132 kWh

ELECTRICITY
1,500,000 kWh

+

NATURAL GAS
737,132 kWh

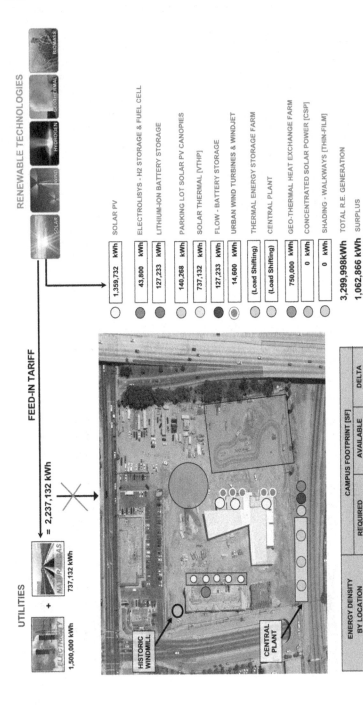

HISTORIC
WINDMILL

CENTRAL
PLANT

ENERGY DENSITY BY LOCATION	CAMPUS FOOTPRINT [SF]		
	REQUIRED	AVAILABLE	DELTA
GROUND LEVEL	45,272	45,272	
ROOFTOP	63,975	46,000	(17,975)

LOS ANGELES NORTHEAST CAMPUS

Section Six:
Financial Structure

Alternate Energy 3rd Party Arithmetic

- Federal Energy Credit – 30%
- Accelerated Depreciation – 22%
- Bonus Depreciation – 5%
- Utility Incentives – 10%
- RECs – 5%
- Bulk Procurement – 10% (?)
- 18 Cents on the Dollar !!!

Section Seven:
Power Purchase Agreement

Power Purchase Agreement

	Annual PV Production (kWh)	Pre PPA Rate	PPA Rate	Potential Utility Dollar Savings under PPA
City College	7,197,952	$0.21	$0.15	$431,877
East LA	9,859,828	$0.21	$0.15	$591,590
Harbor College	3,592,365	$0.21	$0.15	$215,542
Mission College	4,342,258	$0.21	$0.15	$260,535
Pierce College	5,460,089	$0.21	$0.15	$327,605
Southwest College	5,674,813	$0.21	$0.15	$340,489
Trade Tech	3,684,048	$0.21	$0.15	$221,043
Valley College	5,790,966	$0.21	$0.15	$347,458
West LA	4,071,449	$0.21	$0.15	$244,287
Total				**$2,980,426**

* Per Year

Power Purchase Agreement

	Annual PV Production (kWh)	Pre PPA Rate	PPA Rate	Potential Utility Dollar Savings under PPA
City College	7,197,952	$0.21	$0.13	$575,836
East LA	9,859,828	$0.21	$0.13	$788,786
Harbor College	3,592,365	$0.21	$0.13	$287,389
Mission College	4,342,258	$0.21	$0.13	$347,381
Pierce College	5,460,089	$0.21	$0.13	$436,807
Southwest College	5,674,813	$0.21	$0.13	$453,985
Trade Tech	3,684,048	$0.21	$0.13	$294,724
Valley College	5,790,966	$0.21	$0.13	$463,277
West LA	4,071,449	$0.21	$0.13	$325,716
Total				$3,973,901

* Per Year

Section Eight:
Finance Partners

LOS ANGELES
COMMUNITY COLLEGE DISTRICT

RENEWABLE ENERGY PROGRAM
FINANCE PARTNERS

The District facilitated a competitive process to determine the most qualified partners and the most advantageous finance structure for its renewable energy projects. As a result of this course of action, the finance committee developed a short list of entities based on the evaluation criteria set forth by the District.

A sample of the evaluation criteria:

- Financial ratings and guarantees
- Experience with financing renewable energy programs
- Experience with financing non-solar technologies
- Current appetite for federal tax credits
- PPA financing structure and pricing
- A corporate mission for investing in renewable energy projects

The preferred financial entities which the District has elected to partner with are as follows:

1. **Hannon Armstrong**, which will act as the facilitator for the financial syndication as well as provide gap funding if necessary.

2. **HSH Nordbank,** the largest financier of renewable energy programs in the world

3. **Citi Community Capital**, a division of Citi Bank, which has pledged $50 million investment in renewables (nationwide)

4. **Bank of America,** pledged $20 million investment in renewables (nationwide)

5. **Macquarie Power Cook**, which offers a prepay finance model

Section Nine:
Feed-in Tariff Analysis / RECs

Feed-in Tariff at $0.31 per kWh

	FIT $	REC $	Carbon $	Potential Annual Revenue
City College	$899,744	$359,898	$19,532	$1,279,173
East LA	$1,232,478	$492,991	$15,746	$1,741,216
Harbor College	$449,046	$179,618	$9,748	$638,412
Mission College	$542,782	$217,113	$11,783	$771,678
Pierce College	$682,511	$273,004	$14,816	$970,331
Southwest College	$709,352	$283,741	$9,063	$1,002,155
Trade Tech	$460,506	$184,202	$9,997	$654,705
Valley College	$723,871	$289,548	$15,714	$1,029,133
West LA	$508,931	$203,572	$6,502	$719,006
Total	**$6,209,221**	**$2,483,688**	**$112,900**	**$8,805,809**

* Per Year

Feed-in Tariff at $0.21 per kWh

	FIT $	REC $	Carbon $	Potential Annual Revenue
City College	$539,846	$359,898	$19,532	$919,276
East LA	$739,487	$492,991	$15,746	$1,248,225
Harbor College	$269,427	$179,613	$9,748	$458,793
Mission College	$325,669	$217,113	$11,783	$554,565
Pierce College	$409,507	$273,004	$14,816	$697,327
Southwest College	$425,611	$283,741	$9,063	$718,414
Trade Tech	$276,304	$184,202	$9,997	$470,503
Valley College	$434,322	$289,548	$15,714	$739,584
West LA	$305,359	$203,572	$6,502	$515,433
Total	$3,725,532	$2,483,688	$112,900	$6,322,120

* Per Year

Feed-in Tariff at $0.12 per kWh

	FIT $	REC $	Carbon $	Potential Annual Revenue
City College	$215,939	$359,898	$19,532	$595,368
East LA	$295,795	$492,991	$15,746	$804,532
Harbor College	$107,771	$179,618	$9,748	$297,137
Mission College	$130,268	$217,113	$11,783	$359,163
Pierce College	$163,803	$273,004	$14,816	$451,623
Southwest College	$170,244	$283,741	$9,063	$463,048
Trade Tech	$110,521	$184,202	$9,997	$304,721
Valley College	$173,729	$289,548	$15,714	$478,991
West LA	$122,143	$203,572	$6,502	$332,218
Total	**$1,490,213**	**$2,483,688**	**$112,900**	**$4,086,801**

* Per Year

Section Ten:
Education / Jobs

SUSTAINABLE DEVELOPMENT CURRICULUM

- Learn from actual green projects
- Sustainable curriculum
  - Solar, wind, geothermal, etc.
  - Economics, business, life cycle, etc.
  - Operations and maintenance
- Certificates, licenses, and degrees
- Train for green collar jobs
- Climate solutions today

412

Los Angeles Community College District Sustainable Building Program

Leading the Way to a Sustainable Future

"The LACCD has shown true vision and leadership in the field of sustainable design."

Bharat Patel, Chair of the U.S. Green Building Council, Los Angeles chapter

L.A. Southwest College's Child Development Center is LEED™ certified and was honored with a Project Achievement Award.

L.A. Valley College's Allied Health & Science Center's rooftop has a solar energy system that generates approximately 10 percent of the building's electricity needs.

By developing and implementing one of the largest public sector sustainable building efforts in the United States, the Los Angeles Community College District (LACCD) has become a leader in the green industry. The District received numerous prestigious awards for its work in sustainable construction and renewable energy.

In addition, representatives from LACCD have been called upon to serve as **industry experts** at international, national and local sustainability events, and conferences. LACCD is taking that role one step further by **promoting sustainability** and educating others on how they can create similar successful programs. Finally, the District is developing curriculum and training models that will prepare the green workforce of tomorrow.

Leading by Example

In step with its policy to increase energy efficiency and decrease greenhouse gas emissions, the **LACCD became one of the first community college districts in the state to join the California Climate Action Registry (CCAR).** The Registry was established as a non-profit, voluntary entity for businesses and public agencies to track their emissions, which cause global warming. In 2007, the LACCD had an independent firm verify its annual

emissions and made the findings public to become a role model for other institutions of higher education.

LACCD was also honored by the Clinton Climate Initiative (CCI) for its innovative green building program. **LACCD is one of 11 CCI partners** in a pilot program **to dramatically reduce greenhouse gas emissions by upgrading its campus buildings without using their capital budgets or increasing**

ing by Example (continued)

monthly operating expenses. The program will serve as a model for other colleges striving to combat global warming.

The LACCD is also one of more than 600 colleges and universities to sign the **American College and University Presidents Climate Commitment**

(ACUPCC), which provides a framework for America's higher education institutions to go climate neutral. Recently, LACCD Chancellor Dr. Marshall E. Drummond was selected to serve on the ACUPCC Steering Committee, the chief governing body responsible for guidance, policy and direction of the Commitment.

Spreading the Word on Sustainability

The LACCD Board of Trustees and District staff are committed to sharing sustainable building policy ideas with other public entities around the country. Through participation in conferences, summits and industry panels, the District is using its high profile in the academic community

to guide others through the "green" process and foster the growth of the sustainable construction movement.

Thanks to their role in the District's successful sustainable construction and renewable energy programs, LACCD leaders have become

regarded highly by the education and construction industries, the community-at-large and the media. Called upon to share their insight, these "green ambassadors" have addressed key audiences at conferences, summits and seminars throughout the U.S. and on an international scale.

nt Sustainability Awards

2008 CoreNet Global REmmy Corporate Citizens Award
2008 Energy Efficiency Partnership Program Best Practices in New Construction & Sustainable Operations
2008 California Construction Best of 2008 - LAVC Allied Health-Science Center
2007 Clinton's Climate Initiative (CCI)
2007 Governor's Environmental and Economic Leadership
2007 California Construction Owner of the Year - McGraw Hill (California Construction Magazine)
2007 Asian American Architects and Engineers Association Community Enrichment Award
2006 Savings by Design Award (Southern California Gas Company)
2006 USGBC (U.S. Green Building Council) Sustainable Future Award
2006 Construction Management Association of America: Annual Award
2005 & 2006 Flex Your Power by California Energy Efficiency and Conservation
2005 Visionary Leadership Award (US Green Building Council)
2005 Sempra Energy Efficiency Excellence Award
2005 Consulting Engineers and Land Surveyors of California: Client of the Year

Note: Page numbers followed by "f" and "t" refer to figures and tables, respectively

A

Adaptability, building, 169–170, 173–174, 174f
Adaptation, 25
Adders, 122–124
Agile energy system model, 19
Alternating current (AC), 204, 205, 221
Alternative energy public policy, 149
Alternative energy supply sources
  legal mechanisms, 150–161
    energy management agreement by
      community college districts, 150
    energy service contract and facility ground
      lease, by public agencies, 150–153
    photovoltaic system, lease of, 159–161
    power purchase agreement, by government
      agency, 153–159
America, 12, 15–17, 20, 48, 224, 275
American Association of Sustainable Higher
  Education (AASHE), 182–183
American values clarification, through sustainable
  agriculture, 233
  food value, appreciating, 245–246
  intercampus produce exchange,
    244–245
  Murphy apple orchard, 233–242
    beginning, 236–237
    living legacy, 238
    multiple applications and benefits, 239–242
  new beginning, 243–244
  standard American diet (SAD) truth, 242–243
    malnourished values, 243
An Inconvenient Truth, 1, 14, 60f, 273
Annual revenue impact (ARI), 116, 118–119
Aptera, 60, 60f
Asia, 9, 12
Assembly Bill 2660, 153
Automatic responses/emotional responses
  consumption behavior from, 33–34

B

Bacevich, Andrew, 48
Bacterial fuel cell energy generation, 18
Barriers to adoption, 36
Baseline, establishing, 87–88
Benefit/cost ratio (BCR)
  of participant test, 111–113
  of program administrator cost, 128

  of RIM, 116, 118–119
  of TRC, 122, 125–126
BerkeleyFIRST program, 70–71
Biodiesel, 66
Biogas, 256–257, 256f
Bio-generator, 18
Biomass, 17, 76–78, 80
  in Aalborg, 257f
  resources, 256, 257
Biopower, 76–78
Borehole thermal energy storage (BTES), 30–33,
  53–55, 58
Brundltand Report, 1
Budget, balancing, 11–12
Building, natural form of, 171–172
Building adaptability, 169–170, 173–174, 174f
Building independence, 170–171
Building-integrated photovoltaics (BIPV), 79
Building interconnectivity, 172–173, 173f
Building interdependence, 175–176
Building performance, 174–175
Building service, 168, 172, 172f

C

California, 2, 10, 18, 50–51, 67–68, 75, 191
  economic bankruptcy in, 10
  energy efficiency program, 20
California College, 83, 95
  energy conservation efforts, 84–86
California Energy Commission (CEC), 2, 103
California Environmental Quality Act (CEQA),
  154
California Legislature, 149, 150, 205
California Public Utilities Commission (CPUC),
  2, 50–51, 101, 103, 142
California Solar Initiative (CSI), 142, 149,
  162
California Standard Practice Manual (CSPM),
  6, 101
  spreadsheet, 102f
Campus food generation, 243
Canada, 46–47
Carbon capture and storage system, in Denmark,
  248f
Carter, Jimmy, 48
Chicago School of Economics, 27–28
China, 12

CHP (cogeneration of heat and power) plants, 253
see also Strandby CHP plant
Cisco, 228
Clean and green energy, 14
Clean Air Markets Division, 161
Cleantech energy production industry, 220
Climate Action Registry, 2
Climate change, and energy usage, 41
Climate change mitigation, from bottom-up community approach, 247
 Frederikshavn
  current energy system in, 253–255
  energy demand in, 250–252
  energy resources and scenario, 256–259
  energy system integration, 259–263
  public involvement, 263–264
Climate Group, 12
Community college districts
 energy management agreement by, 150
 see also Los Angeles Community College District (LACCD), case study
Community-sponsored agriculture (CSA), 244
Concentrated solar power (CSP), 72
Confirmation bias, 30
Conservation and load management (C & LM) programs, in California, 103
Conservation programs, 104
Consumption and saving energy
 conflicting messages about, 34–35
Coriolis effect, 224
Cost-effectiveness tests, 107t
Cradle to Cradle™, 168

D

Demand-side management categories and program definitions, 104–106
Denmark, 6, 30–31, 247–249
 energy system in, 12, 13f
Digestive process, 17
Direct-combustion steam production, 76
Direct current (DC), 204, 221
District heating, in Denmark, 250–255
Dollar savings, 84, 86
Due diligence procedures, 140

E

Earth Day, 23–24, 234f
"Earthships," 53–55
Economist, The, 14
Electric drive, 61
Electric heat pumps, 45–46, 58

Electric utility rates, 190–191
Electrical energy efficiency, 49–50
Electricité de France, 17
Electricity and energy efficiency retrofits, of existing buildings, 55–56
Emerging Renewables Program, 162
Energy conservation and energy efficiency, 45
 efficient lighting, 56
 electrical energy efficiency, 49–50
 electricity and energy efficiency retrofits, of existing buildings, 55–56
 energy independence and carbon neutrality, key technologies for, 58–59
 green design, 51–53
 heat pumps, 56–58
 near-zero, net-zero, and plus-energy buildings, 53–55
 price signals and energy efficiency, 62
 quality assurance and certification, 59
 transport, energy efficiency in, 59–61
  longer-term measure, 61
  short-term solutions, 60–61
 United States, energy efficiency in, 62
 utility revenue decoupling and energy efficiency, 50–51
Energy conservation facility, 150–151
Energy conservation measures (ECM), in LACC
 energy management services, 200
 HVAC fan and motor upgrades, 200
 interior lighting, 199
 lighting controls installation, 199–200
 solar film application, 200
 sub-metering installation, 201
Energy conservation program, implementation of (case study), 83
 concepts, 45–47
 getting started, 84–86
 installation, 92–95
  comprehensive project, 92
  equipment and subcontractors, selection of, 92
  ongoing training and support, 95
  project cost, 95
  project management, 92
  project savings guarantee, 95
  rebates and incentives, 95
 investment grade audit (IGA), 86–92
  baseline, establishment of, 87–88
  energy conservation measures (ECM), selection of, 88
  savings calculations, 92
 measurement and verification, 95–97

Energy consumption, 24, 27–28, 33–34, 36, 41, 191–196, 202, 217
  rational choices about, 27–28
  versus energy demand, 190
Energy efficiency, 45–47, 49, 59, 104, 227
  in electrical sector, 45–46, 49–50
  of existing buildings, 55–56
  meaning of, 45–47
  and price signals, 62
  in transport, 59–61
  in United States, 62
  and utility revenue decoupling, 50–51
Energy efficiency gap, 26–27
Energy independence and carbon neutrality, key technologies for, 58–59
Energy management agreement
  by community college districts, 150
Energy management services, in LACC, 200
Energy Policy Act, 162
Energy savings versus nonenergy savings, 85
Energy Security and Reliability Act of 2000, 103–104
Energy service contract, and facility ground lease
  by public agencies, 150–153
Energy services company (ESCO), 83–85
  key questions to select, 85–86
Energy usage, 39–40
  and rapid climate change, 41
EnergyPLAN model, 261–262, 262f
ENRON, email surveillance data on, 15f
Enthanol, 66
Environmental attributes, meaning of, 161
Environmental incentives, treatment of, 161–163
Environmental Protection Agency, 161
Equipment lease agreements, 144
Europe, 9, 12, 49
Expected performance-based buy down, 143
Externality values, 109

F

Facility ground lease, and energy service contract
  by public agencies, 150–153
Fee-producing infrastructure project, 153
Feed-in tariffs (FiTs), 4
First-year revenue impact (FRI), 116
Fluorescents, 56
Food Inc., 243
Food system, value of, 242, 246
Food value, appreciating, 245–246

Frederikshavn (in Denmark)
  current energy system in, 253–255
  energy demand in, 250–252
  energy resources and scenario, 256–259
    geothermal resources in, 257–258
    wind resources in, 257
  energy system integration, 259–263
  public involvement, 263–264

G

General Motors, 10
Geo-exchangers, 17
Geothermal energy, 17, 73–76
Geothermal/geoexchange heat pumps, 57–58
Geothermal heat pumps, 17, 73, 75–76
Geothermal plant, in Denmark, 258, 258f
Germany, 12, 14–16
Global fuel reserves and production resources, 11f
Global sustainable future, 267
  "green" careers and businesses, 269
  public policy, 270–272
  sustainable and smart agile communities, 272–275
  third industrial revolution in U.S., 267–268
Google, 228
Government agency
  power purchase agreement by, 153–159
Government Motors, 10
Great Depression (1930), 10
Green design, 51–53
Greenhouse gas (GHG) emission, 24, 36–38, 173, 217
Greenwashing, 38
Grid neutrality, 84
Gross energy savings, 110
Ground source heat pump, 17, 57–58, 57f
Groundwater source heat pumps (GSHPs), 57–58

H

Hannover principles, 166
Heat pumps, 56–58
Heating, ventilation and cooling (HVAC), 195–196
HOMER v2.67 Beta, 202
HVAC fan and motor upgrades, in LACC, 200
Hybrid systems, 80

I

Impact on rate levels test, 115
Incentives, 113, 142–143
Independence, building, 170–171
Induction lighting, 56
Industrial Revolution, 23–25
  see also Second industrial revolution (2IR);
    Third Industrial Revolution (3IR)
Interagency Green Accounting Working Group
    (IGAWG), 101
Intercampus Produce Exchange (ICPE),
    244–245
Interconnectivity, seven principles for
  building adaptability, 173–174, 174f
  building independence, 170–171
  building interconnectivity, 172–173, 173f
  building interdependence, 175–176
  building natural form, 171–172
  building performance, 174–175
  building service, 172, 172f
  sustainable design maxims, 166–168
Interdependence, building, 175–176
Intermittent energy generation, 16
Internal rate of return (IRR), 97, 146
International Performance
    Measurement and Verification
    Protocol, 96
Internet technology, 227
  and grid, 229, 230f
Internet Web model, 226, 227
Inverter unit, 221
Investment grade audit (IGA), 86–92
  baseline, establishment of, 87–88
  energy conservation measures, selection of, 88,
    89f, 90–91t
  savings calculations, 92
Investment Tax Credit (ITC), 142

J

Japan, 12, 49
"Jevon's Paradox," 47

K

Kelly, Walt, 23–24
Kyoto Protocol, 24–25, 161, 224

L

Labor savings, 85
Leadership for Energy and Environmental
    Design (LEED), 1, 52f, 166–167,
    269, 274

LED lamps
  savings calculations for, 93t
LEDs, 56
LEED, *see* Leadership for Energy and
    Environmental Design (LEED)
Life-cycle analysis, projects programs, in
    California, 99
  basic methodology, 103–110
    demand-side management programs,
      104–106
    externality values, 109
    limitations, 109
    policy rules, 110
    tests, balancing, 109
  equations, primary inputs for,
    129–131
  participant test, 110–115
    benefits and costs, 110–111
    definition, 110
    formulas, 112–115
    results, expressing, 111
    strengths, 112
    weaknesses, 112
  program administrator cost test, 126–129
    benefits and costs, 127
    definition, 126–127
    formulas, 128–129
    results expression, 127–128
    strengths, 128
    weaknesses, 128
  ratepayer impact measure test, 115–120
    benefits and costs, 115–116
    definition, 115
    formulas, 118–120
    results, expressing, 116
    strengths, 117
    weaknesses, 117–118
  total resource cost test, 121–126
    definition, 121
    formulas, 125–126
    results expression, 122–124
    strengths, 124–125
    weakness, 125
Life-cycle and cost-benefit analyses, of renewable
    energy, 139
  due diligence procedures, 140
  energy challenges, 139–140
  financing structures, 143–144
    equipment lease agreements, 144
    power purchase agreements, 143
  PV system, financial perspective of, 141–143
    available incentives, 142–143
    costs, 141

PV system, measuring savings from, 144–148
  externality factors, 148
  financial analysis, 144–147
  successful energy plan, 140
Life-cycle revenue impact (LRI), 116, 118–119, 136, 137
Lighting, 56
Lightweight vehicles, 60
Load retention, 106
Local, state, and national policies, 270–272
Loremo, 60
Los Angeles City College (LACC), case study, 194–201
  campus energy consumption and demand, 191–194
  campus growth, 194–197
  demand side management, 197–201
  electric utility rates, 190–191
  energy conservation measures
    energy management services, 200
    HVAC fan and motor upgrades, 200
    installation of sub-metering, 201
    interior lighting, 199
    lighting controls installation, 199–200
    solar film application, 200
  energy demand versus energy consumption, 190
  existing and planned buildings, 196t
  solar insolation, 202–204
Los Angeles Community College District (LACCD), case study, 139–140, 181
  background, 182–184
  current and planned renovations/construction, 212
  current situation of LACC, 188–201
  goals and objectives, 184
  importance, 184–185
  LADWP energy rates, 213
  projected 2015 campus energy demand and consumption, 201–202
  renewable energy options, 185–188
    wind, 186–188
  solar insolation, 202–204
  solar PV array and setup, 204–209
  PV array size, 209
Los Angeles Department of Water and Power, 159–160
Los Angeles Department of Water and Power, 190
Lovins, Amory, 49, 60
Lyle, John, 167–168

M
Make It Right Program, 168
Manure, 256f
Marine, 78
Methane, 76
Microbial fuel cell energy generation, 18
Mitigation, 25
  see also Climate change mitigation, from bottom-up community approach
"Modern Economic Theory," 14
Modified Accelerated Cost Recovery System, 142
Monetary incentives, 142
Murphy apple orchard, 233–242
  beginning, 236–237
  education and environmental sustainability, 234–235
  expansion, 242f
  jobs on campus, 239
  living legacy, 238
  multiple applications and benefits, 239–242
  as symbol of sustainability, 236
Murphy, Jeremiah, 238

N
National Energy Policy Act (2005), 142
National Renewable Energy Laboratory (NREL), 2, 202
Near-zero energy buildings, 45–46, 53
Negawatts, 49–50
Net present value (NPV), 145–146
  of participant test, 111
  of program administrator cost, 127
  of RIM, 116
  of TRC, 122
Net-zero buildings, 53–55
Net-zero houses, 45–46
New York City, 21
New York Times, 30–31
Nitrogen oxides (NO_X), 123
Nonenergy savings and energy savings, 85
Nonparticipant test, 115
Non-profit organizations, 143
  utility cost, calculating, 144, 145t
Nordic countries, 12

O
Obama, Barack, 217–218
Ocean and tidal waves, 17
Ocean current devices, 17
Ocean thermal energy conversion devices, 17

Operational and maintenance savings, 85
Optimization energy plan, 3
Orchard Oversight Committee, 237, 237f
Oregon Institute for Science and Medicine, 30

P

Participant test
  benefits and costs, 110–111
  definition, 110
  formulas, 112–115
  results, expressing, 111
  strengths, 112
  weaknesses, 112
Passive design, 167
Payback period, 111, 146
Payback requirements
  for energy conservation measures, 90–91t
Payment and performance bond, 86
Peak shaving, 190–191
Performance based incentives (PBI), 142–143
Performance contracting, 83–84, 87, 92–96
  comprehensive project, 92
  equipment and subcontractors,
    selection of, 92
  lawsuit/litigation, 86
  measurement and verification, 95–97
  ongoing training and support, 95
  project cost, 95
  project management, 92
  project savings guarantee, 95
  rebates and incentives, 95
Photovoltaic (PV) reaction, 16–17
Photovoltaic (PV) system, *see* PV system,
    life-cycle analysis of
Photovoltaic system, lease of, 159–161
Pinchot, Gifford, 46f
Plastic bags elimination, local campaigns for,
    38
Plug-in hybrid vehicles (PHEVs), 79
Plus-energy buildings, 53–55
"Plus-energy" houses, 45–46
Positive feedback, 36
Poverty and food system, 242
power purchase agreements (PPAs), 4, 142–143,
    154–155
  by government agency, 153–159
Power purchase provisions, 153–154
Price signals, and energy efficiency, 62
Program administrator cost test, 126–129
  benefits and costs, 127
  definition, 126–127
  formulas, 128–129

  results expression, 127–128
  strengths, 128
  weaknesses, 128
Public agencies, 151–152
  energy service contract and facility ground
    lease, 150–153
Public buildings and institutions
  solar power as solution, 149
    alternative energy public policy, 149
    environmental incentives, treatment of,
      161–163
    legal mechanisms, 150–161
Public policy and leadership, 23
  climate change, 28–29
  confirming belief, 29–32
  energy efficiency gap, 26–27
  limited set of factors, focus on, 32
  poor decisions due to fear, 29
  rational choices, about energy consumption,
    27–28
  short-term threats, 32–41
    automatic responses/emotional responses,
      33–34
    conflicting messages, 34–35
    decision making, 35–36
    energy use, 39–40
    following leaders, 36–37
    greenwashing, 38
    people's actions and consequence, direct
      relationship between, 39
    plastic bags elimination, local campaigns
      for, 38
    social networks, 37
    solutions to problems, 41
    theory of moral hazards, 38–39
    tragedy of the commons, 39
Public Resources Code section 25007, 149
Public Resources Code section 25008, 150
PV system, life-cycle analysis of, 141–143
  available incentives, 142–143
  costs, 141–143, 142t
  due diligence procedures, 140
  financing structures, 143–144
    equipment lease agreements, 144
    power purchase agreements, 143
  measuring savings from, 144–148, 145t
    externality factors, 148, 148t
    financial analysis, 144–147, 147t

Q

Qualitative economics, 14
Qualitative externalities, 148

Quality assurance and certification, in energy efficiency, 59

R

Ratepayer impact measure (RIM) test, 115–120
  benefits and costs, 115–116
  definition, 115
  formulas, 118–120
  results, expressing, 116
  strengths, 117
  weaknesses, 117–118
Rational markets, 27–28
Reactive organic gases (ROGs), 123
Reagan, Ronald, 48
Regenerative design, 167–168
Renewable energies and smart grid, transformational relationship of, 217
  power transmission lines and data response, 226–228
  smart grid and market solution, 228–231
  solar electricity system, relationship with grid, 219–223
  wind power, 223–226
Renewable energy, 65
  advanced renewables deployment, 78–80
    hybrid systems, 80
    renewables and buildings, 79
    vehicle-to-grid, 79
  biopower, 76–78
  geothermal, 73–76
  marine, 78
  solar, 70–73
    photovoltaic (PV), 70–72
    thermal, 72–73
  United states, energy use in, 65–67
  wind, 67–69
Renewable energy options, 185–188
Renewable energy power source, 16–21
Renewable energy sources implantation, 140
Renewables and buildings, 79
Reporting rights, 162–163
Reynolds, Mike, 53–55
Rifkin, Jeremy, 2
Rodale, Robert, 167
Rosenfeld, Art, 51f

S

Savings calculations, 89f, 92, 95t
  for LED lamps, 93t
Schneider Electric, 86–88, 92, 95
Schwarzenegger, Arnold, 149
SeaGen tidal power system, 17

Second industrial revolution (2IR), 12, 18, 267, 269
Self-generation, 105
Self-Generation Incentive Program, 141, 162
Short-term threats, focusing on, 32–41
  automatic responses/emotional responses, 33–34
  conflicting messages, 34–35
  decision making, 35–36
  energy use, 39–40
  following leaders, 36–37
  greenwashing, 38
  people's actions and consequence, direct relationship between, 39
  plastic bags elimination, local campaigns for, 38
  social networks, 37
  solutions to problems, 41
  theory of moral hazards, 38–39
  tragedy of commons, 39
Silicon, 70
Skepticism, 30
Smart grid, 21f, 227f, 228–231
Social feedback, 36–37
Social networks, 37
Social norms theory, 36–37
Soft savings, 85
Solar film application, in LACC, 200
Solar generation systems, 16–17
Solar insolation, of LACC, 202–204
Solar photovoltaic (PV) panels, 219, 221
Solar Photovoltaic Incentive Program, 159–160
Solar power, 70–73
  solar photovoltaic (PV), 70–72
  solar thermal, 72–73
Solar PV array and setup, in LACC, 204–209
Solar service agreement, 143
Solar thermal power plant (STPP), 221–222
South Korea, 12
Southern California Edison utility rates, 142–143
Space conditioning/HVAC, 56–57
Spain, 4, 12
Standard accounting mechanisms, 101
Standard American diet (SAD), 242–243
Standard Practices Manual (SPM), 101, 103, 104
State of California General Services Agency, 101
Strandby CHP plant, 253–254, 253f
Sulphur oxides (SOₓ), 123
Sustainability Education for Environmental Development Sessions (SEEDS program), 241

Sustainable and smart agile communities, 272–275
Sustainable agriculture, 233
Sustainable design maxims, 166–168
Syngas, 76

T

Tax shifting, 18
Tests, balancing, 109
Theory of moral hazards, 38–39
Third Industrial Revolution (3IR), 2, 9, 218, 267
   communities and nations, focus of, 12, 13–16
   implementation, 9–10
   renewable energy power source, 16–21
Tidal power devices, 17
Total resource cost test, 121–126
   definition, 121
   formulas, 125–126
   results expression, 122–124
   strengths, 124–125
   weakness, 125
Tragedy of the commons, 39
Transport, energy efficiency in, 59–61
   longer-term measure, 61
   short-term solutions, 60–61
Treasury Cash Grant (TCG), 142
Turbines, in Frederikshavn, 258f

U

UN Climate conference 2009, 30–31
United Nations Framework Convention on Climate Change (UNFCCC), 161

United States, 9, 12, 45–47
   3IR implementation, 9–10
   energy efficiency in, 62
   energy use in, 65–67
   leadership, in 2IR, 12
   oil supplies, 10–12
U.S. Chamber of Commerce, 30–31
U.S. Department of Energy, 10–11, 84
U.S. National Intelligence Council, 24
USGBC (U.S. Green Building Council), 1
Utility cost, calculating, 144
Utility Policy Journal, 15–16
Utility revenue decoupling
   and energy efficiency, 50–51

V

Vehicle efficiency standards and automaker penalties
   versus gas taxes, 60–61
Vehicle-to-grid, 79
Volatile organic compounds (VOCs), 123

W

Waste incineration plant, 253f
Water heating, 56–58
Wave power conversion devices, 17
Waxman-Markley bill, 24–25
Weber, Max, 28
Wheaton College, 233–236, 235f
Wind farms, 67
Wind generation, 16
Wind power, 67–69
Wright, Frank Lloyd, 171

Printed in the United States
By Bookmasters